The Observer's
Basic Military Aircraft Directory

THE OBSERVER'S

BASIC MILITARY AIRCRAFT DIRECTORY

COMPILED BY

WILLIAM GREEN
AND GORDON SWANBOROUGH

FREDERICK WARNE & CO LTD London

FREDERICK WARNE & CO INC New York

Published by Frederick Warne & Co Ltd 1974

Reprinted 1975

LIBRARY OF CONGRESS CATALOG CARD
No 73–70251

ISBN 0 7232 1523 5

Filmset and printed in Great Britain
by BAS Printers Limited, Wallop, Hampshire
170.175

Introduction

First published in 1967 in a smaller format, *The Observer's Basic Military Aircraft Directory* now appears as a comprehensive but compact reference source to the world's most important and most widely used military aircraft, irrespective of their age. Consequently, this work and its companion volume of *Civil Aircraft* supplement but do not replace the annual *Observer's Book of Aircraft* which, each year, surveys the newest aircraft and the latest variants of existing types that have appeared in the preceding twelve months or are expected to appear during the currency of the edition.

Since the purpose of the present volume is to provide a guide to the characteristics of the aircraft in service with air forces and army and naval air arms throughout the world, research types and prototypes with uncertain futures are excluded. All types in production for military users are described, including those which, by the spring of 1974, had reached only the flight test stage but were firmly committed to future production and service. Thus, for example, the Panavia MRCA, perhaps Europe's most important military aircraft of the second half of the present decade and destined to serve with the air forces of the UK, Federal Germany and Italy, is included, although it was scheduled to make its first flight after this volume closed for press, whereas the Rockwell B-1, the USAF's new supersonic bomber which entered flight testing late-1974, is excluded since, at the time of writing, there is uncertainty as to its future production. Similarly, the Kawasaki C-1A, Japan's new military transport, is included but the highly interesting Boeing YC-14 and McDonnell Douglas YC-15 STOL transport prototypes are excluded.

Much of the military equipment still in service around the world comprises aircraft types that are no longer in production. All these types are fully described here, with the exception of a few of the most elderly which may still be found in odd corners of the globe but are of little significance. Some transport, communications and training aircraft used in military guise are basically commercial types bought "off-the-shelf" and such are described and illustrated in the *Observer's Basic Civil Aircraft Directory* which is uniform with this volume.

The arrangement of the content matter is according to the

primary function of the aircraft in question, in 11 categories, the arrangement within each category being alphabetical by manufacturer. This brings into close juxtaposition aircraft types of generally similar characteristics and purpose, and thus facilitates direct comparison. However, it should be remembered that different air forces often classify similar aircraft in different rôles, and in other cases aircraft have a genuine multi-rôle capability—particularly in the case of trainer/light strike types. Consequently, the inclusion of an aircraft type in this volume under a specific rôle heading does not imply that it is not used for other purposes, as is made clear in the "Notes" provided for each type. Dimensions, weights and performance data are given only for the principal production/service versions, but under the "Status" heading, a brief summary is presented in each case of the complete production life of the aircraft, including quantities built of all variants.

All silhouettes are the copyright of Pilot Press Ltd. Credits for the photographs supplied by private individuals appear on page 224.

AERITALIA (FIAT) G.91

The photo above illustrates a G.91R.3 of the Luftwaffe's Waffenschule 50, and the general arrangement silhouette (right) depicts the G.91Y

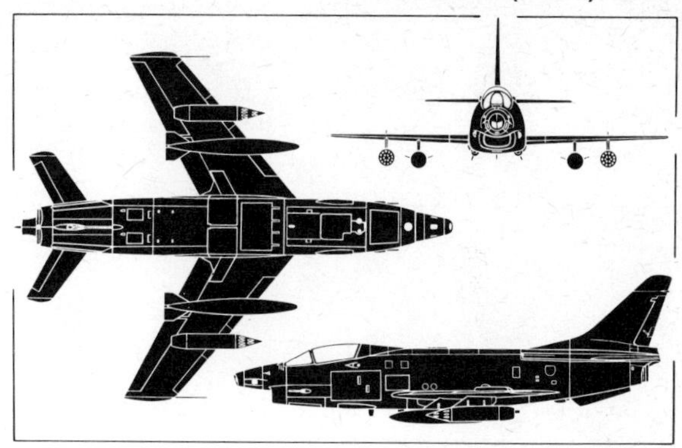

Country of Origin: Italy.
Type: Light fighter-bomber, reconnaissance and trainer.
Power Plant: (G.91R, T) one 5,000 lb st (2 268 kgp) Bristol Siddeley Orpheus 803 or (G.91Y) two 2,725 lb st (1 236 kgp) dry and 4,080 lb st (1 850 kgp) with reheat General Electric J85-GE-13A turbojets.
Performance: (G.91R) Max speed, 668 mph (1 075 km/h) at sea level and 675 mph (1 086 km/h) at 5,000 ft (1 524 m); initial rate of climb, 6,000 ft/min (30,5 m/sec); service ceiling, 43,000 ft (13 100 m); combat radius, 196 mls (315 km); ferry range, 1,150 mls (1 850 km).
Performance: (G.91Y) Max speed, 690 mph (1 110 km/h) at sea level and 670 mph (1 078 km/h) at 32,810 ft (10 000 m); initial rate of climb, 17,000 ft/min (86,36 m/sec); service ceiling, 41,000 ft (12 500 m); radius (lo-lo-lo with 2,910 lb (1 320 kg load) 240 mls (385 km); ferry range, 2,110 mls (3 400 km).
Weights: (G.91R) Empty 6,835 lb (3 100 kg); normal take-off, 11,995 lb (5 440 kg).
Weights: (G.91Y) Empty, 8,598 lb (3 900 kg); normal take-off, 17,196 lb (7 800 kg); max overload, 19,180 lb (8 700 kg).

Dimensions: Span (G.91R, T) 28 ft 1 in (8,56 m), (G.91Y) 29 ft $6\frac{1}{2}$ in (9,01 m); length, (G.91R) 33 ft $9\frac{1}{4}$ in (10,30 m), (G.91T, Y) 38 ft $3\frac{1}{2}$ in (11,67 m); height (G.91R) 13 ft $1\frac{1}{4}$ in (4,00 m), (G.91T) 13 ft $11\frac{1}{4}$ in (4,25 m), (G.91Y) 14 ft $6\frac{1}{4}$ in (4,43 m); wing area (G.91R, T) 176·7 sq ft (16,4 m^2), (G.91Y) 195·15 sq ft (18,13 m^2).
Accommodation: Pilot only or (G.91T) two in tandem.
Armament: (G.91R.1) Four or (G.91T) two 0·50-in (12,7 mm) machine guns or (G.91R.3 and Y) two 30-mm DEFA cannon plus four underwing store stations for max of (G.91R) 2,000 lb (907 kg) or (G.91Y) 4,000 lb (1 814 kg) ordnance.
Status: Prototype first flown 9 August 1956; first pre-production flown on 20 February 1958; first G.91R.3 (German assembly) flown 20 July 1965; first G.91T flown 31 May 1960; first G.91Y flown 27 December 1966. Production totals: prototypes, 3; G.91 pre-production, 27; G.91R.1, 98; G.91R.3, 50 by Fiat and 282 by Dornier; G.91R.4, 50; G.91T.1, 76; G.91T.3, 66; G.91Y prototypes, 2; G.91Y pre-production, 20; G.91Y, 50.
Notes: G.91, R.1, T.1 and Y are Italian Air Force versions; G.91R.3, R.4 and T.3 are *Luftwaffe* versions, of which the R.4s were transferred to Portuguese Air Force.

ARMSTRONG WHITWORTH SEA HAWK

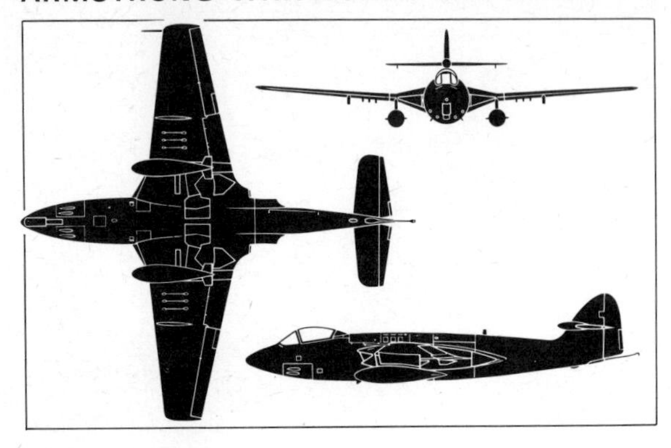

The photo above illustrates a Sea Hawk Mk 3 (brought up to Mk 6 standards) of the Indian Navy's No 300 Sqdn, the general arrangement silhouette (left) depicting the Mk 6

Country of Origin: United Kingdom.
Type: Carrier based fighter.
Power Plant: One Rolls-Royce (F Mk 1, 2, 3, 4) 5,000 lb st (2 270 kgp) Nene 101 or (F Mk 5, 6, 50, 100, 101) 5,400 lb st (2 450 kgp) Nene 103 turbojet.
Performance: Max speed, 599 mph (958 km/h) or Mach 0·79 at sea level, 587 mph (939 km/h) or Mach 0·83 at 20,000 ft (6 100 m) and 530 mph (848 km/h) or Mach 0·80 at 40,000 ft (12 200 m); initial rate of climb, 5,700 ft/min (29,0 m/sec); service ceiling, 44,500 ft (13 560 m); radius of action (clean) 230 mls (368 km); radius of action (two drop tanks), 386 mls (618 km).
Weights: Empty, 9,190 lb (4 165 kg); normal loaded 13,220 lb (6 000 kg); max take-off (drop tanks and bombs), 16,200 lb (7 355 kg).
Dimensions: Span, 39 ft 0 in (11,9 m); length, 39 ft 8 in (12,1 m); height, 8 ft 8 in (2,65 m); wing area, 278 sq ft (25,83 m²).
Accommodation: Pilot only.
Armament: Four 20-mm Hispano cannon in front fuselage plus (FB Mk 3, 5) two wing pylons with 500-lb (227-kg) capacity plus (FGA Mk 4, 6, 50) up to 20 3-in (7,62-cm) rockets.

Status: Prototype (P.1040) first flown on 2 September 1947; prototype (Sea Hawk) flown on 3 September 1948; first production F Mk 1 flown 14 November 1951; first F Mk 2 flown 24 February 1954; first FB Mk 3 flown on 13 March 1954; first F (GA) Mk 4 flown on 26 August 1954. Production quantities: prototypes, 3; F Mk 1, 35 by Hawker, 60 by AWA; F Mk 2, 40; FB Mk 3, 116; F (GA) Mk 4, 97; F (GA) Mk 6, 87; Mk 50, 22; Mk 100, 32; Mk 101, 32. Production completed 1961.
Notes: The Sea Hawk was the production version of Hawker's first jet fighter, identified as the P.1040 in its prototype form. Production was initiated by Hawker and transferred to Armstrong Whitworth, the latter company being responsible for all but the first 35 aircraft. Mks 1 to 4 and 6 were production models for the Royal Navy; Mk 5 was a conversion of the Mk 3. Mk 50 was for the Netherlands and Mks 100/101 for German Navy. Only user now is the Indian Navy, which bought a batch of 24 ex-RN Mk 3s updated to Mk 6 standard by AWA and later acquired 22 more ex-RN Mk 4s and 6s and 28 ex-German Mk 100/101s. They equip No 300 Squadron, the "White Tigers", aboard the INS *Vikrant* and training units, probably continuing in service until 1975–76.

BAC (ENGLISH ELECTRIC) LIGHTNING

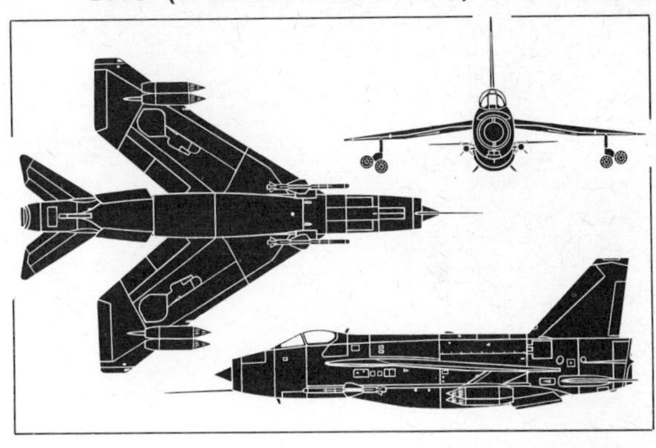

The photo above illustrates a Lightning F Mk 2A of No 92 Sqdn, RAF, and the general arrangement silhouette (right) illustrates the Lightning F Mk 53

Country of Origin: United Kingdom.
Type: Interceptor, strike and reconnaissance fighter.
Power Plant: Two Rolls-Royce (F Mk 1, 1A, T Mk 4, 54) 11,250 lb st (5 103 kgp) dry and 14,430 lb st (6 545 kgp) with reheat Avon 201 or (F Mk 2, 2A, 52) Avon 210 or (F Mk 3, 6, T Mk 5, 55) 11,100 lb st (5 035 kgp) dry and 16,300 lb st (7 393 kgp) with reheat Avon 301 or (F Mk 53) Avon 302C turbojets.
Performance (approximate): Max speed (F Mk 1 and 2), 1,390 mph (2 237 km/h) or Mach 2·1 at 40,000 ft (12 190 m), (F Mk 6) 1,500 mph (2 415 km/h) or Mach 2·27 at 40,000 ft (12 190 m); initial rate of climb, 50,000 ft/min (254 m/sec).
Weights (approx): Normal take-off, 40,000 lb (18 144 kg).
Dimensions: Span, 34 ft 10 in (10,61 m); length (over probe) 55 ft 3 in (16,84 m); height, 19 ft 7 in (5,97 m); wing area (F Mk 1, 1A, 2, 3, 52, T Mk 4), 458·52 sq ft (42,70 m²) (F Mk 2A, 6, 53, T Mk 5, 55) 474·5 sq ft (44,08 m²).
Accommodation: Pilot only or (T Mk 4, 54, 5, 55) pilot and instructor side-by-side.
Armament: (F Mk 1, 2, 53) Two 30-mm Aden guns internally and two Firestreak AAMs on forward fuselage (F Mk 3, 6, 53) two Red Top AAMs or air-to-air rocket packs and optimal ventral pack containing two 30-mm Aden guns and (F Mk 53 only) two underwing pylons with 1,000-lb (454-kg) capacity each.
Status: Prototype (P.1B) first flown 4 April 1957; first development batch F Mk 1 flown on 3 April 1958; first F Mk 1 flown 29 October 1959; first F Mk 2 flown on 11 July 1961; first F Mk 3 flown on 16 June 1962; prototype T Mk 4 flown on 6 May 1959; prototype T Mk 5 flown 29 March 1962; F Mk 6 prototype flown on 17 April 1964; production F Mk 6 flown 16 June 1965. Production totals: P.1B prototypes, 3; pre-production, 2; F Mk 1 and 1A, 48; F Mk 2, 44 (about 30 cvtd to F Mk 2A and 5 to F Mk 52); F Mk 3, 58; F Mk 6, 67; F Mk 53, 35 to Saudi Arabia and 12 to Kuwait; T Mk 4, 20 (2 cvtd to T Mk 54); T Mk 5, 22; T Mk 55, 6 to Saudi Arabia and 2 to Kuwait. Production completed September 1972.
Notes: The Lightning is standard equipment in RAF interceptor squadrons in Germany and the UK. Final production version, the Mk 6, introduced broader wing tips and provision for overwing fuel tanks. The export Mk 53 is similar with additional provision for underwing loads for ground attack rôle.

CONVAIR F-102 DELTA DAGGER

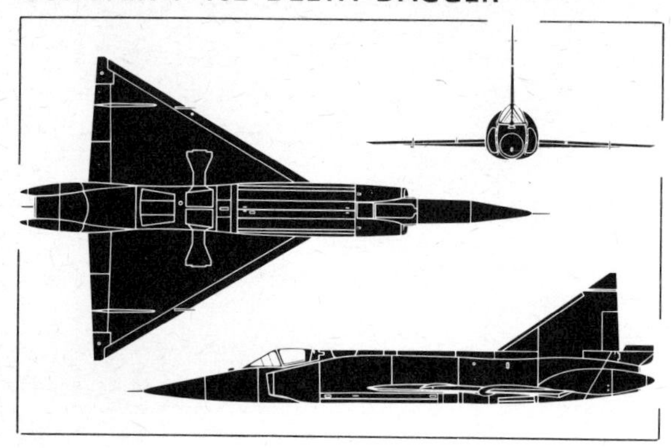

The photo above illustrates an F-102A Delta Dagger of the Hellenic Air Force's 342 Mira, and the general arrangement silhouette (left) also illustrates the F-102A

Country of Origin: USA.
Type: Interceptor fighter and operational trainer.
Power Plant: One 11,700 lb st (5 300 kgp) dry and 17,200 lb st (7 802 kgp) with reheat Pratt & Whitney J57-P-23 turbojet.
Performance: Max speed, 825 mph (1 328 km/h) at 40,000 ft (12 192 m); speed with two drop tanks, 630 mph (1 014 km/h) at 36,000 ft (10 973 m); normal cruising speed, 540 mph (869 km/h) at 35,000 ft (10 668 m); initial rate of climb, 13,000 ft/min (66,0 m/sec); service ceiling, 54,000 ft (16 460 m); max range, 1,350 mls (2 173 km).
Weights: Normal loaded, about 28,000 lb (12 700 kg); max take-off, 31,500 lb (14 288 kg).
Dimensions: Span, 38 ft 1½ in (11,60 m); length over nose probe (F-102A) 68 ft 4⅔ in (20,83 m); (TF-102A) 63 ft 4½ in (19,30 m); height (F-102A) 21 ft 2½ in (6,45 m) (TF-102A) 20 ft 7 in (6,27 m); wing area, 695 sq ft (64,57 m^2).
Accommodation: Pilot only or (TF-102A) two side-by-side, dual controls.
Armament: Three AIM-4C or -4D Falcon and one AIM-26A or -26B Falcon AAMs in internal weapons bay; provision for 24 × 2·75-in rockets in missile bay doors.
Status: Prototypes (YF-102) first flown on 24 October 1953 and 11 January 1954 respectively; YF-102A flown on 20 December 1954; TF-102A first flown on 8 November 1955. Production totals, YF-102, 10; YF-102A, 4; F-102A, 875; TF-102A, 63. Production completed 1958.
Notes: The F-102A was the first delta-wing design to enter service with the USAF, and in its production guise was the first to feature the "Coke-bottle" effect of area-ruling. Used primarily by Air Defence Command in North America but also by units of the USAFE in Europe, the F-102A had been withdrawn from USAF frontline service by 1973 but still equipped nine groups of the Air National Guard. All primary production of F-102As and TF-102As was for USAF, but batches of aircraft released from the final service units in Europe were transferred to the Greek and Turkish Air Forces, each receiving approximately 20 and 40 aircraft respectively for interim use pending the introduction of F-4 Phantoms, these including several TF-102As. The two-seat model of the Delta Dagger retains the weapons capability of the single-seater but lacks the MG-10 fire control and is subsonic.

CONVAIR F-106 DELTA DART

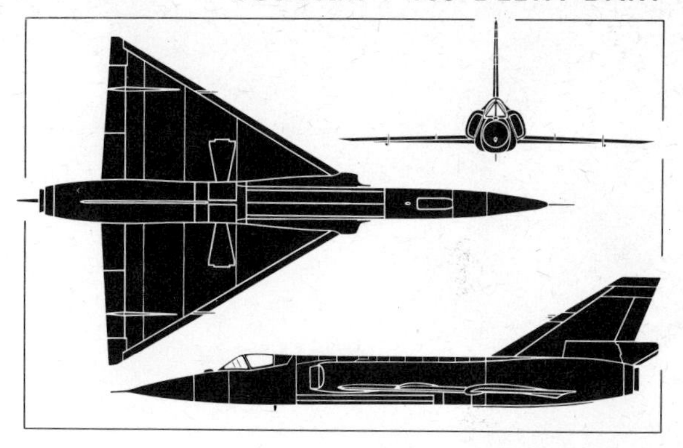

Both photograph above and general arrangement silhouette (right) illustrate the F-106A Delta Dart of the USAF's Aerospace Defence Command

Country of Origin: USA.
Type: All-weather interceptor and (F-106B) operational trainer.
Power Plant: One 17,200 lb st (7 800 kgp) dry or 24,500 lb st (11 130 kgp) with reheat Pratt & Whitney (F-106A) J75-P-17 or (F-106B) J75-P-9 turbojet.
Performance: Max speed, 1,525 mph (2 455 km/h) at 40,000 ft (12 192 m); service ceiling, 57,000 ft (17 374 m); combat radius (clean) 575 mls (925 km); ferry range, 1,500 mls (2 415 km).
Weights: Empty, 23,646 lb (10 725 kg); normal take-off (clean) 35,500 lb (16 100 kg); max overload, 38,250 lb (17 350 kg).
Dimensions: Span, 38 ft $3\frac{1}{2}$ in (11,67 m); length (over probe) 70 ft $8\frac{3}{4}$ in (21,55); height, 20 ft $3\frac{1}{3}$ in (6,18 m); wing area, 697·8 sq ft (64,80 m^2).
Accommodation: Pilot only or (F-106B) pilot and instructor or observer in tandem.
Armament: Internal weapons bay for AAMs; typical load comprises two AIR-2A or AIR-2B Genie unguided rockets and four AIM-4E or AIM-4G Falcon AAMs. One 20-mm M-61 multi-barrel rotary cannon being introduced as standard from 1973.
Status: First trials aircraft flown on 26 December 1956; first F-106B flown on 9 April 1958. Deliveries began in July 1959. Production totals, F-106A, 277; F-106B, 63, completed in 1960.
Notes: The F-106 Delta Dart was an outgrowth of the F-102 (see page 10), based on the same delta wing but with a redesigned fuselage containing the uprated J75 engine and improved systems and equipment. The two production versions differed from each other only in the cockpit, the F-106B having a second seat, with some reduction in fuel capacity, and being used at operational squadron level either as a proficiency trainer or as a two-seat interceptor. The F-106As and Bs entered service with Air (now Aerospace) Defense Command of the USAF to operate within the Semi-Automatic Ground Environment defense system over North America. Since their introduction they have undergone a number of modification programmes; one of these introduced underwing drop tanks and flight refuelling equipment, and another, initiated in 1973, includes installation of an M-61 multi-barrel cannon in the weapons bay in a semi-retractable installation for added dog-fight capability, in conjunction with a new SnapShoot gunsight. Front-line service with the ADC is projected to continue until the late 'seventies.

DASSAULT ETENDARD

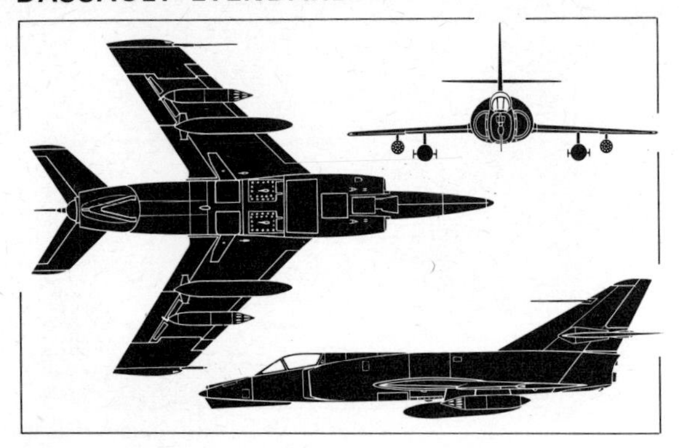

The Etendard IVM, illustrated by the photograph above and the general arrangement silhouette (left) is currently Aéronavale's standard shipboard strike fighter

Country of Origin: France.
Type: Carrier-based fighter and tactical reconnaissance aircraft.
Power Plant: One 9,700 lb st (4 400 kgp) SNECMA Atar 8B or (Super Etendard) approx 11,000 lb st (5 000 kgp) Atar 8K-50 turbojet.
Performance: (IVM) Max speed, 673 mph (1 083 km/h) at 36,090 ft (11 000 m), 683 mph (1 099 km/h) at sea level; initial rate of climb, 19,685 ft/min (100 m/sec); time to reach 42,000 ft (12 800 m) 6 min; service ceiling, 50,850 ft (15 500 m); tactical radius 186 mls (300 km) at sea level and 435 mls (700 km) at 42,000 ft (12 800 m); range with two 132 Imp gal (600 l) tanks, 1,750 mls (2 816 km) at 510 mph (820 km/h) at 36,000 ft (10 970 m).
Weights: Empty, 12,786 lb (5 800 kg); max catapult weight, 19,840 lb (9 000 kg); max take-off, 22,486 lb (10 200 kg).
Dimensions: Span, 31 ft 6 in (9,60 m); length, 47 ft 3 in (14,40 m); height, 14 ft 2 in (4,30 m); wing area, 312 sq ft (29,0 m²).
Accommodation: Pilot only.
Armament: (IVM) Two 30-mm DEFA cannon in fuselage. Four underwing pylons with max capacity of 3,000 lb (1 360 kg) of bombs, rockets, fuel tanks etc.
Status: Prototype Etendard IV-01 first flown 24 July 1956; prototype IVM-01 flown 21 May 1958; first pre-production (IVM-02) flown in 1959; second pre-production (IVB) flown on 2 December 1959; sixth pre-production (IVP-07) flown on 19 November 1960; first production IVM flown in July 1961. Production totals, prototype, 1; pre-production, 6; IVM, 69; IVP, 21; Super Etendard, 100 ordered August 1973.
Notes: The Etendard evolved from a Dassault private venture submission for a NATO tactical strike fighter requirement (won by the Fiat G.91, page 7), and entered production in 1960 to fulfil the French *Aéronavale's* need for a carrier-based strike fighter. For service aboard the *Foch* and *Clémenceau*, *Flotilles* 11F and 17F maintain a front-line strength of 12 Etendard IVMs each, while 15F serves as a training unit and 16F operates the Etendard IVP version with cameras replacing the AIDA nose radar and two cannon. In 1973 *Aéronavale* selected the Super Etendard, with uprated Atar 8K-50 engine of 11,000 lb (5,000 kg) and a new radar, as the Etendard replacement for 1976/77. A modified Etendard commenced trials in October 1974 as a Super Etendard prototype.

DASSAULT MYSTÈRE IVA

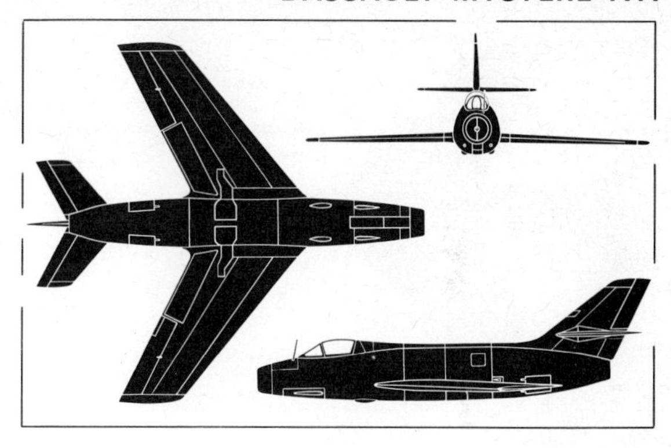

The photo above illustrates a Mystère IVA of the Israeli air arm, Heyl Ha'Avir, and the general arrangement silhouette (right) illustrates the standard production model

Country of Origin: France.
Type: Interceptor fighter.
Power Plant: One 7,716 lb st (3 500 kgp) Hispano-Suiza Verdon 350 turbojet.
Performance: Max speed, 696 mph (1 120 km/h) at sea level and 615 mph (990 km/h) at 39,370 ft (12 000 m); long-range cruising speed 510 mph (820 km/h); initial rate of climb, 8,860 ft/min (45,0 m/sec); service ceiling, 49,200 ft (15 000 m); range (clean), 570 mls (917 km); range with two drop tanks, 820 mls (1 319 km); ferry range, 1,050 mls (1 690 km).
Weights: Empty, 12,950 lb (5 870 kg); normal loaded, 16,530 lb (7 500 kg); max take-off, 20,950 lb (9 500 kg).
Dimensions: Span, 36 ft 5¾ in (11,12 m); length, 42 ft 1¾ in (12,85 m); height, 15 ft 1 in (4,59 m); wing area, 344·5 sq ft (32,0 m²).
Accommodation: Pilot only.
Armament: Two 30-mm DEFA cannon in front fuselage. Provision in fuselage for retractable pack of Matra air-to-air rockets. Four underwing pylons for up to 2,000 lb (900 kg) of bombs or rocket pods.

Status: Prototype (IVA-01) first flown on 28 September 1952; prototype IVB flown on 16 December 1953; prototype IVN flown on 19 July 1954. Production totals: prototype 1; IVA, 421; IVB, three prototypes and 16 pre-production; IVN, 1.
Notes: Evolved, by way of the Mystère I and II, from the straight-wing Ouragan, the Mystère IV was Dassault's first swept-wing fighter built in quantity. It was the subject of an "off-shore" contract placed by the US government to help arm NATO forces, for 225 examples, with additional contracts placed by the French government. Of the total built, 110 were supplied to the Indian Air Force, 60 to Israel and the remainder to *l'Armée de l'Air.* During 1973, examples were flying in the service of all three nations but were close to being finally replaced by more up-to-date types in each case. The first 50 supplied to *l'Armée de l'Air* were powered by the 6,280 lb st (2 850 kgp) Rolls-Royce Tay 250, a switch then being made to the French development of that engine named the Verdon. Final French Air Force units to fly the Mystère IVA were two *Escadrons* of the 7e *Escadre de Chasse,* each comprising two *Escadrilles,* these aircraft having been modified since delivery to have Martin-Baker in place of SNCASO ejection seats.

DASSAULT SUPER-MYSTÈRE B2

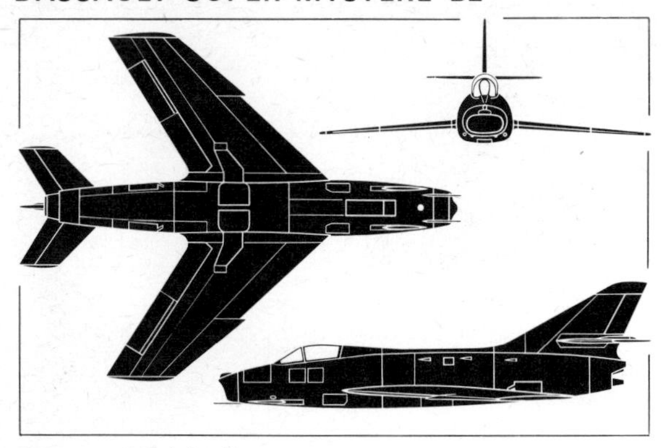

Both photograph (above) and general arrangement silhouette (left) illustrate the Super-Mystère B2, the photo depicting a fighter of this type in service with Israel

Country of Origin: France.
Type: Interceptor and tactical strike fighter.
Power Plant: One 7,495 lb st (3 400 kgp) dry and 9,920 lb st (4 500 kgp) with reheat SNECMA Atar 101G turbojet.
Performance: Max speed, 743 mph (1 200 km/h) at 40,000 ft (12 192 m) and 686 mph (1 104 km/h) at sea level; max cruising speed, 620 mph (998 km/h) at 40,000 ft (12 192 m); speed for best range, 560 mph (901 km/h) at 36,090 ft (11 000 m); initial rate of climb, 17,500 ft/min (89,0 m/sec); service ceiling, 55,750 ft (17 000 m); range (clean) 540 mls (870 km) at 36,000 ft (10 973 m); range with two 180 Imp gal tanks, 730 mls (1 175 km).
Weights: Empty, 15,400 lb (6 985 kg); normal loaded, 19,840 lb (9 000 kg); max overload, 22 046 lb (10 000 kg).
Dimensions: Span, 34 ft 5¾ in (10,50 m); length, 46 ft 1¼ in (14,04 m); height, 14 ft 10¾ in (4,53 m); wing area, 377 sq ft (35,0 m²).
Accommodation: Pilot only.
Armament: Two 30-mm DEFA cannon and internal stowage for 35 SNEB Type 22 68-mm air-to-air rockets in Matra launcher. Two underwing pylons for up to 2,000 lb (908 kg) of bombs, Matra M.116E rocket pods each containing 19 68-mm missiles, two clusters of 5-in (12,7-cm) HVARs, fuel tanks etc.
Status: Prototype (B1) first flown on 2 March 1955; first pre-production (B2) flown on 15 May 1956; first production (B2) flown on 26 February 1957; prototype B4 flown on 9 February 1958. Production, 180 completed in 1959.
Notes: The Super-Mystère B2 derived from the Mystère IVA (page 13), from which it differed primarily in having a more powerful engine and a wing of thinner section, bestowing upon it a level supersonic capability—the first such European design to reach production. The prototype had an Avon RA 7R engine, an Atar being substituted in pre-production and production models. Initial production contracts for the Super-Mystère were cut back owing to the appreciably higher performance offered by the Mirage III. The Super-Mystère B4 had the uprated Atar 9 but did not enter production. Israel acquired 24 Super-Mystère B2s and still had one squadron in use in 1973. In *l'Armée de l'Air*, two *Escadres de Chasse*, the 10e and 12e, each with two *Escadrons*, still flew the Super-Mystère B2 in 1973, although re-equipment with the Mirage F was under way by the end of the year.

DASSAULT-BREGUET MIRAGE III

The photograph above depicts a Mirage III-EBR of the Brazilian Air Force and the general arrangement silhouette (right) depicts the standard Mirage III-E of the Armée de l'Air

Country of Origin: France.
Type: Interceptor and close-support fighter, recce and trainer.
Power Plant: One (III-C,B) 9,370 lb st (4 250 kgp) dry and 13,225 lb st (6 000 kgp) with reheat SNECMA Atar 9B or (III-E, O, R and S) 9,436 lb st (4 280 kgp) dry and 13,670 lb st (6 200 kgp) with reheat Atar 9C turbojet plus provision for one 3,307 lb st (1 500 kgp) SEPR 844 rocket engine.
Performance: (IIIE) Max speed, 850 mph (1 370 km/h) at sea level and 1,460 mph (2 350 km/h) at 39,375 ft (12 000 m) or Mach 2·2; typical cruising speed at 36,090 ft (11 000 m), 593 mph (954 km/h) or Mach 0·9; time to reach 36,090 ft (11 000 m), 3 min; service ceiling at Mach 1·8, 55,775 ft (17 000 m); combat radius, ground attack mission, 745 mls (1 200 km).
Weights: Empty (IIIB) 13,820 lb (6 270 kg) (III-E) 15,540 lb (7 050 kg) (III-R) 14,550 lb (6 600 kg); max take-off (III-B) 26,455 lb (12 000 kg) (III-E,R) 29,760 lb (13 500 kg).
Dimensions: Span, 27 ft 0 in (8,22 m); length (III-E) 49 ft 3½ in (15,03 m) (III-B) 50 ft 6¼ in (15,40 m) (III-R) 50 ft 10¼ in (15,50 m); height, 13 ft 11½ in (4,25 m); wing area 375 sq ft (34,85 m²).
Accommodation: Pilot only or (III-B,D) two in tandem.

Armament: Two 30-mm DEFA cannon in fuselage. Fuselage centre-line pylon carries one AAM or two 1,000-lb (454-kg) bombs or one ASM; two underwing pylons each carry one 1,000-lb (454-kg) bomb or rocket pods, AAMs ASMs or fuel tanks.
Status: Prototype (III-001) first flown 17 November 1956; first III-A flown on 12 May 1958; prototype III-B flown on 20 October 1959; first production III-B flown on 19 July 1962; first production III-C flown on 9 October 1960; prototype III-E on 5 April 1961; prototype III-R flown on 31 October 1961; first Australian assembled III-O flown 16 November 1963; first Swiss-assembled III-S flown 28 October 1965. Three prototypes, 16 pre-production and 1,170 production models built or on order, plus 48 III-OF and 50 III-OA built in Australia and 34 III-S in Switzerland.
Notes: Mirage III-C is basic interceptor, III-E is ground attack version, III-B and III-D are two-seat operational trainers and III-R is reconnaissance version, all in service with French *Armée de l'Air*. Export customers include: Brazil (III-DBR, EBR); Israel (III-BJ, CJ); Lebanon (III-BL, EL); South Africa (III-BZ, CZ, DZ, EZ, RZ); Argentina (III-DA, EA); Pakistan (III-DP, EP, RP); Spain (III-DE, EE); Australia (III-D, O) and Switzerland (III-C, BS, RS and S).

DASSAULT-BREGUET MIRAGE 5

The photograph above depicts a Mirage 5-COA of the Colombian Air Force and the general arrangement silhouette illustrates the standard production Mirage 5

Country of Origin: France.
Type: Ground-attack fighter.
Power Plant: One 9,436 lb st (4 280 kgp) dry and 13,670 lb st (6 200 kgp) with reheat SNECMA Atar 9C turbojet.
Performance: Max speed (clean) 835 mph (1 335 km/h) or Mach 1·1 at sea level, 1,386 mph (2 230 km/h) or Mach 2·1 at 29,370 ft (12 000 m); typical cruising speed, 594 mph (956 km/h) at 36,090 ft (11 000 m); time to climb to 36,090 ft (11 000 m) at Mach 0·9, 3 mins, and to 49,210 ft (15 000 m) at Mach 1·8 6 min 30 sec; combat radius with 2,000 lb (907 kg) bomb load (hi-lo-hi profile) 805 mls (1 300 km).
Weights: Empty equipped, 14,550 lb (6 600 kg); max loaded, 29,760 lb (13 500 kg).
Dimensions: Span, 26 ft $11\frac{1}{2}$ in (8,22 m); length, 51 ft $0\frac{1}{4}$ in (15,55 m); height, 13 ft $11\frac{1}{2}$ in (4,25 m); wing area, 375·12 sq ft (34·85 m^2).
Accommodation: Pilot only or (5D) two in tandem, dual control.
Armament: Two 30-mm DEFA 5-52 cannon in base of intake fairings; seven external ordnance stations (three under fuselage, four under wings) with maximum capacity (weapons and fuel) of 9,260 lb (4 200 kg).
Status: Prototype first flown on 19 May 1967. Milan S-01 flown on 29 May 1970. First 5-BA flown on 6 March 1970. Production total, over 300, including, for Belgium, 27 5-BA, 63 5-BR and 16 5-BD; for Pakistan, 28; for Peru, 22 5-P and two 5-PD; for Colombia, 14 5-COA, COD and COR; for Libya, 100 5-D, 5-DE and 5-DR and 10 5-DD; for Abu Dhabi, 14; for Venezuela, 4 5-V and 2 5-VD and for *l'Armée de l'Air*, 50 5-F (built for Israel but not exported).
Notes: The Mirage 5 is the export version of the Mirage IIIE, optimised for the ground attack rôle and with simplified avionics. It is also available with retractable foreplanes to improve the low-speed performance, one prototype (a converted Mirage IIIE) having been flown with this feature as the Milan S-01, primarily to meet Swiss Air Force requirements. The 106 Belgian Mirage 5s were assembled by SABCA in Belgium, incorporating some components of local manufacture. The two-seat training versions carry a "D" suffix after the country indicator in the designation. The Mirage 5-F equips one *escadron* in each of the 3e and 13e *Escadres* at Nancy and Colmar, the other component *escadrons* being equipped with the Mirage IIIE.

DASSAULT-BREGUET MIRAGE F

Both photograph (above) and general arrangement silhouette (right) depict the standard production Mirage F1 optimised for the intercept rôle for the Armée de l'Air

Country of Origin: France.
Type: Multi purpose fighter.
Power Plant: One 11,023 lb st (5 000 kgp) dry and 15,873 lb (7 200 kgp) with reheat SNECMA Atar 9K-50 turbojet.
Performance: Max speed (clean), 915 mph (1 472 km/h) or Mach 1·2 at sea level, 1,450 mph (2 335 km/h) or Mach 2·2 at 39,370 ft (12 000 m); cruising speed for best range, 550 mph (885 km/h) at 29,530 ft (9 000 m); service ceiling, 65,600 ft (20 000 m); range with max external fuel, 2,050 mls (3 300 km), with max external combat load of 8,818 lb (4 000 kg), 560 mls (900 km) and with external combat load of 4,410 lb (2 000 kg) 1,430 mls (2 300 km).
Weights: Empty, 16,314 lb (7 400 kg); loaded (clean), 24,030 lb (10 900 kg); max take-off, 32,850 lb (14 900 kg).
Dimensions: Span, 27 ft 6¾ in (8,40 m); length, 49 ft 2½ in (15,000 m); height, 14 ft 9 in (4,50 m); wing area, 269·098 sq ft (25 m²).
Accommodation: Pilot only.
Armament: Two 30-mm DEFA cannon in forward fuselage and (intercept rôle) two Matra 550 Magic or AIM-9 Sidewinder AAMs at wingtips and three Matra 530 AAMs on hardpoints under fuselage centre line and wings or (ground attack rôle) up to 14 bombs or various combinations of rocket pods, fuel tanks, etc, on centreline and six wing harpoints.
Status: First prototype flown 23 December 1966. First of three pre-production models flown 20 March 1969. First production model flown 15 February 1973. Initial contracts placed by *Armée de l'Air* for 85 and by Spanish and South African Air Forces for 15 and 16 respectively, the former having options on a further 18 and the latter to receive another 32 (assembled by Atlas).
Notes: The Mirage F1 was evolved on a French government contract as a replacement for the Mirage III in the interceptor rôle in *l'Armée de l'Air*. It entered production in 1969 and deliveries began in mid-1973, with total procurement of 105 planned. Spain has an option on 18 in addition to 15 on order and South Africa holds a licence for production by Atlas Aircraft Corporation following acquisition of an initial batch from France. Under development is an improved version (referred to as the Mirage F1E and F1 International) with the SNECMA M53 engine, rated at 12,380 lb st (5 615 kgp) dry and 18,646 lb (8 458 kgp) with reheat, the first prototype having flown on 22 December 1974.

DE HAVILLAND VENOM

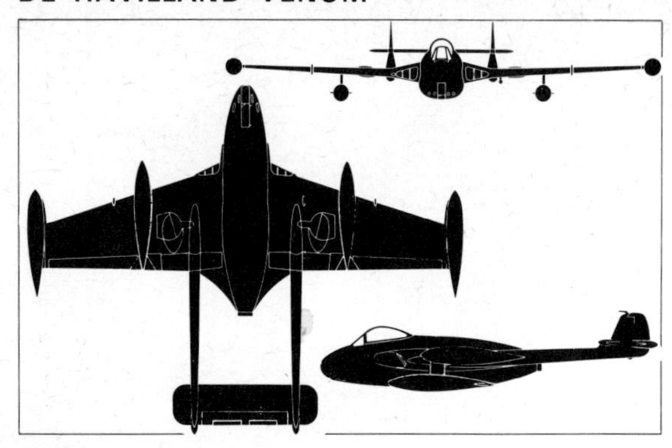

The photograph above depicts a Swiss-manufactured Venom FB Mk 50 (with a Mk 4 tail) and the general arrangement silhouette (left) illustrates the Venom FB Mk 4

Country of Origin: United Kingdom.
Type: Fighter-bomber.
Power Plant: One 4,850 lb st (2 200 kgp) de Havilland Ghost 103 turbojet.
Performance: (FB Mk 4) Max speed, 597 mph (961 km/h) at sea level, 560 mph (901 km/h) at 20,000 ft (6 096 m) and 530 mph (853 km/h) at 45,000 ft (13 716 m); initial rate of climb, 7,230 ft/min (36,7 m/sec); patrol duration (with allowance for take-off, climb, 10-min combat and descent), 1·35 hrs at 202 mph (325 km/h) at sea level, 2·45 hrs at 391 mph (629 km/h) at 40,000 ft (12 192 m).
Weights: Take-off with tip tanks, bombs and rockets, 15,310 lb (6 945 kg).
Dimensions: Span, 41 ft 8 in (12,7 m); length, 31 ft 10 in (9,70 m); height, 6 ft 2 in (1,88 m); wing area, 279·8 sq ft (26,0 m^2).
Accommodation: Pilot only.
Armament: Four 20 mm cannon in nose. Two underwing pylons with 1,000 lb (454 kg) capacity each plus provision for eight 60 lb (27,2 kg) rockets.
Status: Prototype first flown on 2 September 1949; prototype FB Mk 4 first flown on 29 December 1953; Venom 2 prototype flown on 22 August 1950; NF Mk 3 first flown 22 February 1953; first production FAW Mk 20 flown on 27 March 1953; first production FAW Mk 21 flown on 6 June 1957. Production total in the UK, 1,143 including 775 FB Mk 1, NF Mk 2, NF Mk 3 and FB Mk 4 for RAF and 256 FAW Mk 20, 21, 22 and 53 for RN and RAN, 19 FB Mk 50 for Iraq and 22 FB Mk 4 for Venezuela. In addition, Sud-East in France built 117 Aquilon (Sea Venoms) and FFA in Switzerland built 150 FB Mk 1 and 100 FB Mk 4 (all as FB Mk 50).
Notes: By 1973, all the two-seat Venom night fighters and Sea Venoms had been retired, but Switzerland retained a substantial number of the 250 Venom 1s and 4s built under licence between 1951 and 1954, and these are expected to remain in service with eleven ground-attack *Fliegerstaffeln* until the end of the 'seventies. These two types differed significantly only in respect of the tail unit, the Mk 4 having later "kidney-shaped" rudders to improve handling at low speeds. Swiss Venoms have now been fitted with additional avionic equipment to increase their effectiveness in the ground attack rôle. A handful of the British-built Venom FB4s supplied to Venezuela in 1956 were withdrawn by the beginning of the 'seventies.

GENERAL DYNAMICS (CONVAIR) F-111

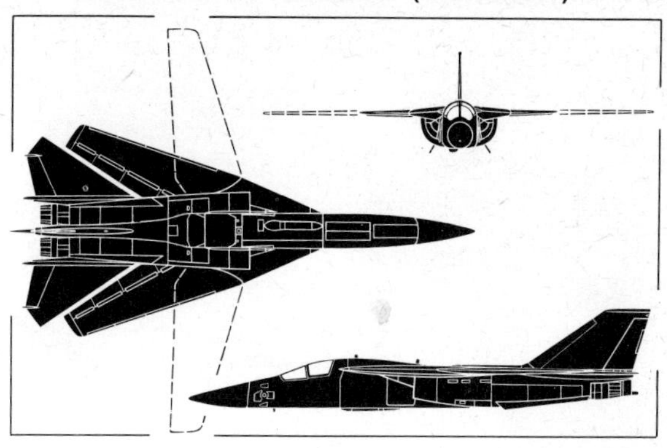

The photograph above depicts an F-111A of the 4410th Special Operations Training Group and the general arrangement silhouette (right) illustrates the F-111E

Country of Origin: USA.
Type: Tactical strike fighter.
Power Plant: Two Pratt & Whitney (F-111A, C) 18,500 lb st (8 390 kgp) with reheat TF30-P-3 (F-111D, E) 19,600 lb st (8 890 kgp) with reheat TF30-P-9 or (F-111F) 25,000 lb st (11 340 kgp) TF30-P-100 turbofans.
Performance: (F-111E) Max speed, 865 mph (1 390 km/h) or Mach 1·2 at sea level, 1,650 mph (2 655 km/h) or Mach 2·5 at 40,000 ft (12 190 m); service ceiling, over 60,000 ft (18 300 m); tactical radius, with 16,000-lb (7 257 kg) combat load for hi-lo-hi mission profile, 1,500 mls (2 415 km); ferry range with max internal fuel, 3,800 mls (6 115 km).
Weights: (F-111E) Empty equipped, 47,500 lb (23 525 kg); normal take-off, 74,000 lb (33 566 kg); max overload, 91,500 lb (41 504 kg).
Dimensions: Span (spread, except F-111C) 63 ft 0 in (19,20 m); (F-111C) 70 ft 0 in (21,34 m); span (swept-back, except F-111C), 31 ft 11⅓ in (9,74 m); (F-111C), 33 ft 11 in (10,34 m); length, 73 ft 6 in (22,40 m); height, 17 ft 1⅓ in (5,22 m).
Accommodation: Pilot and observer side-by-side in emergency escape module.
Armament: One 20-mm M-61A1 multi-barrel cannon internally, optionally replaced with two 750-lb (340-kg) bombs. Four fixed and four pivotting underwing pylons have total capacity of about 30,000 lb (13 608 kg) bombs, ASMs, rockets or fuel tanks.
Status: First development batch F-111A flown on 21 December 1964; first development batch F-111B (cancelled US Navy variant) flown on 18 May 1965; first production F-111B flown on 29 June 1968; RF-111A prototype flown on 17 December 1967. Production totals, F-111A development batch, 18 (including RF-111A prototypes); F-111A production, 141; F-111B development batch, 5; F-111B production, 2; F-111C, 24; F-111D, 96; F-111E, 94; F-111F, 82; F-111K/YF-111A, 2. Production completed end 1973.
Notes: The world's first swing-wing aircraft to enter service, the F-111 is used by four wings of Tactical Air Force (one each with A, D, E and F models); model differences are concerned with intake design, engine thrust and avionics. The F-111C for RAAF has bigger wing originally designed for abortive F-111B programme for US Navy and used also by FB-111 (see page 66). F-111K was projected version for RAF; only two built before cancellation.

GRUMMAN F-14A TOMCAT

The photograph above depicts an F-14A Tomcat of US Navy squadron VF 124 and the general arrangement silhouette (left) illustrates the standard production Tomcat

Country of Origin: USA.
Type: Ship-borne multi-purpose fighter.
Power Plant: Two (approx) 20,600 lb st (9 344 kgp) with reheat Pratt & Whitney TF30-P-412 turbofans.
Performance: (Estimated) Max speed with four AIM-7 Sparrow missiles for intercept missions at approx 57,300 lb (25 990 kg), 910 mph (1 470 km/h) or Mach 1·2 at sea level, 1,564 mph (2 517 km/h) or Mach 2·34 at 40,000 ft (12 190 m).
Weights: (Estimated) Empty, 37,500 lb (17 010 kg); normal take-off (intercept mission), 55,000 lb (24 948 kg); max take-off, 66,200 lb (30 028 kg).
Dimensions: Span (spread) 64 ft 1½ in (19,55 m); span (swept-back), 37 ft 7 in (11,45 m); length, 61 ft 11.9 in (18,90 m), height 16 ft 0 in (4,88 m).
Accommodation: Pilot and observer in tandem.
Armament: One 20 mm M-61A1 rotary cannon and (intercept mission) four AIM-7 Sparrow and four AIM-9 Sidewinder AAMs or six AIM-54 Phoenix and two AIM-9 AAMs.
Status: First of 12 research and development aircraft began flight trials on 21 December 1970, followed by second on 24 May 1971. Deliveries to training squadron VF 124 began mid-1972 and the first two operational squadrons, VF-1 and VF-2, were activated in October 1972. Total projected US Navy procurement, 313, of which 134 had been ordered in five batches up to mid-1973. Prototype F-14B (No 7 F-14A development airframe converted) first flown 12 September 1973. Total of 80 to be procured by Iran.
Notes: The F-14A Tomcat was chosen by the US Navy in January 1969 as its new carrier-based fighter to take the place of the proposed F-111B programme. Total planned procurement was 463 aircraft but increased production costs may cause a revision of the programme beyond the 134 ordered up to 1973. First Navy squadron to fly the Tomcat was VF-124, assigned the task of training crews for operational units, the first two of which, VF-1 and VF-2, were working up in 1973. The seventh of 12 development F-14As was re-engined during 1973 with Pratt & Whitney F401-P-400 turbofans, becoming the prototype F-14B. The Tomcat fire control system allows it to track and engage six independent targets at the same time, and to launch one of its Phoenix missiles at each target simultaneously. Variants proposed during 1973 included the F-14D with simplified avionics and weapons system.

HAL GNAT

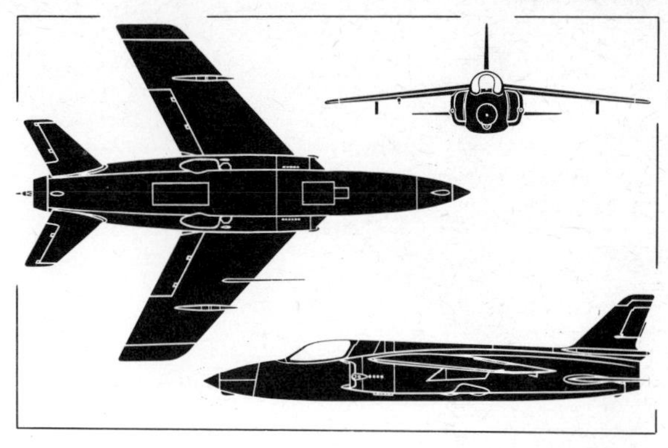

The Hindustan Aeronautics-built Gnat, illustrated by the photograph (above) and the general arrangement silhouette (right), currently equips eight squadrons of the Indian Air Force

Country of Origin: United Kingdom.
Type: Day interceptor and (T Mk 1) advanced trainer.
Power Plant: One 4,520 lb st (2 050 kgp) Bristol Siddeley (HAL-built) Orpheus 701 or (T Mk 1) 4,230 lb st (1 920 kgp) Orpheus 101 turbojet.
Performance: (Gnat 1) Max speed, 695 mph (1 118 km/h) at 20,000 ft (6 096 m); time to climb to 45,000 ft (13 700 m), 5·25 min; service ceiling, 50,000 ft (15 000 m); tactical radius with two drop tanks, 500 mls (805 km).
Performance: (Gnat T Mk 1) Max speed, 636 mph (1 026 km/h) at 31,000 ft (9 450 m); initial rate of climb, 9,850 ft/min (50,0 m/sec); service ceiling, 48,000 ft (14 600 m); range with drop tanks, 1,180 mls (1 900 km).
Weights: Basic weight (T Mk 1), 5,613 lb (2 546 kg); normal loaded (T Mk 1), 8,250 lb (3 742 kg); max take-off (T Mk 1), 9,350 lb (4 240 kg), (interceptor), 6,650 lb (3 010 kg), (tactical fighter), 8,885 lb (4 020 kg).
Dimensions: Span, 22 ft 2 in (6,75 m), (T Mk 1) 24 ft 0 in (7,32 m); length, 29 ft 9 in (9,06 m), (T Mk 1) 31 ft 9 in (9,65 m) over nose probe; height, 8 ft 10 in (2,69 m), (T Mk 1) 10 ft 6 in (3,2 m); wing area, 136·6 sq ft (12,69 m²), (T Mk 1) 175 sq ft (16,26 m²).
Accommodation: Pilot only or (T Mk 1) two in tandem, dual control.
Armament: Two 30-mm Aden guns and wing strong points for two 500-lb (227-kg) bombs or drop tanks.
Status: Prototype Gnat fighter first flown on 18 July 1955; first production aircraft flown 16 May 1956; first HAL-built Gnat flown on 18 November 1959; first T Mk 1 flown on 31 August 1959. Production totals (fighters) development batch, 6; Finland, 12, Yugoslavia, 2, India, 25 plus 15 for local assembly by HAL followed by 200 licence-built. Programme completed mid-1973; T Mk 1, 105.
Notes: Designed by WEW Petter for production by Folland Aircraft Ltd, the Gnat was a lightweight fighter offering high performance at low cost. The Finnish Air Force bought one squadron and India picked the design for local production by Hindustan Aeronautics. Total of 200 Mk 1s had been built by mid-1973, when Mk 2 was under development with fuel tanks in the wings, a re-designed longitudinal control system, updated avionics and uprated Orpheus engine. The slightly longer Gnat T Mk 1 is standard training equipment in the RAF and will remain until 1977.

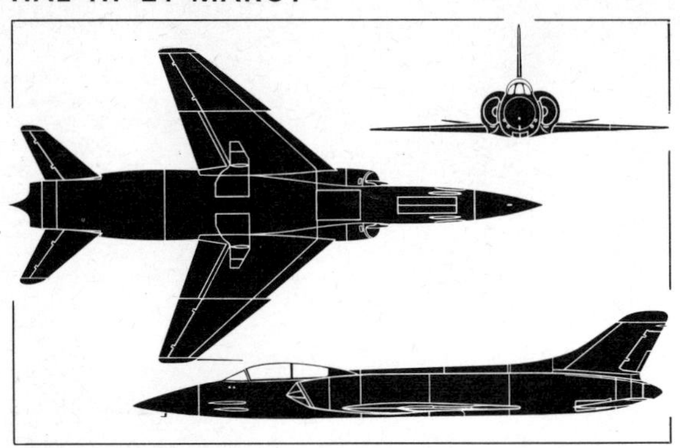

The photograph above depicts a Marut Mk 1 of the Indian Air Force's No 10 "Dagger" Squadron and the general arrangement silhouette (left) illustrates the two-seat Marut Mk 1T

Country of Origin: India.
Type: Fighter and light strike aircraft and (Mk 1T) operational trainer.
Power Plant: Two 4,850 lb st (2 200 kgp) HAL-built Rolls-Royce Bristol Orpheus 703 turbojets.
Performance: Max level speed, 673 mph (1 083 km/h) or Mach 1·02 at 40,000 ft (12 192 m).
Weights: Empty equipped, 13,658 lb (6 195 kg); take-off (clean) 19,734 lb (8 951 kg); max take-off, 24,048 lb (10 908 kg).
Dimensions: Span, 29 ft 6½ in (9,00 m); length, 52 ft 0¾ in (15,87 m); height, 11 ft 9¾ in (3,60 m); wing area, 306·8 sq ft (28,50 m^2).
Accommodation: Pilot or (Mk 1T) two in tandem, dual controls.
Armament: Four 30-mm Aden Mk 2 cannon in forward fuselage and retractable Matra 103 launcher pack for 50 68-mm unguided rockets in lower fuselage. Four wing pylons each with capacity for one 1,000-lb (454-kg) bomb, rocket pods, air-to-ground rocket clusters or similar stores.
Status: First and second prototypes flown on 17 June 1961 and 4 October 1962 respectively; first pre-production HF-24 Mk 1 flown in April 1963; first full production HF-24 Mk 1 flown on 15 November 1967; Mk 1BX EI-300 engine test-bed first flown 29 March 1967; Mk 1A with reheat Orpheus flown September 1966; Mk 1T prototype first flown on 30 April 1970 and second prototype in March 1971. Production quantities include two prototypes, 15 pre-production Mk 1s, approx 100 Mk 1s with production continuing. Pre-production batch of four Mk 1Rs with re-heat Orpheus engines ordered 1973, and delivery of 10 Mk 1Ts to commence 1974.
Notes: The HF-24 is India's first indigenous combat aircraft, design work having begun in June 1957 under the direction of Dr Kurt Tank. A two-seat glider representing the aerodynamic shape of the HF-24 was flown behind a C-47, making 78 flights between 1 April 1959 and 24 March 1960 to help finalise the design. One pre-production HF-24 became the Mk 1BX with a Brandner EI-300 turbojet replacing one of the Orpheus engines and three others were fitted with re-heat systems on their Orpheus engines, one as a Mk 1A and two as Mk 1Rs. The two-seat Mk 1T entered production in 1973 and development of a Mk 3 HF-24 fighter, the HF-73 with Rolls-Royce/Turbo Union RB.199 engines, was projected at beginning of 1975 while production of the Mk 1 continued.

The Hunter, illustrated above and right in its FGA Mk 9 form, still serves with three RAF squadrons for continuation training and more than a dozen foreign air forces

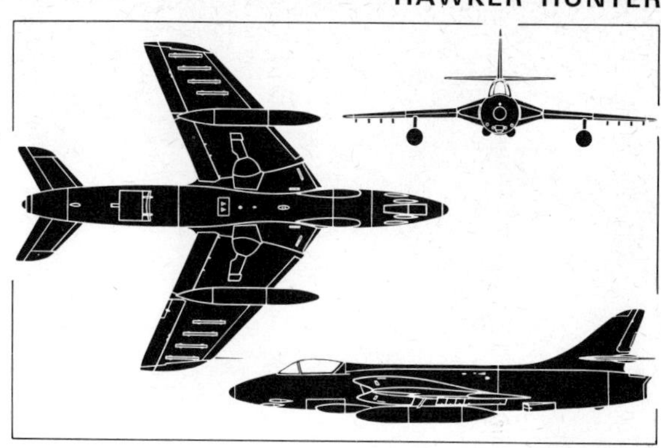

Country of Origin: United Kingdom.
Type: Interceptor, ground attack, reconnaissance fighter and trainer.
Power Plant: One (Mks 7, 8, 11, 12, 50, 51, 52, 53 and 62) 7,575 lb st (3 435 kgp) Rolls-Royce Avon 113, 115, 119, 121A or 122 or (Mks 6, 9, 10, 57, 58, 59, 66, 67, 69, 70–79) 10,150 lb st (4 600 kgp) Avon 203 or 207 turbojet.
Performance: (FGA Mk 9) Max speed, 710 mph (1 144 km/h) at sea level, 620 mph (978 km/h) at 36,000 ft (10 973 m); best range cruise, 460 mph (740 km/h); initial rate of climb, approx 8,000 ft/min (40,7 m/sec); service ceiling, about 50,000 ft (15 250 m); range (clean), 490 mls (789 km); range with four drop tanks, 1,840 mls (2 965 km).
Weights: (FGA Mk 9) Empty, 13,270 lb (6 020 kg); max take-off, 24,000 lb (10 885 kg).
Dimensions: Span, 33 ft 8 in (10,26 m); length, (single-seaters) 45 ft 10½ in (13,98 m) (two-seaters), 48 ft 10½ in (14,90 m); height, 13 ft 2 in (4,26 m); wing area, 349 sq ft (32,43 m²).
Accommodation: (Mks 6, 9, 10, 11, 50, 51, 52, 57, 58, 59, 70, 71, 73, 74, 76, 78) Pilot only or (Mks 7, 8, 12, 53, 62, 66, 67, 69, 72, 75, 77, 79) two side-by-side, dual controls.
Armament: Four (single-seaters) or two (two-seaters) 30 mm Aden cannon in front fuselage; two inner wing pylons with 1,000-lb (454-kg) capacity; two outer wing strongpoints for up to 24 3-in (7,6 cm) rockets.
Status: Hunter prototype first flown 20 June 1951; first production Mk 1 flown 16 May 1953; Mk 4 flown 20 October 1954; Mk 6 flown 25 March 1955; Mk 7 flown 11 October 1957. Production totals, UK single-seaters, 1,426, UK two-seaters, 101, Belgian/Dutch Mk 4, 208 and Mk 6, 237.
Notes: Of the UK production totals, 429 were for direct export; in addition over 520 used Hunters have been refurbished by HSA for export. User nations include RAF (two squadrons Mk 9 and about 30 Mk 7); Rhodesia (Mk 9); Peru (Mk 52 and 62); India (Mk 56, 66, 66D, 66E); Switzerland (Mk 58, 58A); Iraq (Mk 59, 59A, 59B, 69); Jordan (Mk 6, 66B and 73); Lebanon (Mk 66C and 70); Kuwait (Mk 57, 67); Chile (Mk 71, 72); Singapore (Mk 74, 74A, 74B, 75); Abu Dhabi (Mk 76, 76A, 77) and Qatar (Mk 78, 79). Deliveries of refurbished Hunters to India, Switzerland and other countries were continuing in 1974.

LOCKHEED F-104 STARFIGHTER

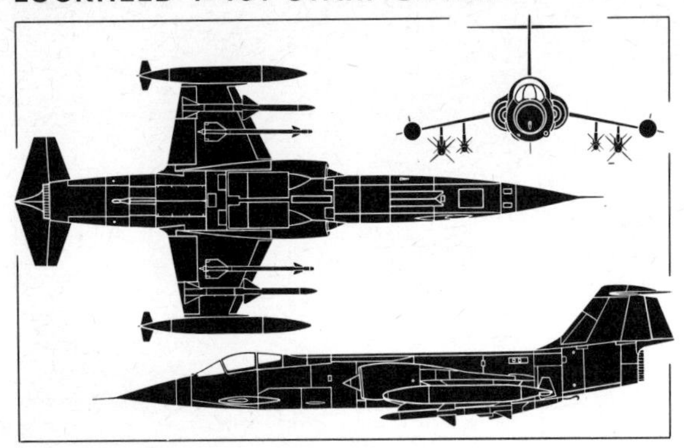

The only current production version of the Starfighter is the F-104S for the Italian Air Force, illustrated above and left, which is being licence-built in Italy by Aeritalia

Country of Origin: USA.
Type: Interceptor, all-weather strike and reconnaissance fighter and operational trainer.
Power Plant: One General Electric (F-104A, B) 14,800 lb st (6 713 kgp) with reheat J79-GE-3B, (F-104C, D, F) 15,800 lb st (7 166 kgp) with reheat J79-GE-7 or (F-104G, 104-DJ, CF-104) J79-GE-11A, (F-104S) 17,900 lb st (8 120 kgp) with reheat J79-GE-19 turbojet.
Performance: (F-104G) Max speed, 1,320 mph (2 124 km/h) at 40,000 ft (12 192 m) and 915 mph (1 473 km/h) at 1,000 ft (305 m); initial rate of climb over 50,000 ft/min (254 m/sec); service ceiling, about 55,000 ft (16 764 m); tactical radius (with four drop tanks) 690 mls (1 110 km) at 610 mph (982 km/h).
Weights: (F-104G) Empty, 14,082 lb (6 390 kg); normal take-off (clean), 19,841 lb (9 000 kg); maximum overload, 28,780 lb (13 000 kg).
Dimensions: Span, 21 ft 11 in (6,68 m); length, 54 ft 9 in (16,69 m); height, 13 ft 6 in (4,11 m); wing area, 196·1 sq ft (18,22 m²).
Accommodation: Pilot only or (F-104B, D and F, TF-104G, CF-104D, F-104DJ) two in tandem, dual controls.
Armament: One M-61 rotary barrel cannon in forward fuselage and (according to rôle) one AAM at each wing-tip and/or four underwing pylons with total load of 4,000 lb (1 815 kg).
Status: Prototype XF-104 flown on 7 February 1954; F-104A flown on 17 February 1956; F-104B flown on 7 February 1957; F-104G flown on 5 October 1960; CF-104 flown on 28 March 1961; CF-104D flown on 14 June 1961; F-104J flown on 30 June 1961; prototype F-104S flown in December 1966 and first production F-104S on 30 December 1968. Production totals, XF-104, 2; YF-104A, 15; F-104A, 153; F-104B, 26; F-104C, 77; F-104D, 21; F-104F, 30; F-104G, 1,266; TF-104G, 181; F-104J, 210 (by Mitsubishi), F-104DJ, 20 (by Mitsubishi), CF-104 and D, 200 (by Canadair), F-104S, 205 (by Fiat with completion 1975).
Notes: F-104A used by Jordan and Pakistan, F-104G and B by Chinese Nationalist Air Force, F-104J and DJ by JASDF, CF-104 by Canadian Armed Forces. F-104G production includes 977 built by European consortium and 110 by Canadair; user nations are Germany, Netherlands, Italy, Belgium, Denmark, Norway, Turkey and Greece. F-104S production for Italian Air Force to end 1975.

McDONNELL F-101 VOODOO

The photograph above depicts a CF-101B of No 416 Squadron, Canadian Armed Forces, this being externally similar to the F-101B illustrated by the general arrangement silhouette

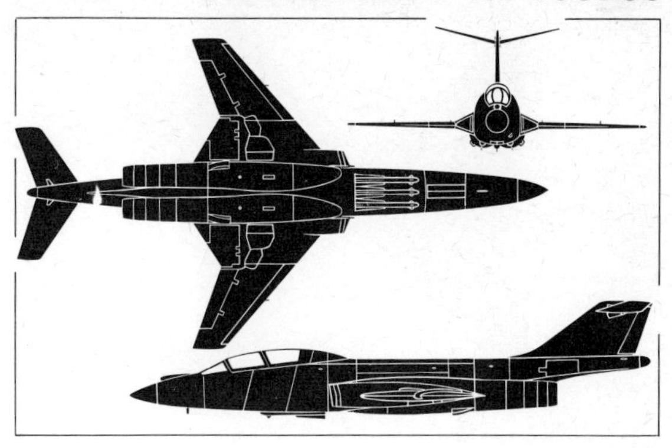

Country of Origin: USA.
Type: All-weather interceptor and reconnaissance aircraft.
Power Plant: Two (F-101B) 11,990 lb st (5 440 kgp) dry and 14,990 lb st (6 800 kgp) with reheat Pratt & Whitney J57-P-53 or -55 or (RF-101C) 10,100 lb st (4 580 kgp) and 14,880 lb st (6 750 kgp) with reheat J57-P-13 turbojets.
Performance: (F-101B) Max speed, 1,220 mph (1 963 km/h) at 40,000 ft (12 192 m) and 716 mph (1 152 km/h) at sea level; initial rate of climb, 17,000 ft/min (5 182 m/sec); service ceiling, 52,000 ft (15 850 m); max range, 1,550 mls (2 500 km/h) clean, 2,200 mls (3 540 km) with two drop tanks.
Weights: (F-101B) Normal loaded, 39,900 lb (18 100 kg); max take-off, 46,700 lb (21 180 kg).
Dimensions: Span, 39 ft 8 in (12,09 m); length (F-101B), 67 ft $4\frac{3}{4}$ in (20,55 m), (RF-101C) 69 ft 3 in (21,10 m); height, 18 ft 0 in (5,49 m); wing area, 368 sq ft (34,2 m^2).
Accommodation: (F-101B) Pilot and observer (TF-101B, TF-101F) two pilots in tandem and (RF-101C) pilot only.
Armament: (F-101B, F-101F) Internal bay accommodating three AIM-4D Falcon AAMs and two AIR-2A Genie missiles under fuselage (F-101C) Four 20 mm cannon in forward fuselage, (RF-101 versions unarmed.)
Status: First F-101A flown on 29 September 1954; first of two YRF-101A prototypes flown on 10 May 1956; first RF-101C flown on 12 July 1957; first F-101B flown on 27 March 1957.
Production quantities: F-101A, 77; F-101C, 47; YRF-101A, 2; RF-101A, 35; RF-101C, 166; F-101B, 480. Production completed 1961.
Notes: F-101A and improved F-101C served with Tactical Air Command; most were converted to RF-101G and RF-101H reconnaissance configuration for Air National Guard, which had three TR Groups so equipped in 1973. RF-101A and RF-101C were camera-equipped production versions and some RF-101Cs served with the Chinese Nationalist Air Force. Two-seat F-101B interceptor has been a mainstay of US continental air defence since 1961 and is used by six ANG Fighter Groups in ADC, plus three Canadian squadrons. The latter used 56 CF-101Bs and 10 dual-control CF-101Fs. From 1970, the 58 surviving Canadian Voodoos were exchanged for 66 late-model USAF F-101Bs with more advanced MG-13 fire control and MB-5 autopilot.

McDONNELL DOUGLAS F-4 PHANTOM II

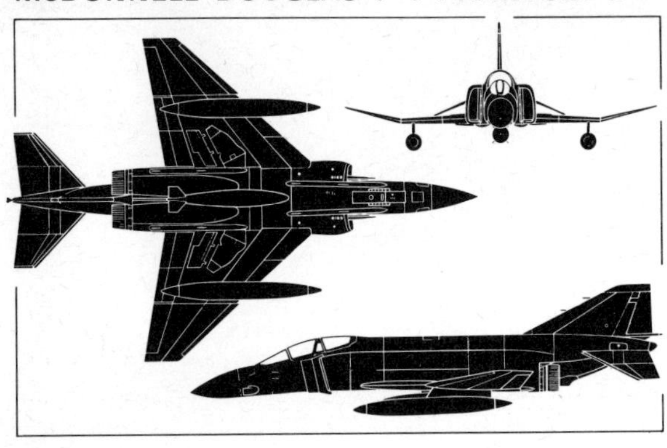

The photograph above depicts a Phantom FGR Mk 2 (F-4M) of No 14 Sqdn, RAF, this version of the Phantom also being illustrated by the general arrangement silhouette (left)

Country of Origin: USA.
Type: Tactical strike fighter and reconnaissance aircraft.
Power Plant: Two (F-4B, G, N, RF-4B) 17,000 lb st (7 710 kgp) with reheat J79-GE-8 or (F-4C, D, RF-4C) J79-GE-15, (F-4E, EJ, F, RF-4E) 17,900 lb st (8 120 kgp) J79-GE-17 or (F-4J) J79-GE-10 turbojets; (Phantom FG Mk 1, FGR Mk 2) 20,315 lb st (9 300 kgp) with reheat Rolls-Royce Spey 202 turbofans.
Performance: (F-4E) Max speed, with external stores, 910 mph (1 464 km/h) or Mach 1·2 at 1,000 ft (305 m), 1,500 mph (2 414 km/h) or Mach 2·27 at 40,000 ft (12 190 m); initial rate of climb, about 30,000 ft/min (152 m/sec); service ceiling, over 60,000 ft (19 685 m); tactical radius (with eight AAMs) 140 mls (225 km); (with four 1,000-lb/454-kg bombs, four AAMs and one drop tank) 656 mls (1 056 km).
Weights: Empty equipped, 30,425 lb (13 800 kg); typical loaded, 51,810 lb (21 500 kg); fully-loaded (eight AAMS, one drop tank) 58,000 lb (26 308 kg); max overload, 60,630 lb (27 502 kg).
Dimensions: Span, 38 ft 4¾ in (11,70 m); length (F-4A, B, C, D, G, J, N), 58 ft 1½ in (17,76 m), (RF-4B, C, E, F-4E), 62 ft 10½ in (19,19 m); (Phantom 1 and 2), 57 ft 7 in (17,55 m); height, 16 ft 3⅓ in (4,96 m); wing area, 530 sq ft (49,2 m²).
Accommodation: Two pilots or pilot and observer.
Armament: Semi-recessed installation of four AIM-7 Sparrow AAMs under fuselage and four wing pylons for four AIM-9 Sidewinder or up to 16,000 lb (7 257 kg) of external stores plus (F-4E, F-4EJ, F-4F) one 20-mm M61-1 rotary cannon in nose.
Status: Prototype (XF4H-1) first flown 27 May 1958; RF-4B flown 12 March 1965; F-4C on 27 May 1963; YRF-4C 9 August 1963, production RF-4C on 18 May 1964; F-4D on 8 December 1965; F-4E on 30 June 1967; F-4EJ on 14 January 1971; F-4J in May 1966; YF-4K on 27 June 1966; YF-4M on 17 February 1967. Production total about 5,000 by mid-1973 including F-4A, 47; F-4B, 635; F-4C, 583; F-4D, 825; RF-4C, over 500; RF-4E (*Luftwaffe*) 88; plus 175 F-4F on order for *Luftwaffe* and 128 F-4EJ for JASDF (produced by Mitsubishi).
Notes: Twelve F-4G, 178 F-4N and some QF-4B drones converted from F-4Bs for USN. USAF supplied 18 F-4Es to South Korea and 36 F-4Cs to Spain. Other exports include 168 F-4E/RF-4E to Israel, 64 F-4Ds to Iran, 36 F-4E to Greece, 40 F-4E to Turkey and 14 RF-4EJ to Japan (for delivery 1974–75).

McDONNELL DOUGLAS F-15 EAGLE

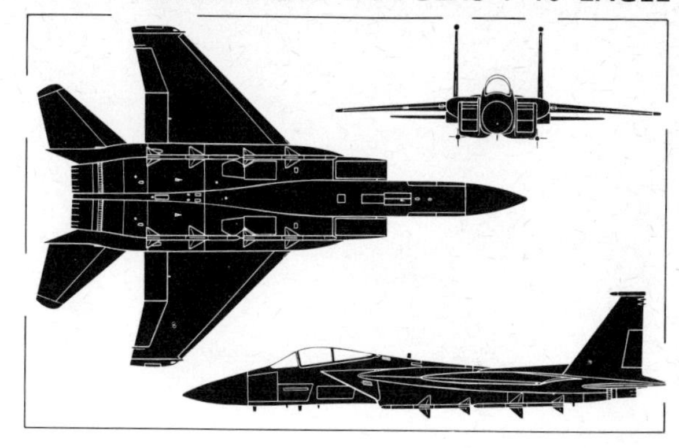

Both the photograph (above) and the general arrangement silhouette (right) depict the pre-production F-15A Eagle participating in the manufacturer's test programme late 1973

Country of Origin: USA.
Type: Air superiority fighter.
Power Plant: Two (approx) 19,000 lb st (8 618 kgp) dry and 27,000 lb st (12 247 kgp) with reheat Pratt & Whitney F100-PW-100 turbofans.
Performance: Max sustained speed (approx) 1,520 mph (2 446 km/h) or Mach 2·3 above 36,000 ft (10 975 m); max short-period dash speed, 1,650 mph (2 655 km/h) or Mach 2·5; max low altitude speed (approx) 915 mph (1 470 km/h) or Mach 1·2 at 1,000 ft (305 m); initial rate of climb, over 50 000 ft/min (254 m/sec).
Weights: Approx max loaded weight (air superiority mission), 40,000 lb (18 144 kg); max take-off weight, 56,000 lb (25 400 kgp).
Dimensions: Span, 42 ft 9½ in (13,04 m); length, 63 ft 9½ in (19,44 m); height, 18 ft 7¼ in (5,67 m).
Accommodation: Pilot only or (TF-15) two in tandem.
Armament: One (initial production models) 20-mm M-61A-1 rotary cannon or (ultimate version) 25-mm Philco-Ford GAU-7/A rotary cannon using caseless ammunition. Provision in semi-recessed housings under fuselage for four Raytheon AIM-7F Sparrow and four AIM-9L Sidewinder AAMs.
Status: First of 20 development and test F-15As flown on 27 July 1972; second aircraft flown on 26 September 1972 and third aircraft on 4 November 1972. Eight of test batch for use in service evaluation and 12 in contractor's test programme. Production aircraft for delivery starting in 1975. USAF procurement plans for total of 729 aircraft over six years; first production batch of 30 ordered in FY 73 and second batch of 77 in FY 74. First TF-15 flown 7 July 1973.
Notes: The F-15A Eagle was developed to provide USAF's Tactical Air Command with an air superiority fighter for service in the period 1975-1985. The thrust-to-weight ratio of approximately 2 to 1 allows the Eagle to climb supersonically and to accelerate from subsonic cruise to Mach 1·5 in less than one minute. Production plans include a quantity of TF-15 two-seat trainers, with two prototypes of this version in the initial production batch of 20 aircraft. The Eagle's structure comprises 35·5 per cent aluminium, 26·7 per cent titanium and 37·8 per cent composites and other materials. It is allegedly the most highly manoeuvrable warplane in its performance category, and its missions include fighter sweep, escort and combat air patrol.

MIKOYAN-GUREVICH MIG-17F

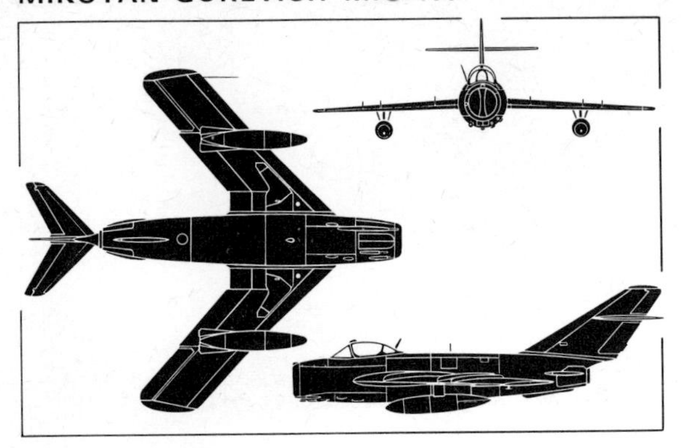

Although obsolescent for a number of years, the MiG-17 remains in service with a number of air arms in its -17F version illustrated above (in Nigerian service) and left

Country of Origin: USSR.
NATO Code Name: *Fresco-C.*
Type: Interceptor fighter.
Power Plant: One 4,732 lb (2 600 kg) dry and 7,452 lb (3 380 kg) with reheat Klimov VK-1F turbojet.
Performance: Max speed (clean), 711 mph (1 145 km/h) at 9,840 ft (3 000 m), 702 mph (1 130 km/h) at 16,400 ft (5 000 m), 626 mph (1 071 km/h) at 32,810 ft (10 000 m); max climb rate, 12,795 ft/min (65 m/sec); ceiling (without reheat), 49,540 ft (15 100 m), (with reheat), 54,460 ft (16 600 m); range (internal fuel), 422 mls (680 km) at 16,400 ft (5 000 m), 603 mls (970 km) at 32,810 ft (10 000 m); range with two 88 Imp gal (400 l) drop tanks, 640 mls (1 030 km) at 16,400 ft (5 000 m), 913 mls (1 470 km) at 32,810 ft (10 000 m).
Weights: Loaded (clean), 11,773 lb (5 340 kg); take-off with drop tanks, 13,380 lb (6 069 kg).
Accommodation: Pilot only.
Armament: Three 23-mm NR-23 cannon.
Status: First prototype flown January 1950. Licence manufacture undertaken in Poland as LIM-5P and in China as F-4. Production completed in Soviet Union in early 'sixties.
Notes: The MiG-17, which entered service with the Soviet Air Forces in 1952, resulted from the extensive redesign of the MiG-15-bis, a lengthened aft fuselage and an entirely new wing being adopted. The initial production MiG-17 day fighter (*Fresco-A*) retained the cannon armament of the MiG-15bis (i.e., one 37-mm N-37 and two 23-mm NR-23s) as did also the first limited all-weather variant, the MiG-17P (*Fresco-B*) with nose-mounted Izumrud AI radar, power being provided by the non-afterburning VK-1 engine of 5,952 lb (2 700 kg) thrust. Production of both variants quickly gave place to the MiG-17F (*Fresco-C*) and MiG-17PF (*Fresco-D*) day and limited all-weather versions respectively, these being powered by the afterburning VK-1F. The final production derivative, the limited all-weather MiG-17PFU (*Fresco-E*), had cannon armament replaced by a quartet of small beam-riding Alkali missiles. The Polish production version of the MiG-17F, the LIM-5P, was eventually adapted for the fighter-bomber rôle as the LIM-5M with large braking chute housing beneath the vertical tail surfaces, RATOG and additional ordnance pylons. The MiG-17F remains in widespread service.

MIKOYAN MIG-19

The photograph above depicts a Chinese built MiG-19SF in Pakistani service, the -19SF version of the fighter, also illustrated by the silhouette (right), still serving in numbers

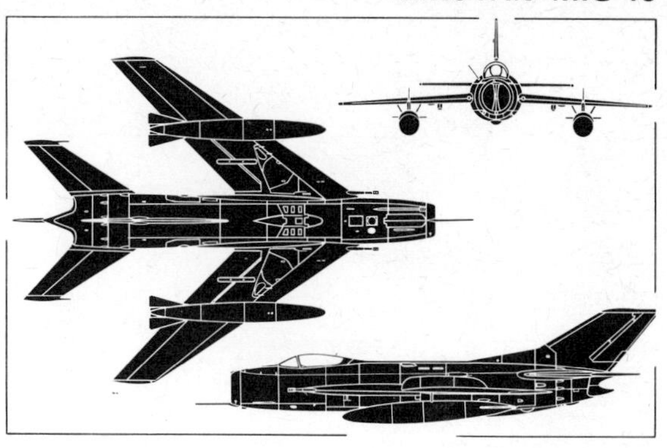

Country of Origin: USSR.
NATO Code Name: *Farmer-C.*
Type: Interceptor fighter.
Power Plant: Two 5,732 lb (2 600 kg) dry and 7,165 lb (3 250 kg) Tumansky RD-9B turbojets.
Performance: Max speed, 902 mph (1 452 km/h) at 32,810 ft (10 000 m); range cruise, 590 mph (950 km/h) at 32,810 ft (10 000 m); initial climb, 22,640 ft/min (115 m/sec); service ceiling, 58,725 ft (17 900 m); normal range, 864 mls (1 390 km) at 45,930 ft (14 000 m); max range (with two 176 Imp gal/800 l drop tanks), 1,367 mls (2 200 km).
Weights: Empty, 12,698 lb (5 760 kg); normal loaded, 16,755 lb (7 600 kg); max take-off, 19,180 lb (8 700 kg).
Dimensions: Span, 29 ft $6\frac{1}{2}$ in (9,00 m); length, 42 ft $11\frac{1}{3}$ in (13,09 m); height, 13 ft $2\frac{1}{4}$ in (4,02 m); wing area, 269·098 sq ft (25,0 m^2).
Accommodation: Pilot only.
Armament: Three 30-mm NR-30 cannon and provision for two 550-lb (250-kg) bombs or two 212-mm ARS-212 rocket missiles.
Status: Prototypes flown late 1953 with initial production deliveries to Soviet Air Forces commencing early 1955. Manufacture undertaken in China as F-6. Production completed in Soviet Union in early 'sixties and probably in China in late 'sixties.
Notes: The first Soviet production fighter capable of supersonic speed in level flight, the MiG-19 remains in service in some numbers with the Soviet Air Forces, and the Chinese copy of the SF version serves with the Sino-Communist air arm and the air forces of Pakistan and North Vietnam. The initial production model of the MiG-19 (*Farmer-A*) was powered by two Mikulin AM-5 turbojets rated at 4,850 lb (2 200 kg) and 6,700 lb (3 040 kg) with reheat and had an armament of one 37-mm N-37 and two 23-mm NR-23 cannon. This was supplanted by the MiG-19S (*Farmer-B*) with an all-moving slab-type tailplane, a ventral air brake and trio of NR-30 cannon. The version built in the largest numbers, the MiG-19SF, switched to the RD-9B engine, a limited all-weather version produced simultaneously, the MiG-19PF (*Farmer-D*) introduced Izumrud (*Emerald*) AI radar, the final production development being the MiG-19PM which differed from the PF primarily in having cannon armament deleted and provision for four Alkali beam-riding missiles beneath the wings.

MIKOYAN MIG-21

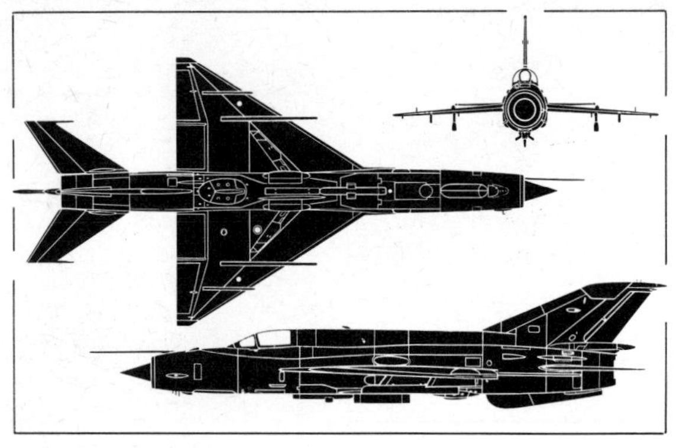

The photograph above depicts a MiG-21F in Yugoslav service and the general arrangement silhouette (left) illustrates the multi-purpose MiG-21MF

Country of Origin: USSR.
NATO Code Name: *Fishbed-J.*
Type: Multi-purpose fighter.
Power Plant: One 11,244 lb (5 100 kg) dry and 14,550 lb (6 600 kg) with reheat Tumansky R-13 turbojet.
Performance: Max speed, 808 mph (1 300 km/h) or Mach 1·06 at 1,000 ft (305 m), 1,386 mph (2 230 km/h) or Mach 2·1 above 36,090 ft (11 000 m); range with max internal fuel, 683 mls (1 100 km); ferry range with max external fuel, 1,118 mls (1 800 km); service ceiling, 59,055 ft (18 000 m).
Weights: Normal loaded (with four K-13A AAMs), 18,078 lb (8 200 kg), (with two K-13As and two 108 Imp gal/490 1 drop tanks), 19,731 lb (8 950 kg); max take-off, 21,605 lb (9 800 kg).
Dimensions: Span, 23 ft 5½ in (7,15 m); length (including probe), 51 ft 8½ in (15,76 m), (without probe), 44 ft 2 in (13,46 m); wing area, 247·57 sq ft (23,0 m²).
Accommodation: Pilot only or (MiG-21U) two with dual controls.
Armament: Twin-barrel 23-mm GSH-23 cannon in fuselage and two or four K-13A *Atoll* infra-red AAMs on wing pylons, or two 1,100-lb (500-kg) or four 550-lb (250-kg) bombs, or four BY-57B pods each housing 16 57-mm rockets.
Status: The MiG-21MF is a multi-purpose derivative of the MiG-21PFM (*Fishbed-F*) interceptor and entered service with the Soviet Air Forces in the late 'sixties. Licence manufacture in India by Hindustan Aeronautics with deliveries of interim MiG-21M 1973.
Notes: The MiG-21MF is current production version of basic MiG-21 first flown in 1955. Initial quantity production model, the MiG-21F (*Fishbed-C*), was simple day fighter with single 30-mm NR-30 cannon with radar ranging, parallel production being undertaken of the limited all-weather MiG-21PF (*Fishbed-D*) with R1L search-and-track radar, the cannon being deleted. Late production MiG-21PF introduced broader-chord vertical tail surfaces, a bullet-type parabrake housing, improved R2L radar and an uprated R-11-F2S-300 turbojet of 13,670 lb (6 200 kg) with reheat. This version was licence-built in India as the MiG-21FL. Progressive development resulted in the MiG-21PFM (*Fishbed-F*) with cockpit quarter-lights, side-hinging canopy and zero-zero ejection seat. Late production MiG-21F received broad-chord vertical tail (*Fishbed-E*), a two-seat version is designated MiG-21U (*Mongol*), and a Tac-R variant of the MiG-21MF is in service (*Fishbed-H*).

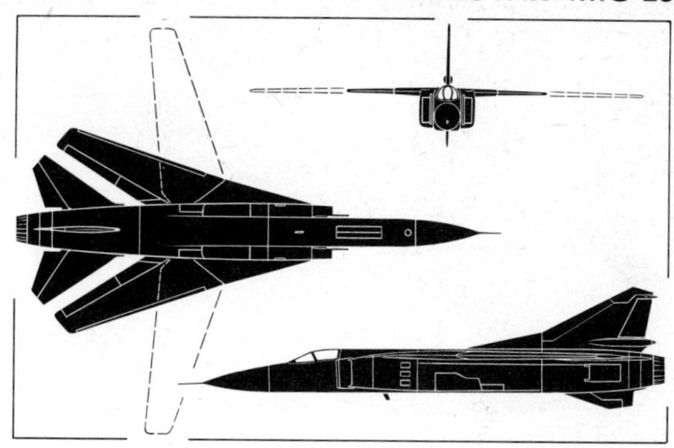

The MiG-23, illustrated by the photograph (above) and general arrangement silhouette (right) is one of the most recent fighters to enter the inventory of the Soviet Air Forces

Country of Origin: USSR.
NATO Code Name: *Flogger.*
Type: Interceptor fighter.
Power Plant: One (approx) 28,000 lb (12 700 kg) reheat Lyulka turbojet.
Performance: (Estimated) Max speed, 865 mph (1 390 km/h) or Mach 1·2 at sea level, 1,520 mph (2 446 km/h) or Mach 2·3 at 39,370 ft (12 000 m), in high-drag configuration (eg, two AAMs of advanced Anab type on fuselage stations and two AAMs on wing root stations), 1,120 mph (1 800 km/h) of Mach 1·7 at 39,370 ft (12 000 m); combat radius (with twin drop tanks on fuselage stations), 700 mls (1 126 km); service ceiling, 50,000 ft (15 250 m).
Weights: (Estimated) Normal take-off (with two AAMs), 30,000–32,000 lb (13 600–14 515 kg).
Dimensions: (Estimated) Span (minimum sweep), 48 ft 0 in (14,63 m), (maximum sweep), 24 ft 0 in (7,31 m); length (including probe), 60 ft 0 in (18,29 m).
Accommodation: Pilot only.
Armament: Twin-barrel 23-mm GSH-23 cannon in fuselage and up to four radar-guided AAMs on two fuselage and two wing root stations or mix of two infra-red and two radar-guided AAMs.
Status: Prototypes reportedly flown initially in 1967 with production deliveries to the Soviet Air Forces commencing in 1971.
Notes: The first Soviet combat aircraft designed from the outset for variable-geometry to achieve production and service status, the MiG-23 is roughly equivalent to the projected interceptor version of the Anglo-German-Italian multi-rôle combat aircraft (MRCA). Optimised for the air superiority rôle in the form currently serving with the Soviet Air Forces but presumably amenable to development for the multi-purpose fighter category, the MiG-23 possesses, according to US intelligence sources, avionics comparable with those of the F-4J Phantom and had achieved limited operational status by mid-1973. The shoulder-positioned wing translates from fully-forward to full-swept position in approximately four seconds, but unlike that of the MRCA the movable portion of the wing has no ordnance pylons, these being confined to the fuselage and to the short fixed inner portion of the wing. The MiG-23 reportedly offers excellent short-field performance and possesses long subsonic range and endurance in maximum fuel configuration.

MIKOYAN MIG-25

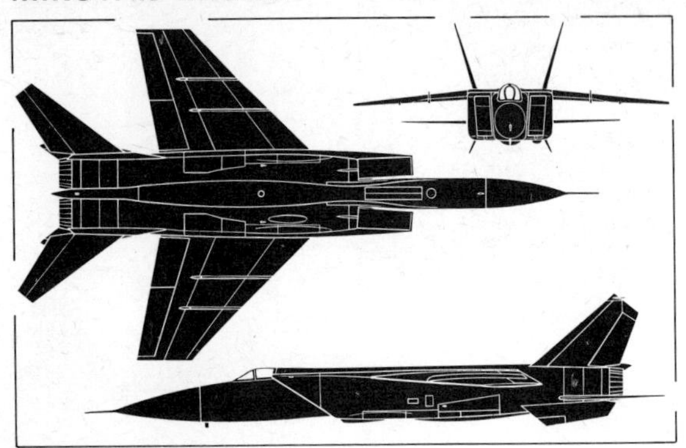

The MiG-25, illustrated by the photograph (above) and general arrangement silhouette (left) is a dual-rôle aircraft, being employed for both intercept and reconnaissance missions

Country of Origin: USSR.
NATO Code Name: *Foxbat.*
Type: Interceptor and reconnaissance fighter.
Power Plant: Two (approx) 24,250 lb (11 000 kg) reheat Tumansky turbojets.
Performance: (Estimated) Max short-period dash speed, 2,100 mph (3 380 km/h) or Mach 3·2 at 39,370 ft (12 000 m), max sustained speed, 1,780 mph (2 865 km/h) of Mach 2·7 at 39,370 ft (12 000 m), 975 mph (1 570 km/h) of Mach 1·3 at 4,920 ft (1 500 m); normal combat radius, 700 mls (1 125 km); time to 36,000 ft (10 970 m), 2·5 min.
Weights: (Estimated) Empty equipped, 34,000 lb (15 420 kg); normal loaded, 50,000–55,000 lb (22 680–24 950 kg); max take-off, 64,200 lb (29 120 kg).
Dimensions: (Estimated) Span, 41 ft 0 in (12,5 m); length, 70 ft 0 in (21,33 m).
Accommodation: Pilot only.
Armament: Four wing stations for radar homing AAMs for the intercept rôle. Internal bay in the fore-part of each air intake trunk, forward of the wheel well, capable of housing either 23-mm or 30-mm cannon or reconnaissance equipment.
Status: Believed flown in prototype form 1963–64 with service deliveries following from 1970–71.
Notes: First reported to be operational with Soviet Air Force units based in Egypt early in 1971, the MiG-25 has established a number of FAI-recognised records since 1965 under the designation Ye-266. These have included a record of 1,852·61 mph (2 981,5 km/h) or Mach 2·8 over a 310-mile (500-km) closed circuit on 5 October 1967, followed, on 27 October, by a 1,841·8 mph (2 920,67 km/h) or Mach 2·7 record over a 621-mile (1 000 km) circuit. In April 1973, the Ye-266 achieved a speed of 1,616 mph (2 600 km/h) over a 62-mile (100-km) closed circuit. The MiG-25 was apparently designed primarily for the interception of fast strike aircraft and it may be assumed that this fighter will eventually be equipped with a look-down/shoot-down radar/missile system, enabling maximum use to be made of the exceptional altitude capability offered by the MiG-25, operational ceiling reportedly being in excess of 80,000 ft (24 385 m). During the short period in which the MiG-25 was flown from Cairo West airfield (Autumn 1971–Spring 1972) it was used primarily for high-speed high-altitude reconnaissance missions.

NORTH AMERICAN F-86 SABRE

The photograph above depicts an F-86F of the Japanese Air Self-Defence Force's 4th Wing, this Sabre variant also being illustrated by the general arrangement silhouette (right)

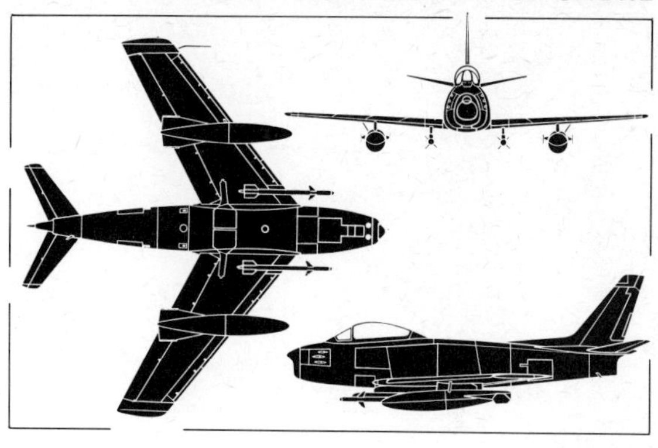

Country of Origin: USA.
Type: Interceptor and fighter-bomber.
Power Plant: One (F-86E) 5,200 lb st (2 358 kgp) General Electric J47-GE-13 or (F-86F) 5,970 lb st (2 710 kgp) J47-GE-27 turbojet.
Performance: (F-86F) Max speeds, 678 mph (1 091 km/h) at sea level, 599 mph (964 km/h) at 35,000 ft (10 670 m); initial rate of climb, 10,000 ft/min (50,8 m/sec); service ceiling, 50,000 ft (15 240 m); combat radius (with two drop tanks), 463 mls (745 km).
Weights: Empty, 11,125 lb (5 045 kg); normal loaded, 15,198 lb (6 895 kg); max take-off, 20,611 (9 350 kg).
Dimensions: Span, 39 ft 1 in (11,91 m); 37 ft 6 in (11,43 m); height, 14 ft 8¾ in (4,47 m); wing area, 313 sq ft (29,08 m²).
Accommodation: Pilot only.
Armament: Six 0·50-in (12,7-mm) Colt-Browning M-3 machine guns and underwing provision for two AAMs or bombs up to 1,000-lb (454-kg) each or eight rockets.
Status: Prototype (XP-86) first flown on 1 October 1947; first F-86A flown 18 May 1948; YF-86D flown on 22 December 1949; F-86E flown on 23 September 1950; F-86F flown 19 March 1952; F-86H flown 30 April 1953; YF-86K flown 15 July 1954; Fiat-built F-86K flown 23 May 1955; TF-86F flown 14 December 1953; first Canadian CL-13 Sabre 1 flown on 9 August 1950; prototype Australian CA-26 flown 3 August 1953; (Shipboard derivatives) XFJ-2B Fury prototype flown 27 December 1951; FJ-3 prototype flown 3 July 1953; FJ-4 prototype flown 28 October 1954.
Production totals: XP-86, 3; F-86A, 554; F-86D, 2,504; F-86E, 336 plus 120 by Canadair; F-86F, 1,794; F-86F assembled in Japan, 300; F-86H, 475; F-86K, 120 plus 221 assembled in Italy; Canadair Sabre 1, 1; Sabre 2, 350; Sabre 3, 1; Sabre 4, 438; Sabre 5, 370; Sabre 6, 655; Commonwealth Sabre CA-26, 1; Sabre 30, 21; Sabre 31, 21; Sabre 32, 69; XFJ-2, 3; FJ-2, 200; FJ-3, 538; FJ-4, 152; FJ-4B, 222.
Notes: Many nations still use Sabres, including Argentina, Burma, Nationalist China, Colombia, Ethiopia, Honduras, Indonesia, Iran, Italy, Japan, South Korea, Pakistan, Peru, the Philippines, Portugal, Saudi Arabia, South Africa, Thailand, Tunisia, Venezuela and Yugoslavia. The most widely-used model is the F-86F, but the F-86K is employed by the last two listed countries.

NORTH AMERICAN F-100 SUPER SABRE

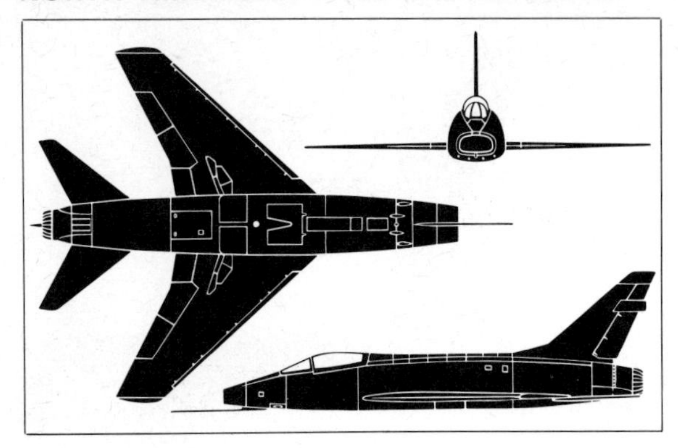

The photograph above depicts an F-100D of the 11e Escadre of the Armée de l'Air, the general arrangement silhouette (left) also illustrating this version of the Super Sabre

Country of Origin: USA.
Type: Tactical fighter-bomber (and operational trainer).
Power Plant: One (F-100A) 9,700 lb st (4 400 kgp) Pratt & Whitney J57-P-7 or (F-100C or F-100D) 11,700 lb st (5 310 kgp) dry and 16,950 lb st (7 690 kgp) with reheat J57-P-21A turbojet.
Performance: (F-100C, D and F) Max speed, 864 mph (1 390 km/h) at 35,000 ft (10 670 m); typical cruising speed, 565 mph (910 km/h) at 36,000 ft (10 970 m); initial rate of climb, 16,000 ft/min (81,3 m/sec); max range, 1,500 mls (2 415 km).
Weights: Empty (F-100D), 21,000 lb (9 525 kg) (F-100F) 22,300 lb (10 115 kg); normal loaded (F-100D), 29,762 lb (13 500 kg), (F-100F) 30,700 lb (13 925 kg), max take-off (F-100D), 34,832 lb (15 800 kg).
Dimensions: Span, 38 ft $9\frac{1}{3}$ in (11,81 m); length (F-100D), 49 ft 6 in (15,09 m), (F-100F), 52 ft 6 in (16,00 m); height, 16 ft $2\frac{2}{3}$ in (4,95 m); wing area, 385·2 sq ft (35,77 m^2).
Accommodation: (F-100C and D) Pilot only; (F-100F) Two in tandem, full dual controls.
Armament: (F-100C and) Four or (F-100F) two 20-mm M-39E cannon and (F-100D) 7,500 lb (3 400 kg) or (F-100F) 6,000 lb (2 720 kg) of external ordnance on eight underwing pylons.
Status: Prototypes (YF-100A) first flown on 25 May and 14 October 1953 respectively; first production F-100A flown on 29 October 1953; first production F-100C (Los Angeles) flown on 17 January 1955 and first F-100C (Columbus) on 8 September 1955; first F-100D (Los Angeles) flown on 24 January 1956 and first F-100D (Columbus) flown on 12 June 1956; prototype two-seat TF-100C flown on 6 August 1956; first production F-100F flown on 7 March 1957. Production totals, YF-100A, 2; F-100A, 203; F-100C, 451 at Los Angeles and 25 at Columbus; F-100D, 940 at Los Angeles and 334 at Columbus; F-100F, 339. Production completed October 1959.
Notes: F-100 was world's first operational fighter with a true supersonic performance in level flight. Successive models saw extensive service with the USAF. By 1973, the Super Sabre had been retired from TAC but equipped 19 Tactical Fighter Groups of the Air National Guard. Through MAP, several European nations received F-100s and the type is still used by the French *Armée de l'Air,* the Turkish Air Force, and the Royal Danish Air Force, as well as the Chinese Nationalist Air Force in Taiwan.

NORTHROP F-5

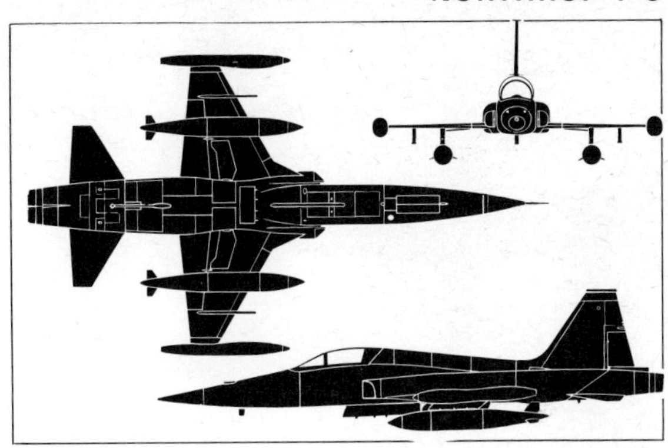

The photograph above illustrates an NF-5A of No 315 Squadron of the Royal Netherlands Air Force, the general arrangement silhouette depicting the standard F-5A version

Country of Origin: USA.
Type: Single-seat air-superiority fighter.
Power Plant: Two (F-5A, F-5B) 2,720 lb st (1 233 kgp) dry and 4,080 lb st (1 850 kgp) with reheat General Electric J85-GE-13 or (F-5E) 3,500 lb st (1 588 kgp) dry and 5,000 lb st (2 268 kgp) with reheat J85-GE-21 turbojets.
Performance: (NF-5A) Max speed, 790 mph (1 271 km/h) at sea level or Mach 1·04, 977 mph (1 572 km/h) at 36,000 ft (10 973 m) or Mach 1·48; initial rate of climb (clean), 33,000 ft/min (167,6 m/sec); tactical radius with 1,500 lb (680 kg) ordnance plus external fuel, hi-lo-hi profile, 575 mls (925 km); ferry range, 1,580 mls (2 543 km).
Weights: (F-5A) Empty equipped, 10,380 lb (4 710 kg); normal take-off (F-5A), 14,150 lb (6 420 kg), (F-5E), 15,400 lb (6 985 kg); max overload (F-5A), 20,390 lb (9 250 kg), (F-5E), 24,083 lb (10 924 kg).
Dimensions: Span (F-5A), 25 ft 3 in (7,70 m), (F-5E), 26 ft 8 in (8,13 m); length (F-5A), 47 ft 2 in (14,38 m), (F-5B), 46 ft 4 in (14,12 m), (F-5E), 48 ft 3¾ in (14,73 m); height, (F-5A) 13 ft 2 in (4,01 m), (F-5E), 13 ft 4½ in (4,08 m).

Accommodation: (F-5A, F-5E) Pilot only; (F-5B, F-5F) Two in tandem, dual controls.
Armament: Two 20-mm M-39 cannon in forward fuselage; provision for AAM at each wing tip and five external strong points for maximum of (F-5A) 6,200 lb (2 812 kg) or (F-5E) 7,000 lb (3 175 kg) of ordnance.
Status: First prototype (N-156) flown 30 July 1959; first production F-5A flown 19 May 1964; first F-5B flown 24 February 1964; YF-5-21 development aircraft flown 28 March 1969; first F-5E flown 11 August 1972. Production totals (completed or planned) include 621 F-5A, 134 F-5B, 89 CF-5A, 26 CF-5D, 75 NF-5A, 30 NF-5B, 35 SF-5A, 34 SF-5B and 500-plus F-5E.
Notes: F-5A (single-seat) and F-5B (two-seat) Freedom Fighter produced under US government or foreign contracts for Iran, Taiwan, Greece, South Korea, Philippines, Turkey, Ethiopia, Morocco, Norway, Thailand, South Vietnam and Libya. Both types built in Canada for CAF and R Neth AF, and in Spain. Improved F-5E in production for South Korea, South Vietnam, Brazil, Taiwan, Thailand, Jordan, Iran, Saudi Arabia and Malaysia. The F-5E Tiger II has been optimised for the intercept-air superiority rôle.

REPUBLIC F-84 THUNDERJET

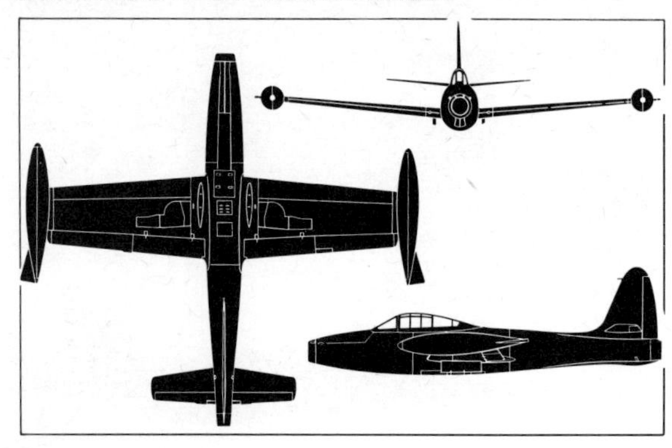

The obsolete F-84G Thunderjet remains in the inventories of only the Portuguese and Yugoslav air arms, an example serving with the former being illustrated above

Country of Origin: USA.
Type: Fighter-bomber.
Power Plant: One 5,600 lb st (2 540 kgp) Allison J35-A-29 turbojet.
Performance: (F-84G) Max speed, 622 mph (1 000 km/h) at sea level; cruising speed, 483 mph (777 km/h); time to climb to 35,000 ft (10 668 m), 9·4 min; service ceiling, 40,500 ft (12 344 m); range, 2,000 mls (3 218 km).
Weights: Empty, 11, 095 lb (5 030 kg); loaded, 23,525 lb (10 670 kg).
Dimensions: Span, 36 ft 5 in (11,09 m); length, 38 ft 1 in (11,61 m); height, 12 ft 7 in (3,84 m); wing area, 260 sq ft (24,15 m²).
Accommodation: Pilot only.
Armament: Six 0·50-in (12,7-mm) Colt-Browning M-3 machine guns (four in the front fuselage and two in the wings), and up to 4,000 lb (1 814 kg) of ordnance carried externally, including provision for tactical nuclear bomb.
Status: First of three XP-84 prototypes flown on 28 February 1946. Production deliveries of P-84B began in summer of 1947. Production totals: XP-84, 3; YP-84A, 15; P-84B, 226; P-84C, 191; F-84D, 154; F-84E, 843; F-84G, 3,025.
Notes: The F-84 Thunderjet was among the first generation of jet fighters to serve with the USAF, entering service after the P-80 Shooting Star and becoming the last of the straight-wing subsonic fighter-bombers to see operational service. Deployed to Korea in December 1950, F-84Ds flew escort missions for B-29s but were later used, together with F-84Es, primarily for ground attack duties, for which they were found highly satisfactory. The F-84G had the added capability of carrying a tactical nuclear bomb and was deployed in Europe from 1953 onwards. Of the total production quantities given above, 100 F-84Es and 1,936 F-84Gs were acquired through the USAF for MAP supply to NATO Air Forces. Only one of the latter still operated F-84Gs in 1973, this being Portugal, which retained about 30 out of an original force more than twice that size, for use by *Esquadra de Caca*/BA-9 operating against insurgents and guerillas from its base at Luanda in Angola. Only other air force still using the type is that of Yugoslavia, which originally equipped nine squadrons using aircraft declared surplus by NATO nations, but has been progressively replacing this aged type in recent years with the indigenous SOKO Jastreb (see page 199).

REPUBLIC F-84F THUNDERSTREAK

The photograph above and the general arrangement silhouette (right) illustrate the F-84F Thunderstreak, the former depicting an example currently serving with the Hellenic Air Force

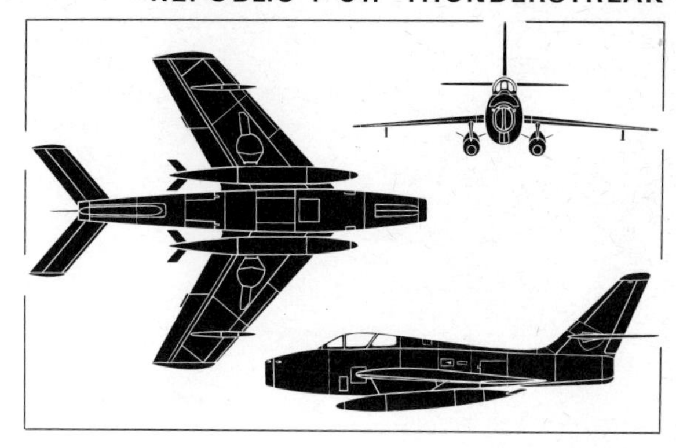

Country of Origin: USA.
Type: (F-84F) Tactical fighter-bomber and (RF-84F) reconnaissance fighter.
Power Plant: One (F-84F) 7,220 lb st (3 275 kgp) Wright J65-W-3 or (RF-84F) 7,800 lb st (3 538 kgp) J65-W-7 turbojet.
Performance: (F-84F) Max speed, 695 mph (1 118 km/h) at sea level, 658 mph (1 059 km/h) at 20,000 ft (6 096 m); initial rate of climb, 8,200 ft/min (41,6 m/sec); service ceiling, 46,000 ft (1 402 m); combat radius, 450 mls (724 km) clean, 810 mls (1 305 km) with two drop tanks; ferry range, 2,140 mls (3 440 km).
Performance: (RF-84F) Max speed, 679 mph (1 093 km/h) at sea level, 607 mph (976 km/h) at 35,000 ft (10 668 m); initial rate of climb, 7,900 ft/min (40,1 m/sec); service ceiling, 46,000 ft (14 020 m); range, up to 2,200 mls (3 540 km).
Weights: Loaded (clean), 19,340 lb (8 755 kg); (RF-84F) normal loaded, 26,800 lb (12 156 kg); max overload, 28,000 lb (12 700 kg).
Dimensions: Span, 33 ft 7$\frac{1}{4}$ in (10,24 m); length (F-84F), 43 ft 4$\frac{3}{4}$ in (13,22 m), (RF-84F), 47 ft 7$\frac{3}{4}$ in (14,51 m); height (F-84F), 14 ft 4$\frac{3}{4}$ in (4,38 m), (RF-84F), 15 ft 0 in (4,57 m); wing area, 325 sq ft (30,19 m^2).

Accommodation: Pilot only.
Armament: (F-84F) Six 0·5-in (12,7-mm) Colt-Browning M-3 machine guns in forward fuselage and wings, and up to a maximum of 6,000 lb (2 722 kg) of external ordnance on four underwing strong points. (RF-84F) Four 0·5-in (12,7-mm) M-3 guns.
Status: Prototype (YF-84F) first flown on 3 June 1950 with XJ35-A-25 engine; second YF-84F flown on 14 February 1951 with YJ65-W-1 engine; first production F-84F flown on 22 November 1952; YRF-84F prototype flown in February 1952. Production deliveries began in 1954; production totals, YF-84F, 3; F-84F, 2,713; RF-84F, 715; completed 1958.
Notes: The F-84F was derived from the straight-wing F-84 Thunderjet, and the RF-84F was a further development with wing-root intakes and cameras in the nose. Of the total production, 1,301 F-84F and 386 RF-84F were purchased through USAF for supply to NATO forces, but by 1973 these aircraft remained in first-line only with the air forces of Greece and Turkey both countries having three F-84F squadrons and the former having one and the latter two squadrons of RF-84Fs. Only odd examples remain in the USAF active inventory.

REPUBLIC F-105 THUNDERCHIEF

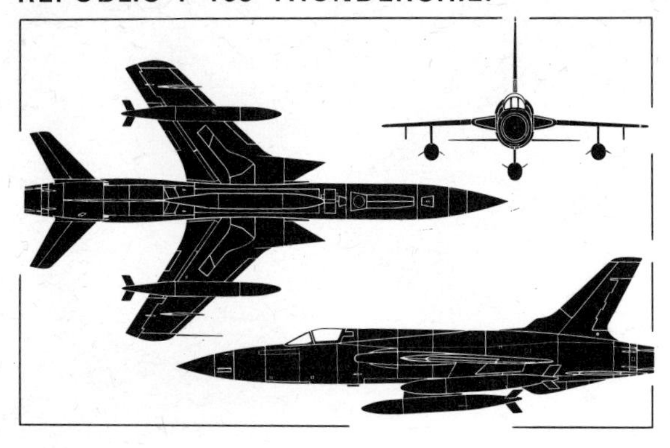

The photograph above depicts a two-seat F-105F which now serves with the Air National Guard and the general arrangement silhouette illustrates the single-seat F-105D

Country of Origin: USA.
Type: Long-range fighter-bomber.
Power Plant: One (F-105B) 23,500 lb st (10 660 kgp) with reheat Pratt & Whitney J75-P-5 or (F-105D, F and G) 17,200 lb st (7 800 kgp) dry and 24,500 lb st (11 113 kgp) with reheat J75-P-19W turbojet.
Performance: (F-105D) Max speed, 855 mph (1 375 km/h) at sea level, 1,390 mph (2 237 km/h) at 36,000 ft (10 973 m) or Mach 2·1; 1,122 mph (1 805 km/h) at 50,000 ft (15 240 m) or Mach 1·7; initial rate of climb, 34,500 ft/min (175,2 m/sec); service ceiling, 52,000 ft (15 850 m); tactical radius (three drop tanks) 920 mls (1 480 km); ferry range, 2,390 mls (3 846 km).
Weights: Empty, 27,500 lb (12 474 kg); normal loaded (F-105D) 38,034 lb (17 250 kg) (F-105F) 40,073 lb (18 175 kg); maximum overload (F-105D), 52,546 lb (23 834 kg), (F-105F), 54,027 lb (24 500 kg).
Dimensions: Span, 34 ft $11\frac{1}{4}$ in (10,65 m); length (F-105D), 64 ft 3 in (19,58 m), (F-105F), 69 ft $7\frac{1}{2}$ in (21,21 m); height (F-105D), 19 ft 8 in (5,99 m), (F-105F), 20 ft 2 in (6,15 m); wing area, 385 sq ft (35,76 m^2).

Accommodation: (F-105D) Pilot only; (F-105F) two in tandem, dual controls; (F-105G) pilot and observer in tandem.
Armament: One 20-mm M-61 multi-barrel cannon. Internal weapons bay with capacity for up to 8,000 lb (3 628 kg) of bombs plus 6,000 lb (2 722 kg) externally on fuselage and wing points.
Status: First of two YF-105A prototypes flown on 22 October 1955; first F-105B development aircraft flown on 24 May 1957; first F-105D flown on 9 June 1959; first F-105F flown on 11 June 1963; prototype T-Stick II conversion flown on 9 August 1969. Production deliveries of F-105B began May 1958. Production totals: YF-105A, 2; JF-105B, 3; F-105B, 75; F-105D, 600; F-105F, 143, completed 1964.
Notes: In addition to the single-seat F-105B and F-105D, the two-seat F-105F was used operationally in Vietnam, with an observer in the rear seat, and about 30 were converted to F-105G with "Wild Weasel" radar homing and warning system. Thirty F-105Ds were modified to T-Stick II configuration with improved all-weather bombing capability, and had saddle-back fairing from cockpit to fin. After withdrawal from Vietnam, F-105Bs and Ds were issued to four Air National Guard Wings in 1972.

SAAB 35 DRAKEN

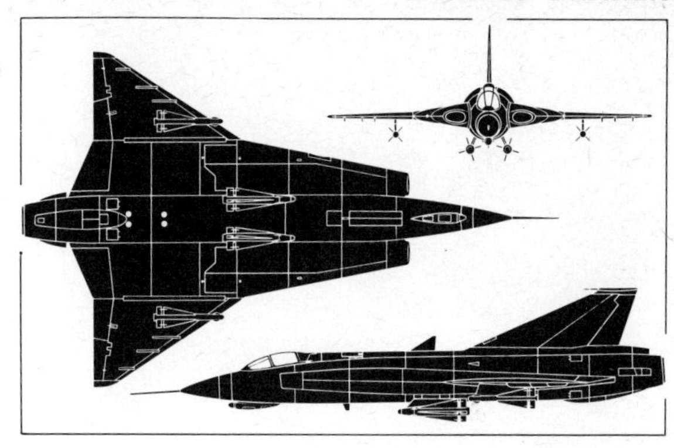

The photograph above depicts the F-35 version of the Saab 35X Draken serving with the Danish Air Force and the general arrangement silhouette illustrates the J 35F Draken

Country of Origin: Sweden.
Type: All-weather fighter, reconnaissance and attack aircraft.
Power Plant: One (J 35A, J 35B and SK 35C) 15,000 lb st (6 803 kgp) with reheat Volvo Flygmotor RM6B (Avon 200) or (J 35D, S 35E, J 35F and Saab 35X) 12,790 lb st (5 800 kgp) dry and 17,635 lb st (8 000 kgp) with reheat RM6C (Avon 300).
Performance: (Saab 35X) Max speed, 924 mph (1 487 km/h) or Mach 1·4 at 36,090 ft (11 000 m); initial rate of climb, 22,650 ft/min (11,5 m/sec); radius of action (clean), hi-lo-hi profile, 395 mls (635 km); radius of action (two drop tanks and two 1,000 lb/454 kg bombs) hi-lo-hi profile, 621 mls (1 000 km); ferry range, 2,020 mls (3 250 km).
Weights: Take-off clean, 25,132 lb (11 400 kg); take-off (two 1,000 lb/454 kg bombs and two drop tanks), 32,165 lb (14 590 kg); max take-off, 33,070 lb (15 000 kg); max overload, 35,275 lb (16 000 kg).
Dimensions: Span, 30 ft 10 in (9,40 m); length (including nose probe), 50 ft 4 in (15,4 m); height, 12 ft 9 in (3,9 m); wing area, 529·6 sq ft (49,2 m^2).
Accommodation: Pilot or (SK 35C) two with dual controls.

Armament: (J 35F) One or (Saab 35X) two 30-mm Aden cannon in wings (optional in place of extra fuel); nine external stores stations (three beneath fuselage and three under each wing) with capacity of 1,000 lb (454 kg) each. Primary air-to-air armament comprises two or four RB 24 Sidewinders.
Status: First of three prototypes flown on 25 October 1955; second in March 1956, third in June 1956. First production J 35A flown on 15 February 1958; first J 35B flown on 29 November 1959; first SK 35C flown on 30 December 1959; first J 35D flown on 27 December 1960; first S 35E flown on 27 June 1963; first Saab 35XD flown on 29 January 1970. Total production over 600, completed 1972, including 20 F-35, 20 RF 35 and 6 TF-35 for R. Danish Air Force, 12 Saab 35s for Finnish Air Force (assembled by Valmet OY) and remainder for Swedish Air Force.
Notes: The delta-winged Saab 35 Draken is the primary front-line fighter equipment of the Swedish Air Force, pending introduction of the JA 37 Viggen (see page 40). J 35F with Saab S7B collision-course fire-control system has replaced earlier fighter version, and is complemented by S 35E reconnaissance model and two-seat SK 35C.

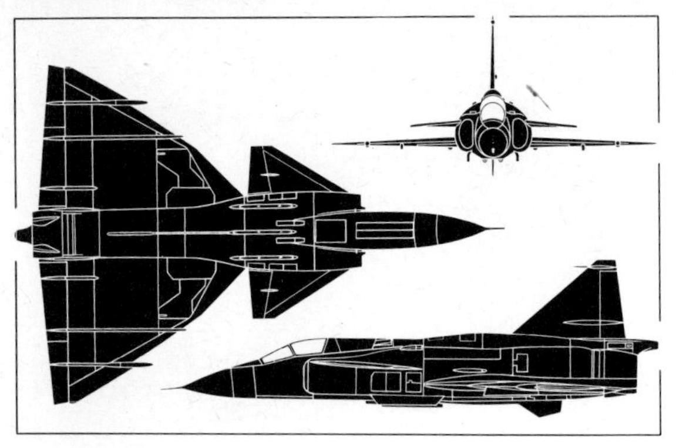

Both the photograph (above) and the general arrangement silhouette (left) depict the AJ 37 version of the Viggen, the former illustrating an example serving with F 7

Country of Origin: Sweden.
Type: Multi-purpose fighter and two-seat operational trainer.
Power Plant: One 14,700 lb st (6 667 kgp) dry and 26,450 lb st (12 000 kgp) with reheat Volvo Flygmotor RM 8 (Pratt & Whitney JT8D-22) turbofan.
Performance: (Estimated) Max speed without external stores, 1,320 mph (2 125 km/h) or Mach 2·0 at 36,090 ft (11 000 m), 875 mph (1 410 km/h) or Mach 1·15 at 305 ft (100 m); tactical radius with typical external ordnance load for hi-lo-hi mission profile, 620 mls (1 000 km), for lo-lo-lo mission profile, 310 mls (500 km); time to 36,090 ft (11 000 m), 2 min.
Weights: Normal take-off weight, 35,275 lb (16 000 kg).
Dimensions: Span, 34 ft 9¼ in (10,60 m); length, 50 ft 8¼ in (15,45 m); length (including nose probe), 53 ft 5¾ in (16,30 m); height, 18 ft 4½ in (5,60 m).
Accommodation: (AJ 37, JA 37, SF 37, SH 37) Pilot only, (SK 37) Two in tandem, dual controls.
Armament: All ordnance carried on seven external stores stations (four beneath wings and three under fuselage), primary armament being RB 04E or RB 05A ASMs for the attack rôle, or RB 24 (Sidewinder) or RB 28 (Falcon) AAMs for the intercept rôle.
Status: First of six single-seat prototypes flown 8 February 1967 and the sixth by April 1969. Two-seat SK 37 prototype flown 2 July 1970. First production AJ 37 flown 23 February 1971, deliveries to Swedish Air Force began 21 June 1971. First SF 37 flown 21 May 1973. Production orders placed for 150 AJ 37, 25 SK 37 and a quantity of SF 37/SH 37 reconnaissance versions.
Notes: Initial production version of the double-delta Viggen is the AJ 37, a single-seat all-weather attack version with secondary interceptor capability. Deliveries began in 1971 and the first *Division* (Squadron) of F7 at Satenäs was declared operational early in 1973, with conversion of F15 at Söderhamn and F6 at Karlsborg to be initiated in 1974. For operational and conversion training, the SK 37 has a second cockpit within a fuselage of the same overall length, and deliveries began in mid-1972. The reconnaissance version is in production in two variants—SH 37 for overland tactical reconnaissance and SF 37 for sea surveillance—with different equipment. The JA 37 is a planned single-seat interceptor for eventual procurement by the Swedish Air Force to replace the J 35, eight squadrons of this model having been requested in 1973.

SUKHOI SU-11

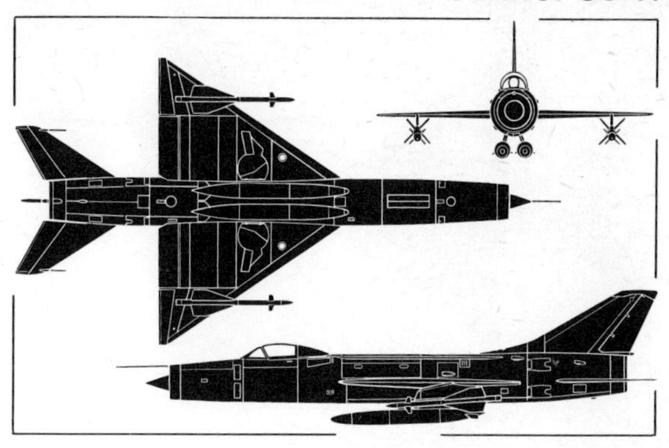

The Sukhoi Su-11, illustrated by the photograph (above) and general arrangement silhouette (right) is currently one of the major fighters in the inventory of the Soviet Air Forces

Country of Origin: USSR.
NATO Code Name: *Fishpot.*
Type: All-weather interceptor fighter.
Power Plant: One 14,200 lb (6 440 kg) dry and 22,046 lb (10 000 kg) with reheat Lyulka AL-7F-1 turbojet.
Performance: (Estimated) Max speed without external stores, 720 mph (1 160 km/h) or Mach 0·95 at 1,000 ft (305 m), 1,190 mph (1 915 km/h) or Mach 1·8 at 40,000 ft (12 190 m); max speed in high-drag configuration (eg, two Anab AAMs and two 132 Imp gal/600 l drop tanks), 790 mph (1 270 km/h) or Mach 1·2 at 40,000 ft (12 190 m); initial climb, 27,000 ft/min (137 m/sec); service ceiling, 55,000 ft (16 765 m).
Weights: (Estimated) Empty, 20,000 lb (9 072 kg); normal loaded, 27,000 lb (12 247 kg); max take-off, 30,000 lb (13 608 kg).
Dimensions: (Estimated) Span, 27 ft 0 in (8,23 m); length (including probe), 57 ft 0 in (17,37 m); height, 16 ft 0 in (4,88 m); wing area, 425 sq ft (39,48 m²).
Accommodation: Pilot only.
Armament: Two advanced *Anab* (one infra-red homing and one radar homing) missiles beneath wings.

Status: Prototypes of the Su-11 were initially flown in the mid 'fifties at which time the aircraft possessed some component commonality with the Su-7, the fuselages, power plants and tail assemblies of the two aircraft being similar. The initial production version entered service with the Soviet Air Forces during the early 'sixties, and the type has since been manufactured in progressively improved versions.
Notes: More advanced avionics, more sophisticated missile armament and more powerful versions of the Lyulka AL-7 engine have been progressively introduced by the Su-11 since its service introduction but few external changes have taken place, apart from the adoption of a longer, less-tapered fuselage nose with a larger centre-body. The Su-11 is currently one of the major fighters in the inventory of the interceptor component of the P-VO Strany which is responsible for the aerial defence of the main centres of industry and population. A tandem two-seat conversion training variant of the basic design is currently in service as the Su-11U (*Maiden*). Production of the Su-11 series is believed to have terminated in favour of the Su-15 in 1969–70, but is unlikely to be phased out before the late 'seventies.

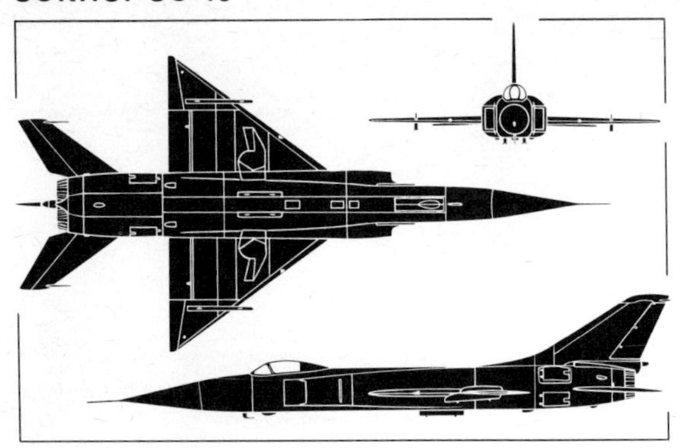

The Sukhoi Su-15, illustrated by the photograph (above) and the general arrangement silhouette (left), is progressively replacing the Su-11 illustrated on the previous page

Country of Origin: USSR.
NATO Code Name: *Flagon-A.*
Type: All-weather interceptor fighter.
Power Plant: Two (approx) 25,000 lb (11 340 kg) reheat Lyulka AL-9 turbojets.
Performance: (Estimated) Max speed (clean), 1,650 mph (2 655 km/h) of Mach 2·5 at 39,370 ft (12 000 m), 910 mph (1 465 km/h) of Mach 1·2 at 1,000 ft (305 m); max speed in high-drag configuration (eg, two AAMs on wing stations and two drop tanks on fuselage stations), 1,120 mph (1 800 km/h) or Mach 1·7 at 39,370 ft (12 000 m); range at subsonic cruise (with max external fuel), 1,500 mls (2 415 km).
Weights: (Estimated) Normal take-off, 35,000–40,000 lb (15 875–18 145 kg).
Dimensions: (Estimated) Span, 31 ft 3 in (9,50 m); length, 70 ft 6 in (21 50 m); height, 16 ft 6 in (5,00 m).
Accommodation: Pilot only.
Armament: Basic armament for intercept mission reportedly comprises two AAMs of the advanced Anab type on wing stations.
Status: The Su-15 is believed to have flown in prototype form during 1964–65 with production deliveries to the Soviet Air Forces commencing 1969 and currently continuing. According to US official sources, some 400 Su-15s were in service with the Soviet IAP-VO Strany by mid-1971 when production was continuing at a rate of some 15 aircraft monthly.
Notes: Optimised for the intercept rôle as a successor to the Su-11 and possessing an obvious family resemblance to its predecessor, the Su-15 reportedly now possesses primary responsibility for the air defence of Moscow and other major Soviet centres, and was deployed in small numbers to Egypt in 1971 for the air defence of Cairo but was withdrawn during the following year. A short-take-off-and-landing (STOL) version of the Su-15 (*Flagon-B*) with three direct lift engines mounted vertically in the centre fuselage and extended wingtips was demonstrated publicly in 1967 but is currently of uncertain status, although unofficial reports have suggested that it is in operational service on a limited scale. Adaptation of the Su-15 for the reconnaissance-strike rôle has also been reported as has a tandem two-seat conversion trainer variant. It is believed that the Su-15 made up 15–20 per cent of the operational fighter inventory of the Soviet Air Forces by mid-1973.

TUPOLEV TU-28P

The Tupolev Tu-28, illustrated by the photograph (above) and the general arrangement silhouette (right) is the largest and heaviest interceptor currently in operational service

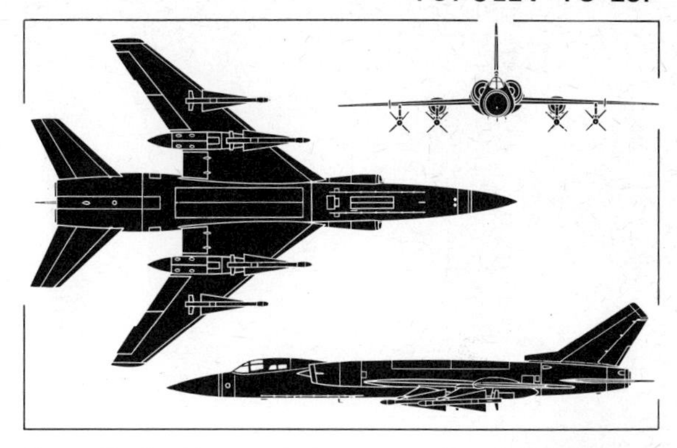

Country of Origin: USSR.
NATO Code Name: *Fiddler.*
Type: Long-range all-weather interceptor fighter.
Power Plant: Two (approx) 25,000 lb (11 340 kg) reheat turbojets.
Performance: (Estimated) Max speed (clean), 1,085 mph (1 745 km/h) or Mach 1·65 at 39,370 ft (12 000 m); max speed in high-drag configuration (eg, four *Ash* AAMs on wing stations), 925 mph (1 490 km/h) or Mach 1·4 at 39,370 ft (12 000 m); tactical radius for high-altitude patrol mission, 900–1,000 mls (1 450–1 770 km).
Weights: (Estimated) Normal loaded, 78,000 lb (35 380 kg); max take-off, 96,000 lb (43 545 kg).
Dimensions: (Estimated) Span, 65 ft 0 in (19,80 m); length, 90 ft 0 in (27,43 m).
Accommodation: Pilot and radar operator in tandem.
Armament: Mix of two infra-red homing and two radar homing *Ash* AAMs on wing stations.
Status: The Tu-28 is believed to have flown in prototype form in 1957 and to have entered the inventory of the Soviet Air Forces in its Tu-28P interceptor version in the early 'sixties.

Notes: The Tu-28, currently the largest and heaviest interceptor in operational service, is believed to have been evolved originally to fulfil a requirement for a long-range reconnaissance-strike aircraft, and its design was biased towards economical high-altitude operation. One version has been seen with a central pack that may be presumed to house early warning radar, and it is probable that an internal weapons bay is standard for the reconnaissance-strike rôle. The Tu-28P is known to be serving with the Soviet Air Forces in some numbers for patrolling the sections of the periphery of the Soviet Union unprotected by surface-to-air missile screens, and presumably operates in conjunction with the *Moss* AWACS aircraft. The Tu-28P is likely to have a normal endurance in the patrol-interceptor rôle of some 3·5 hours and this possibly exceeds 5·5 hours with overload tanks. Apart from its extreme size, unusual features of the Tu-28P include the use of four-wheel main undercarriage bogies retracting into bulged housings beneath the wings and extending aft of the trailing edges. Progressively more powerful turbojets were installed in the Tu-28P during its production life, and this extremely large fighter was in process of being phased out in 1975 in favour of an interceptor version of the Tu-22 bomber.

VOUGHT F-8 CRUSADER

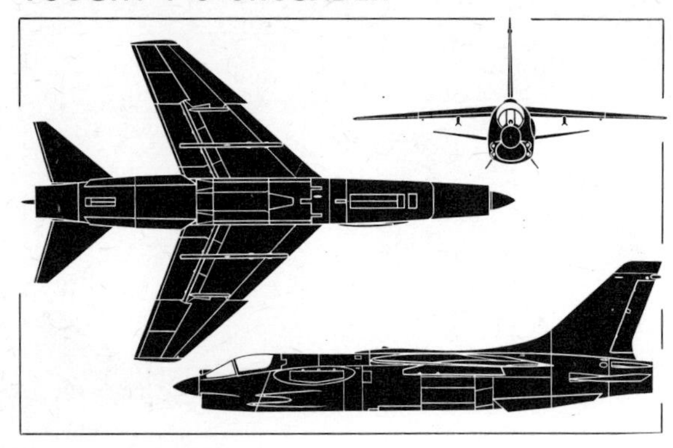

The F-8J Crusader, illustrated by the photograph (above) and the general arrangement silhouette (left) is a remanufactured F-8E for US Navy Reserve units

Country of Origin: USA.
Type: Carrier-borne fighter.
Power Plant: One (F-8L) 16,200 lb st (7 327 kgp) Pratt & Whitney J57-P-12 or (F-8K) 16,900 lb st (7 665 kgp) J57-P-16 or (F-8E (FN), F-8H and F-8J) 18,000 lb st (8 165 kgp) J57-P-20 or -20A turbojet.
Performance: (F-8J) Max speed, 1,120 mph (1 800 km/h) at 40,000 ft (12 190 m); cruising speed for best range, 560 mph (900 km/h) at 36,000 ft (10 970 m); time to climb to 57,000 ft (17 373 m), 6·5 min; service ceiling, 58,000 ft (17 680 m); combat radius, 600 mls (966 km); max range (clean) 1,400 mls (2 253 km).
Weights: Normal loaded, approx 29,000 lb (13 155 kg); max take-off, 34,000 lb (15 420 kg).
Dimensions: Span, 35 ft 8 in (10,87 m); length (F-8H, K and L), 54 ft 3 in (16,54 m); length (F-8E (FN) and F-8J), 54 ft 6 in (16,61 m); height, 15 ft 9 in (4,80 m); wing area, 375 sq ft (34,84 m^2).
Accommodation: Pilot only.
Armament: Four 20-mm Mk 12 cannon in fuselage nose. Four fuselage stations to carry bombs or ASMs or AAMs.

Status: Two XF-8A prototypes (originally XF8U-1) flown on 25 March and 30 September 1955 respectively. Deliveries of production F-8A began March 1957. YTF-8A (two-seat trainer) prototype flown on 6 February 1962. RF-8A first flown on 17 December 1956; F-8B first flown on 3 September 1958; F-8C first flown December 1957 and first production example 20 August 1958; first F-8D flown 16 February 1960; first F-8E flown on 30 June 1961; prototype F-8E (FN) flown 27 February 1964 and first production 26 June 1964; first F-8H flown 17 July 1967; first F-8J flown 31 January 1968. Production totals: XF-8A prototypes, 2; F-8A, 318; F-8B, 130; F-8C, 187; F-8D, 152; F-8E, 286; F-8E (FN), 42; RF-8A, 144, making grand total of 1,261.
Notes: Total of 446 Crusaders were cycled through conversion/updating programmes between 1967 and 1970 to extend their useful life with US Navy reserve units. Through these programmes, 73 RF-8As became RF-8G, 61 F-8Bs became F-8L, 87 F-8Cs became F-8K, 136 F-8Es became F-8J and 89 F-8D became F-8H. France acquired the F-8F (FN) version in 1964/65, with blown flaps and improved low speed control. They have been modernised to F-8J standard and serve with *Flotilles* 12F and 14F.

YAKOLEV YAK-25F

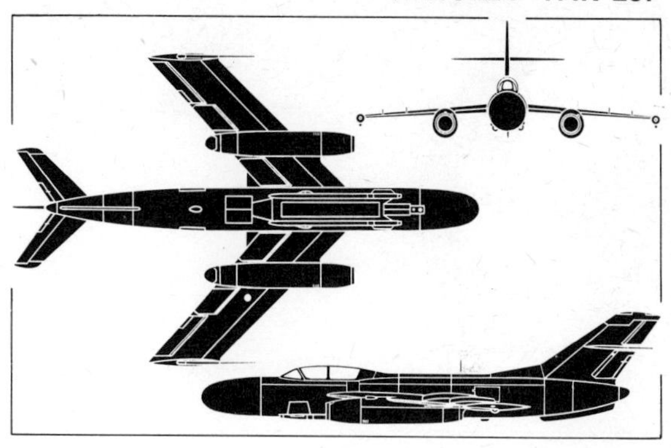

Despite its obsolescence, the Yak-25F, illustrated by the photograph (above) and the general arrangement silhouette (right) remains in service with the Soviet Air Forces

Country of Origin: USSR.
NATO Code Name: *Flashlight-A.*
Type: All-weather interceptor fighter.
Power Plant: Two 5,730 lb (2 600 kg) Tumansky RD-9 tubojets.
Performance: Max speed (clean), 630 mph (1 015 km/h) or Mach 0·83 at sea level, 594 mph (956 km/h) or Mach 0·9 at 36,090 ft (11 000 m); normal endurance, 2·5 hrs; tactical radius (internal fuel), 600 mls (965 km) at 36,090 ft (11 000 m); ferry range with overload fuel, 1,865 mls (3 000 km); initial climb, 9,800 ft/min (49,8 m/sec).
Weights: Normal loaded, 20,282 lb (9 200 kg); max take-off, 23,148 lb (10 500 kg).
Dimensions: Span, 36 ft 1 in (11,00 m); length, 51 ft 4½ in (15,66 m); height, 12 ft 6 in (3,80 m); wing area, 311·51 sq ft (28,94 m²).
Accommodation: Pilot and radar operator in tandem.
Armament: Two 37-mm N-37 cannon and two (one infra-red homing and one radar guided) AAMs.
Status: Prototype flown for the first time in 1953, production deliveries to the Soviet Air Force commencing in 1955 and continuing into early 'sixties. Although obsolescent, the Yak-25F is still included in some numbers in the Soviet fighter inventory.
Notes: The first genuine all-weather interceptor of Soviet design to have attained service status with the Soviet Air Forces, the Yak-25 was designed to meet the requirements of a specification issued in November 1951, its most unusual feature being a novel zero-track tricycle undercarriage arrangement in which the fighter is virtually balanced on the twin-wheel aft member with outrigger units at the wingtips. The initial production version was powered by two 4,850 lb (2 200 kg) Mikulin AM-5 turbojets but these were supplanted by Tumansky RD-9s from 1957, the re-engined model being designated Yak-25F. A parallel development was the Yak-25R (*Flashlight-B*) tactical reconnaissance aircraft which, basically similar to the fighter, featured a redesigned cockpit for a single occupant, the second crew member being accommodated in a revised forward fuselage embodying a pointed, glazed nose. Progressive development of the basic design resulted in the Yak-27R (*Mangrove*) tactical reconnaissance aircraft and Yak-27P (*Flashlight-C*) all-weather interceptor embodying aerodynamic refinements and afterburning engines (see page 101).

YAKOVLEV YAK-28P

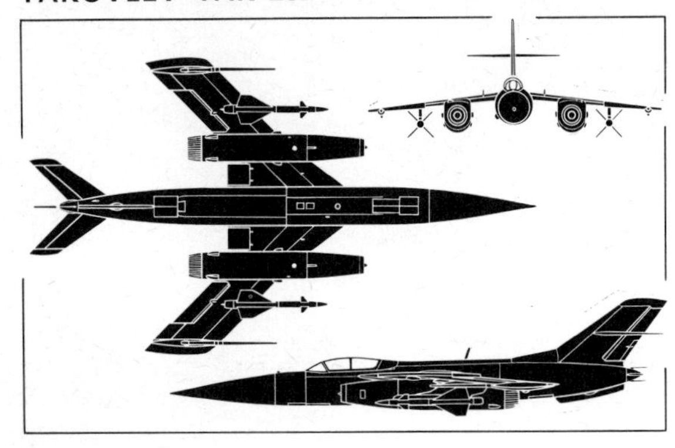

The Yak-28P currently serves in several versions, that illustrated above having the original short nose cone and the general arrangement silhouette (left) showing the more pointed cone

Country of Origin: USSR.
NATO Code Name: *Firebar.*
Type: All-weather interceptor fighter.
Power Plant: Two 10,140 lb (4 600 kg) dry and 13,670 lb (6 200 kg) with reheat Tumansky R-11 turbojets.
Performance: (Estimated) Max speed (clean), 760 mph (1 225 km/h) or Mach 1·15 at 39,370 ft (12 000 m); max speed with two Anab AAMs, 695 mph (1 120 km/h) or Mach 1·05; normal cruise, 560 mph (900 km/h) or Mach 0·9; tactical radius (high-altitude patrol mission), 550 mls (885 km); initial climb, 28,000 ft/min (142,2 m/sec); service ceiling, 55,000 ft (16 765 m).
Weights: (Estimated) Normal loaded, 37,480 lb (17 000 kg); max take-off, 40,785 lb (18 500 kg).
Dimensions: (Estimated) Span, 41 ft 0 in (12,5 m); length, 72 ft 0 in (21,95 m); height, 13 ft 0 in (3,96 m).
Accommodation: Pilot and radar operator in tandem.
Armament: Standard armament comprises two Anab AAMs with alternative semi-active radar or infra-red homing heads. Some examples have been seen with four wing stores stations accommodating two radar-homing Anab and two infra-red homing Atoll air-to-air missiles.
Status: Flown in prototype form in 1960 with production deliveries to the Soviet Air Forces commencing 1963–64. Currently one of the primary interceptors of the IAP-VO Strany.
Notes: The Yak-28P was evolved in parallel with the Yak-28 (*Brewer*) tactical strike-recce aircraft (see page 62), differing from the latter primarily in featuring a di-electric nose cone, tandem seating for the two crew members with the windscreen approximately 2 ft 6 in (76 cm) further forward and the internal weapons bay deleted, this space presumably being occupied by fuel tankage. The Yak-28P bears only a superficial family resemblance to the earlier Yak-25 and Yak-27 (see pages 45 and 101). Features include a widely-spaced "bicycle" undercarriage with outrigger stabilizing members, and an exceptionally long and pointed di-electric nose cone would now seem to have been adopted as standard. The Yak-28P has been fitted with progressively more powerful turbojets and updated avionics and missiles since its service introduction in the mid 'sixties, series production reportedly having been phased out in 1969–70. It is equipped with so-called Skip Spin AI radar with an effective range of about 25 miles (40 km).

DOUGLAS A-1 SKYRAIDER

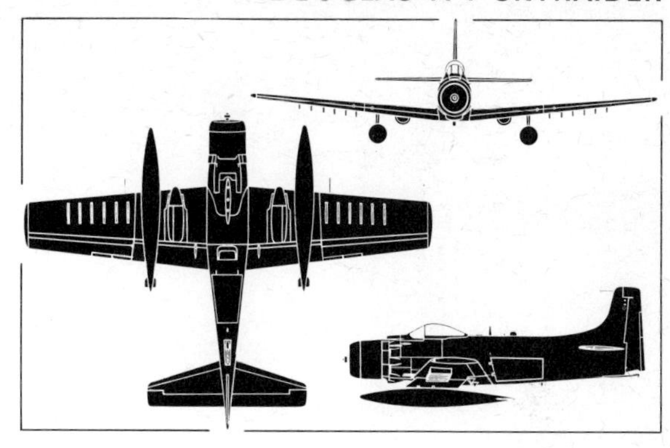

The photograph above depicts a two-seat A-1E of the Republic of Vietnam Air Force and the general arrangement silhouette (right) illustrates the single-seat A-1J

Country of Origin: USA.
Type: Tactical attack and strike aircraft.
Power Plant: One 3,020 hp Wright R-3350-26W or 3,050 hp R-3350-26WB piston radial engine.
Performance: (A-1E) Max speed, 311 mph (501 km/h) at 18,000 ft (5 486 m); normal cruise, 200 mph (322 km/h); initial rate of climb 2,300 ft/min (11,7 m/sec); service ceiling, 32,000 ft (9 753 m); tactical radius, about 450 mls (724 km); max range with external tanks, 3,000 mls (4 828 km).
Weights: Empty, 12,313 lb (5 585 kg); normal loaded, 18,799 lb (8 527 kg); max overload, 25,000 lb (11 340 kg).
Dimensions: Span, 50 ft 9 in (15,47 m); length (A-1E), 40 ft 1 in (12,19 m) (A-1J) 38 ft 10 m (11,84 m); height, 15 ft 10 in (4,83 m) (A-1J) 15 ft $8\frac{1}{4}$ in (4,77 m); wing area, 400·33 sq ft (37,2 m²).
Accommodation: Pilot only or (A-1E) Pilot and observer in tandem.
Armament: Four 20 mm cannon and underwing pylons for up to 8,000 lb (3 629 kg) of bombs, rockets, etc.
Status: Prototype (XBT2D-1) first flown on 18 March 1945; A-1E first flown on 17 August 1951; first A-1J flown in August 1956. Production totals: XBT2D-1 prototypes, 25; AD-1, 242; AD-1Q, 35; AD-2, 156; AD-2Q, 21; AD-2QU, 1; AD-3, 125; AD-3N, 15; AD-3Q, 23; AD-3W, 31; AD-4, 372; AD-4B, 165; AD-4N, 307; AD-4Q, 39; AD-4W, 168; A-1E (AD-5) 239; EA-1E (AD-5W), 218; AD-5S, 1; A-1H (AD-6), 713; A-1J (AD-7), 72; total production, 3,180, completed February 1957.
Notes: The Skyraider, built in single and two-seat versions and a multitude of rôle variants, was withdrawn from US Navy service in April 1968 but the USAF acquired 50 A-1Es and a quantity of A-1Hs and A-1Js which were extensively used in Vietnam and from bases in Thailand up to the end of 1972. Over 100 were supplied to the Republic of Vietnam Air Force before the armistice, about 60 being in service in 1973 with three squadrons. France maintained A-1-equipped units in Madagascar and French Somaliland as late as 1973, and has transferred other A-1s to the air forces of Khmer, the Central African Republic and Chad. These are of the A-1D (AD-4) model, 88 examples of which were originally supplied to the *Armée de l'Air* for use in Algeria. The A-1E, which succeeded the -1D in production, embodied extensive redesign while retaining much of the earlier model's structure.

DOUGLAS A-26/B-26 INVADER

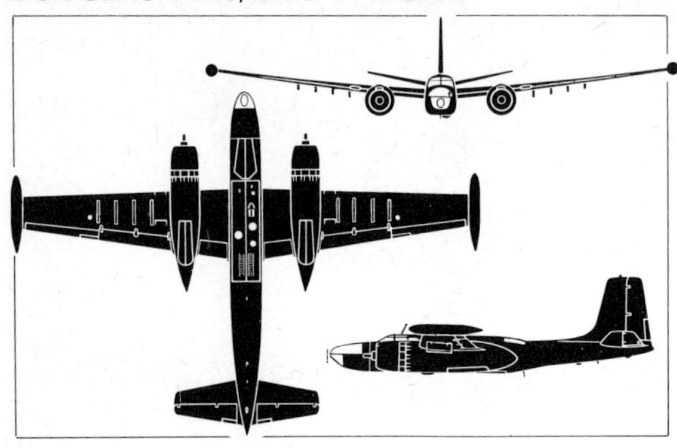

The photograph above depicts a refurbished B-26C of the Brazilian Air Force and the general arrangement silhouette (left) illustrates the B-26K with permanent wingtip tanks

Country of Origin: USA.
Type: Light tactical bomber.
Power Plant: Two 2,000 hp Pratt & Whitney R-2800-27 or -79 radial piston engines.
Performance: Max speed, 355 mph (571 km/h) ; cruising speed, 284 mph (457 km/h) ; time to climb to 10,000 ft (3 048 m), 8 min : service ceiling, 22,100 ft (6 736 m) ; range, 1,400 mls (2 253 km).
Weights: Empty, 22,370 lb (10 145 kg) ; normal loaded, 35,000 lb (15 870 kg) ; max take-off, 38,500 lb (17 460 kg).
Dimensions: Span, 70 ft 0 in (21 34 m) ; length, 50 ft 0 in (15,24 m) ; height, 18 ft 6 in (5,64 m) ; wing area, 540 sq ft (50,16 m²).
Accommodation: Pilot and navigator/bombardier side-by-side ; dorsal gunner.
Armament: Six 0·30-in (7,62-mm) machine guns in wings and optional provision for eight similar guns in nose in lieu of bomb-aiming position. Internal stowage for up to 4,000 lb (1 814 kg) of bombs. Underwing provision for bombs or rockets up to a maximum load (B-26K) of 8,000 lb (3 629 kg).
Status: Prototype (XA-26) first flown on 10 July 1942. Prototype YB-26K flown early 1963. Production totals (1944/45) : XA-26, 3 ; A-26B, 1,355 ; A-26C, 1,091 ; XA-26D, 1.
Notes: The original A-26 Invader was a World War II product designed to succeed the same company's A-20 Havoc and Boston family. It was produced in two versions with "solid" (A-26B) and bombardier (A-26C) noses but operational use was curtailed by the end of the war. The designation was changed from A-26 to B-26, and several squadrons remained in USAF inventory to be called into action in Vietnam in the counter-insurgency rôle. Some were converted to B-26K configuration with 2,500 hp R-2800-103W engines, increased ordnance loads and better take-off performance, and this version was re-designated A-26A during its Vietnam service. Examples of the B-26B, B-26C and B-26K also serve with the air forces of Brazil, Chile, Colombia, Guatamala, Indonesia, Laos, Nicaragua and Peru. Of 30 B-26Bs acquired by Brazil, 18 were later modernized to B-26K Counter-Invader standard, this variant also being operated in the counter-insurgency rôle with the air arms of Brazil and Laos. Apart from more powerful engines, the B-26K, which was re-manufactured by On Mark Engineering, features wingtip tanks and eight underwing ordnance pylons. Day and night photo-reconnaissance capability is built in.

FAIRCHILD A-10A

The photograph (above) and the general arrangement silhouette (right) illustrate the prototype version of the A-10A which differs in minor respects from the proposed production model

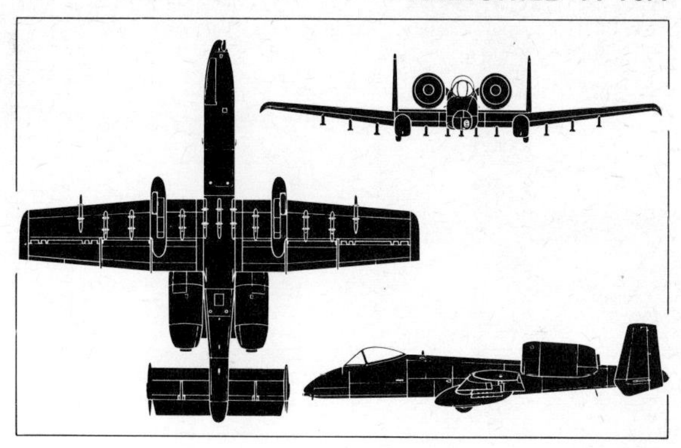

Country of Origin: USA.
Type: Close-support attack aircraft.
Power Plant: Two 9,275 lb st (4 207 kgp) General Electric TF34-GE-100 turbofans.
Performance: (Estimated) Max speed, at 25,500 lb (11 567 kg) 500 mph (805 km/h) at sea level; approx mission radius (with 9,500 lb (4 309 kg) useful payload and including 2 hrs loiter) 300 mls (480 km).
Weights: Approx operational empty weight, 21,300 lb (9 660 kg); approx max take-off 45,825 lb (20 786 kg).
Dimensions: Span, 55 ft 0 in (16,76 m); length, 52 ft 7 in (16,03 m); height, 14 ft $5\frac{1}{2}$ in (4,41 m); wing area, 488 sq ft (45,13 m^2).
Accommodation: Pilot only.
Armament: One (initial standard) 20-mm M-61 rotary cannon or (final standard) 30-mm GAU-8/A cannon and 11 external pylons (three under fuselage, four under each wing) with combined total capacity of 18,500 lb (8 392 kg) of ordnance. Typical possible loads include 24 Mk 82 500-lb (227 kg) bombs, 16 M-117 750-lb (340-kg) bombs, four Mk 84 2,000-lb (907-kg) bombs, 20 Rockeye 11 cluster bombs or nine AGM-65 Maverick missiles.

Status: Two prototypes for USAF evaluation and fly-off first flown respectively on 10 May and 21 July 1972. Declared winner of contest on 18 January 1973 and initial production batch of 32 ordered by USAF, for delivery 1975/1977.
Notes: The A-10A was one of two designs chosen by the USAF for full-scale evaluation to meet a requirement (known as A-X) for a specialised ground-attack aircraft. Two prototypes were built for a fly-off against the two Northrop A-9As, this involving about 125 hrs of flight testing (by each type) in a 60-day period commencing late October 1972. The data collected during these flights was rapidly assessed, leading to selection of the A-10A as the winning design in January 1973. Outline plans were completed by USAF for acquisition of up to 600 A-10As, with an order being placed for a pre-production batch of six to be delivered in 1975. The Philco-Ford and General Electric company held development contracts for the new GAU-8/A 30-mm cannon using caseless ammunition, and the GEC version was selected for use in the future production A-10As. The unusual positioning of the turbofans has been selected to reduce vulnerability to foreign object ingestion when operating from short battlefield-area airstrips.

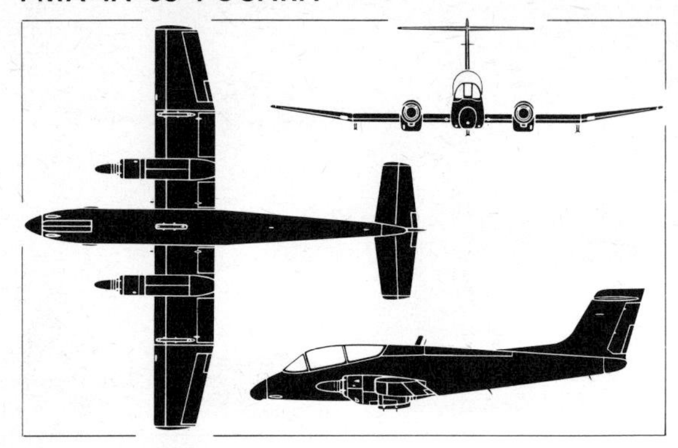

Both photograph (above) and general arrangement silhouette (left) illustrate the definitive second prototype of the Pucará counter-insurgency aircraft

Country of Origin: Argentina.
Type: Counter-insurgency aircraft.
Power Plant: Two 1,022 eshp Turboméca Astazou XV1G turbo-props.
Performance: Max speed, 323 mph (520 km/h) at 9,840 ft (3 000 m); max cruising speed, 301 mph (485 km/h) at 9,840 ft (3 000 m); economical cruising speed, 267 mph (430 km/h); initial rate of climb, 3,543 ft/min (18,0 m/sec); service ceiling, 27,165 ft (8 280 m); range with two 66 Imp gal (300 l) auxiliary tanks, 1,890 mls (3 040 km) at 16,400 ft (5 000 m).
Weights: (First prototype) Basic operating, 7,826 lb (3 550 kg); max take-off, 13,668 lb (6 200 kg); max landing, 11,023 lb (5 000 kg).
Dimensions: Span, 47 ft 6¾ in (14,50 m); length, 45 ft 7¼ in (13,90 m); height, 17 ft 2¼ in (5,24 m); wing area, 326·1 sq ft (30,30 m²).
Accommodation: Pilot and observer in tandem.
Armament: Two 20-mm Hispano cannon and four 0·30-in (7,62 mm) FN machine guns in forward fuselage. One hardpoint on fuselage centre line and one under each wing for various ordnance loads including bombs, rockets or weapon pods.
Status: Aerodynamic prototype (unpowered) first flown on 26 December 1967; first powered prototype flown 20 August 1969, second prototype flown 6 September 1970. In production for *Fuerza Aérea Argentina* (the Argentine Air Force) against initial order for 50 aircraft.
Notes: The Pucará—originally known as the Delfin—was designed at the *Fabrica Militar de Aviones* at Cordoba to meet Argentine Air Force requirements for a light strike aircraft. The design was proved in flight testing of a glider version before construction of the prototypes began. Designated AX-01, the first prototype was powered by two AiResearch TPE 331-U-303 engines, being followed by the AX-02 with Astazou XV1G engines, as specified for production aircraft. Production was under way in 1973 against Argentine Air Force requirements for about 80 Pucarás, to replace the indigenous Huanquero (see page 171) and some other types in service for light attack, reconnaissance and counter-insurgency duties. Several variants of the Pucará were being projected in 1973, including a navalised version for the anti-shipping rôle, with folding wings, arrester gear and catapult spools.

GRUMMAN A-6 INTRUDER

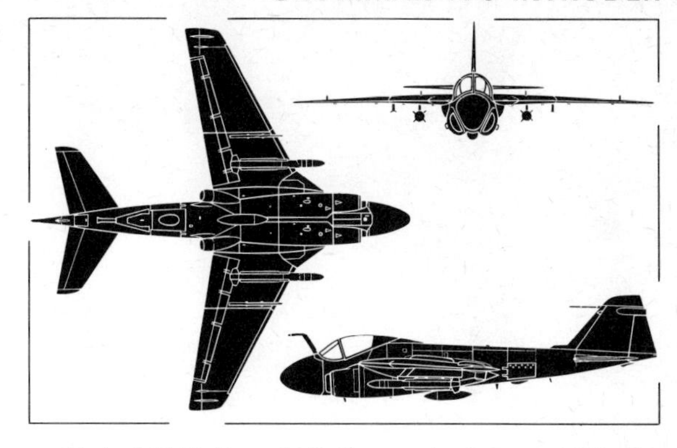

The photograph above depicts an A-6A of US Navy Squadron VA-42 and the general arrangement silhouette (right) illustrates a Standard ARM-equipped A-6B

Country of Origin: USA.
Type: Carrier-borne low-level strike aircraft.
Power Plant: Two 9,300 lb st (4 218 kgp) Pratt & Whitney J52-P-8A turbojets.
Performance: (A-6A) Max speed at 36,655 lb (16 626 kg) in clean condition, 685 mph (1 102 km/h) or Mach 0·9 at sea level, 625 mph (1 006 km/h) or Mach 0·94 at 36,000 ft (10 970 m); average cruise, 480 mph (772 km/h) at 32,750–43,800 ft (9 980–13 350 m); range with max internal fuel and four Bullpup ASMs, 1,920 mls (3 090 km), with single store and four 250 Imp gal (1 136 l) external tanks, 3,040 mls (4 890 km).
Weights: Empty, 25,684 lb (11 650 kg); loaded (clean), 37,116 lb (16 836 kg); max overload take-off, 60,280 lb (27 343 kg).
Dimensions: Span, 53 ft 0 in (16,15 m); length, 54 ft 7 in (16,64 m); height, 15 ft 7 in (4,75 m); wing area, 529 sq ft (49,15 m²).
Accommodation: Pilot and observer side-by-side.
Armament: One strongpoint on fuselage centre line and four underwing pylons having maximum combined capacity of 15,000 lb (6 804 kg) and individual capacity of 3,600 lb (1 633 kg) each.
Status: First of eight test and development aircraft (A2F-1) flown on 19 April 1960. Entered US Navy service February 1963. First KA-6D flown on 23 May 1966; first A-6E flown on 27 February 1970. Total production, over 500 by 1973.
Notes: The Intruder (Grumman G-128) was chosen by the US Navy in 1957 to meet a requirement for a long-range strike aircraft operating at low level from carrier decks. An unusual feature of the initial design was that the jet pipes could be swivelled downwards to shorten the take-off run, but later production aircraft had fixed tail pipes. Three of the development batch of A-6As were later converted to NA-6As as flight refuelling tankers for the F-14A flight test programme. An operational tanker for US Navy use was developed as the KA-6D and was produced in limited numbers, together with the A-6B, an A-6A conversion to carry Standard ARM anti-radar missiles and the A-6C which was fitted with FLIR and LLTV equipment for night target acquisition in Vietnam operations. The A-6E succeeded the A-6A in production in 1970 and has an advanced IBM computer and a multi-mode radar; 192 A-6As were scheduled to be updated to this standard in due course plus procurement of 192 new A-6Es. The EA-6A and EA-6B specialised electronic warfare aircraft are separately described on page 93.

HAWKER SIDDELEY BUCCANEER

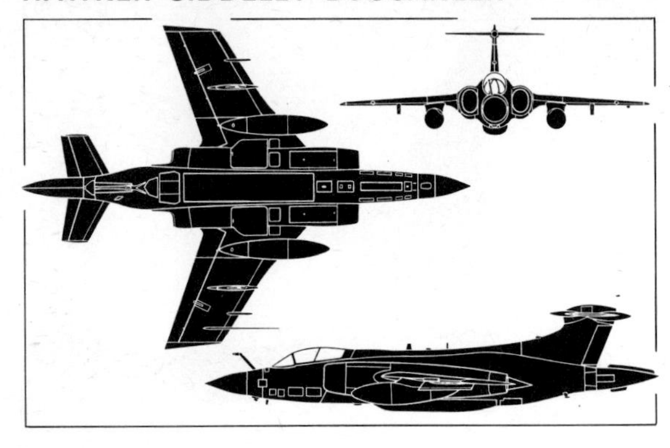

The photograph (above) and the general arrangement silhouette (left) illustrate the Buccaneer S Mk 2B, the former depicting an aircraft of No 16 Squadron, RAF

Country of Origin: United Kingdom.
Type: Strike and reconnaissance aircraft.
Power Plant: Two (S Mk 1) 7,100 lb st (3 220 kgp) Bristol Siddeley Gyron Junior 101 turbojets or (S Mk 2 and 50) 11,100 lb st (5 035 kgp) Rolls-Royce Spey RB.168-1A Mk 101 turbofans plus (S Mk 50 only) 8,000 lb st (3 630 kgp) Bristol Siddeley BS.605 rocket engine.
Performance: (S Mk 2, estimated): Max speed, 645 mph (1 040 km/h) or Mach 0·85 at 250 ft (75 m), 620 mph (998 km/h) or Mach 0·92 at 30,000 ft (9 145 m); typical low-level cruise, 570 mph (917 km/h) or Mach 0·75 at 3,000 ft (915 m); tactical radius (hi-lo-lo-hi) with standard fuel, 500–600 mls.
Weights: Typical loaded 46,000 lb (20 865 kg); max take-off, 62,000 lb (28 123 kg).
Dimensions: Span, 44 ft 0 in (13,41 m); length, 63 ft 5 in (19,33 m); height, 16 ft 3 in (4,95 m); wing area, 514·7 sq ft (47·82 m²).
Accommodation: Pilot and observer in tandem.
Armament: Internal weapons bay (with rotary bomb door), capacity up to 4,000 lb (1 814 kg) and four wing stations each with 3,000 lb (1 360 kg) capacity, carrying Bullpup or Martel ASMs, bombs or rocket pods.
Status: First of 20 development batch aircraft flown on 30 April 1958; first production S Mk 1 flown on 23 January 1962; prototype S Mk 2 flown on 17 May 1963; first production S Mk 2 flown on 5 June 1964; first RAF S Mk 2B flown on 8 January 1970. Operational use of S Mk 1 began July 1962, S Mk 2 in October 1965 (RN) and July 1970 (RAF). Production quantities, development batch, 20; S Mk 1, 40; S Mk 2, 84; S Mk 2B, 42; S Mk 50, 16.
Notes: Buccaneer S Mk 1 and original S Mk 2s were delivered to RN, the S Mk 2B being ordered new for the RAF. Most S Mk 2s were transferred to RAF when relinquished by RN, becoming S Mk 2A with RAF equipment and S Mk 2B with provision for Martel ASMs; unmodified aircraft without and with Martel provision became S Mk 2C and 2D for continuing service with RN. RAF force comprises five squadrons including Nos 12, 15 and 16; SAAF has 13 rocket-boosted S Mk 50s in service with No 24 Squadron. These are essentially similar to the S Mk 2 but are fitted with a retractable Rolls-Royce Bristol BS.605 twin-chamber rocket engine to boost take-off performance under hot-and-high conditions.

HAWKER SIDDELEY HARRIER

The photograph (above) depicts a development Harrier with ECM fairing on tail fin and aerodynamic shape of nose-mounted laser rangefinder. Silhouette (right) depicts Harrier GR Mk 1A

Country of Origin: United Kingdom.
Type: V/STOL strike and reconnaissance fighter.
Power Plant: One (GR 1, T 2) 19,000 lb st (8 620 kgp) Rolls-Royce Pegasus 101 or (GR 1A, T 2A) 20,000 lb st (9 071 kgp) Pegasus 102 or (GR 3, T 4, Mk 52A AV-8A, TAV-8A) 21,500 lb st (9 752 kgp) Pegasus 103 vectored-thrust turbofan.
Performance: Max speed, over 720 mph (1 160 km/h) or Mach 0·95 at 1,000 ft (305 m), speed with typical external loads, 640-660 mph (1 030–1 060 km/h) or Mach 0·85–0·87 at 1,000 ft (305 m); cruising speed, 560 mph (900 km/h) or Mach 0·8 at 20,000 ft (6 096 m); tactical radius (hi-lo-hi) 260 mls (418 km) without external fuel; operational ceiling, over 50,000 ft (15 240 m); ferry range, 2,070 mls (3 330 km).
Weights: Empty, 12,400 lb (5 624 kg); max take-off (VTO), 18,000 lb (8 165 kg); max take-off (STOL) over 23,000 lb (10 435 kg); max take-off, about 26,000 lb (11 793 kg).
Dimensions: Span, 25 ft 3 in (7,70 m); span with ferry tips, 29 ft 8 in (9,04 m); length, (single-seat) 45 ft 6 in (13,87 m); length (two-seat), 55 ft 9½ in (17,00 m); wing area, 201·1 sq ft (18,68 m²); wing area with ferry tips, 216 sq ft (20·1 m²).

Accommodation: (GR 1, 1A, 3, AV-8A) Pilot only (T2, 2A, 4, 52, TAV-8A) Two in tandem, full dual controls.
Armament: Provision for two 30-mm Aden gun pods under fuselage. Four wing hardpoints with combined capacity of 5,300 lb (2 400 kg) for bombs, Martel ASMs (AV-8A) Sidewinder ASMs etc.
Status: Prototype P.1127 first hovered 21 October 1960; first Kestrel prototype flown on 7 March 1964; first development Harrier flown on 31 August 1966; first production GR Mk 1 flown on 28 December 1967; first T Mk 2 flown on 24 April 1969; Mk 52 first flown on 16 September 1971. Production and orders (late-1973): development batch, 6; GR Mk 1/1A/3, 92; T Mk 2/2A/4, 13; AV-8A, 112; AV-8A (Spain), 6; TAV-8A, 8; TAV-8A (Spain), 2; Mk 52, 1.
Notes: Harrier is world's first V/STOL fighter. Initial orders are for four RAF Squadrons in UK and Germany, three USMC squadrons in USA. Versions differ primarily in engine rating; first 10 AV-8As were delivered with F402-RR-400s (equivalent to Pegasus 102), later modified to have F402-RR-401s (Pegasus 103) as fitted in later AV-8As. Mk 52 is company demonstrator. Projected AV-16A has new wing and 25,000 lb st (11 340 kgp) Pegasus 15.

McDONNELL DOUGLAS A-4 SKYHAWK

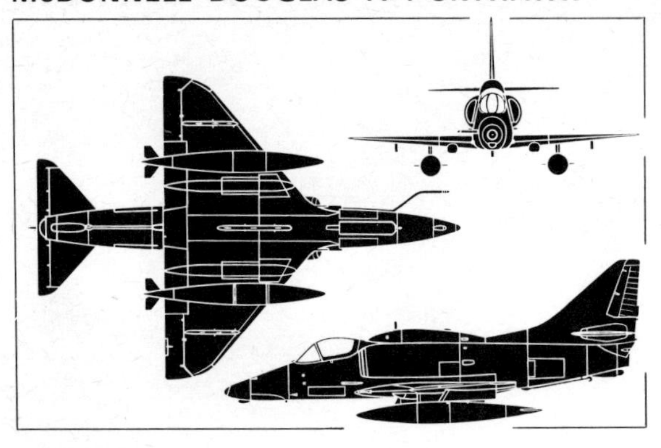

The photograph (above) depicts an A-4M Skyhawk II of the USMC and the general arrangement silhouette (left) illustrates the essentially similar export A-4N

Country of Origin: USA.
Type: Carrier or land-based attack bomber.
Power Plant: One (A-4B, C, L, P, Q) 7,700 lb st (3 493 kgp) Wright J65-W-16A or (A-4EJ) 8,500 lb st (3 856 kgp) J52-P-6 or (A-4F, G, H, K) 9,300 lb st (4 218 kgp) J52-P-8A or (A-4M, N) 11,200 lb st (5 443 kgp) J52-P-408A turbojet.
Performance: (A-4M) Maximum speed, clean, 670 mph (1 078 km/h) at sea level, Mach 0·94 at 25,000 ft (7 620 m); initial rate of climb, 8,440 ft/min (43,0 m/sec); tactical radius, close support mission with 4,000 lb (1 814 kg) bomb load, 340 mls (538 km); ferry range, 2,050 mls (3 307 km).
Weights: (A-4M) Empty, 10,465 lb (4 747 kg); basic operating, 12,250 lb (5 570 kg); max take-off, 24,500 lb (11 113 kg).
Dimensions: Span, 27 ft 6 in (8,38 m); length overall (excluding refuelling probe) (A-4E, F, G, H, K), 40 ft 1½ in (12,22 m), (A-4M, N), 40 ft 3¾ in (12,27 m), (TA-4 versions), 42 ft 7¼ in (12,98 m); height overall (single-seaters), 15 ft 0 in (4,57 m), (two-seaters), 15 ft 3 in (4,66 m); wing area, 260 sq ft (24,16 m²).
Accommodation: (A-4 series) Pilot only or (TA-4 series) two in tandem, dual controls.

Armament: Two (all versions except H and N) 20-mm MK-12 cannon or (A-4H, TA-4H, A-4N) 30-mm DEFA cannon in wing roots. One centreline strong point and four wing pylons, combined capacity, 9,155 lb (4 153 kg).
Status: Prototype (XA4D-1) first flown on 22 June 1954; A-4B (A4D-2) flown 26 March 1956; A-4C (A4D-2N) flown 21 August 1958; A-4E (A4D-5) flown 12 July 1961; TA-4E flown 30 June 1965; first production TA-4F flown April 1966; A-4F flown 31 August 1966; TA-4J flown May 1969; A-4M Skyhawk II flown 10 April 1970; A-4N flown 12 June 1972. Production totals include A-4A and prototypes, 166; A-4B, 542; A-4C, 638; A-4E, 499; A-4F, 154; TA-4F, about 200; A-4M, over 100; TA-4J, unknown; A-4K, 10 and TA-4K, 4 (for RNZAF); A-4G, 8 and TA-4G, 2 (for RAN); A-4H, 90, TA-4H, 10 and A-4N, 104 (for Israel).
Notes: Refurbished A-4Cs serve with Naval Air Reserve as A-4Ls. Argentine Navy has 16 A-4Qs and Argentine Air Force has 50 A-4Ps, these also being refurbished A-4Bs, and Singapore has purchased 40 surplus A-4Bs. The 2,700th A-4 Skyhawk was delivered on 14 June 1973, and production is expected to continue into 1975.

NORTH AMERICAN F-51 MUSTANG

The photograph above illustrates a Cavalier-refurbished F-51D Mustang of the Dominican Air Force and the general arrangement silhouette (right) depicts the standard F-51D Mustang

Country of Origin: USA.
Type: Fighter-bomber.
Power Plant: (F-51D) One 1,490 hp Packard (Rolls-Royce Merlin) V-1650-7 12-cylinder V-in-line piston engine.
Performance: (F-51D) Maximum speed, 437 mph (703 km/h) at 25,000 ft (7 620 m); cruising speed, 362 mph (583 km/h); time to climb to 30,000 ft (9 145 m), 13 minutes; service ceiling, 41,900 ft (12 770 m); range, 950 mls (1 530 km).
Weights: (F-51D) Empty, 7,125 lb (3 230 kg); normal loaded, 10,100 lb (4 580 kg); max take-off, 11,600 lb (5 260 kg).
Dimensions: Span, 37 ft $0\frac{1}{2}$ in (11,29 m); length, 32 ft $2\frac{1}{2}$ in (9,81 m); height, 12 ft $2\frac{1}{2}$ in (3,72 m); wing area, 272·3 sq ft (25,29 m²).
Accommodation: Pilot or (TF-51D) two in tandem, optional dual controls.
Armament: (F-51D re-work) Six 0·50-in (12,7-mm) machine guns in wings and eight underwing hardpoints, two of which each carry 1,000-lb (454-kg) bomb and others carry a 5-in (12,7 cm) rocket.
Status: Prototype (NA-73X) first flown 26 October 1940; first production Mustang I flown on 1 May 1941; first Merlin-engined Mustang flown on 13 October 1942; first Commonwealth CA-17 Mustang XX flown on 29 April 1945; Cavalier Mustang II flown December 1967; Turbo-Mustang III flown in 1968, Piper Enforcer flown on 29 April 1971. Production ended 1946; production quantities were: NA-73X, 1; Mustang I, 620; XP-51, 2; P-51A, 310; A-36A, 500; Mustang IA, 150; P-51B, 1,988; P-51C, 1,750; P-51D, 6,502 at Inglewood and 1,454 at Dallas; P-51K, 1,337; P-51H, 555; Commonwealth Mustang XX, 80; Mustang 21, 40; Mustang 22, 14; Mustang 23, 66.
Notes: One of World War II's most successful fighters, the Mustang flies on in the 'seventies in service with half-a-dozen nations—including Bolivia, Dominica, Haiti, Indonesia, and Salvador. Some are original wartime production, but they have been supplemented more recently by a batch of completely overhauled and modified F-51Ds supplied through MAP by Cavalier Aircraft. These late-delivered models have larger one-piece canopies, provision for a second occupant (with dual controls optional), taller fin and strengthened wings with more weapon strong-points. Cavalier developed an experimental turboprop version and Piper undertook a similar programme as the Enforcer.

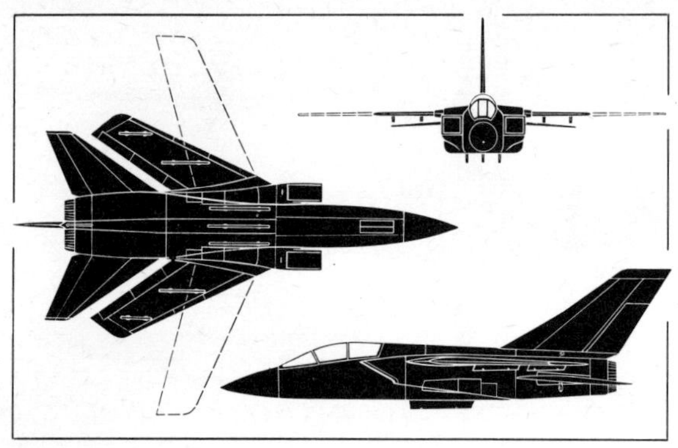

The photograph (above) and the general arrangement silhouette (left) depict the initial version of the MRCA (Multi-Rôle Combat Aircraft)

Country of Origin: United Kingdom/Germany/Italy.
Type: Two-seat multi-purpose fighter.
Power Plant: Two 8,500 lb st (3 855 kgp) dry and 14,500 lb (6 577 kgp) with reheat Turbo-Union RB.199-34R turbofans.
Performance: Max speed, over 1,320 mph (2 125 km/h) or over Mach 2 above 36,000 ft (10 970 km); approx 910 mph (1 465 km/h) at low altitude; combat endurance on internal fuel, about 70–80 minutes.
Weights (Estimated): Empty equipped, 22,000–23,000 lb (9 980–10 430 kg); loaded, 38,000–40,000 lb (17 240–18 145 kg).
Dimensions: Span (max), 45 ft 7 in (13,90 m), (min), 28 ft 2½ in (8,60 m); length, 54 ft 10 in (16,70 m); height, 18 ft 8½ in (5,70 m).
Accommodation: Two pilots or pilot and systems manager.
Armament: Two 27-mm Mauser cannon internally. Three pylons on fuselage and four on wings carry a variety of air-to-ground weapons—bombs, rockets, guided missiles, etc.
Status: First of six prototypes flown on 14 August 1974. Test programme embraces nine prototype and pre-production aircraft, with production deliveries scheduled to begin in 1977–78. Planned production commitments are 385 for the RAF, 322 for the Federal German *Luftwaffe* and the *Marineflieger* and 100 for the Italian *Aeronautica Militare*.
Notes: The MRCA (Multi-Rôle Combat Aircraft) is the product of a multi-national European industrial company, Panavia Aircraft GmbH, set up to handle this programme and combining the resources of the British Aircraft Corporation, Messerschmitt-Bölkow-Blohm and Aeritalia. The design, development and production programmes are shared between the three countries in the appropriate proportion of their respective purchases of production quantities. Thus, four of the nine prototypes are being built in the UK, three in Germany and two in Italy. The two last-named countries will use the MRCA to replace F-104s and Fiat G91s; in the RAF, the type will initially replace Vulcan and Buccaneer bombers but about 165 of the aircraft for the RAF will be built in a specially-equipped version for the intercept rôle and will replace F-4K Phantoms. In *Luftwaffe* service the MRCA will perform conventional all-weather fighter missions, low-level penetration, battlefield interdiction, counterair operations and close air support for land forces.

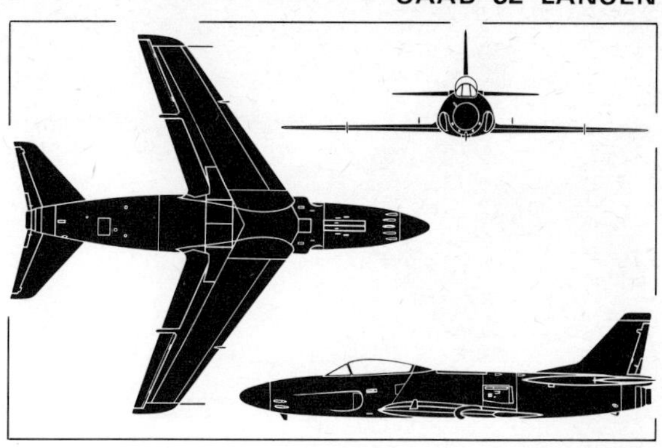

The photograph above depicts an S 32C Lansen of F 11 and the general arrangement silhouette (right) illustrates the A 32A which is now in process of replacement by the AJ 37 Viggen

Country of Origin: Sweden.
Type: All-weather attack aircraft.
Power Plant: One 7,628 lb st (3 460 kgp) dry and 10,362 lb st (4 700 kgp) with reheat Svenska Flygmotor RM5A2 (Rolls-Royce Avon 100) turbojet.
Performance: (A 32A) Max speed, 692 mph (1 114 km/h) or Mach 0·91 at sea level; normal cruising speed, 608 mph (978 km/h) or Mach 0·8; initial rate of climb, 11,810 ft/min (60,0 m/sec); service ceiling, 49,200 ft (15 000 m); max range with external fuel, 2,000 mls (3 219 km).
Weights: Empty equipped, 16,398 lb (7 438 kg); normal take-off, 22,956 lb (10 413 kg); maximum loaded, 28,660 lb (13 600 kg).
Dimensions: Span, 42 ft 7¾ in (13,00 m); length, 49 ft 0¾ in (14,94 m); height, 15 ft 3 in (4,65 m); wing area, 4,026 sq ft (37,40 m²).
Accommodation: Crew of two in tandem.
Armament: Four 20-mm cannon in forward fuselage. Underwing provision for up to 24 rockets, two 1,320 lb (600 kg) bombs, four 550 lb (250 kg) bombs or two RB 04C anti-shipping missiles.
Status: First of four prototypes flown on 3 November 1952. First production A 32As delivered to Swedish Air Force in December 1955. Prototype S 32C flown on 26 March 1957 and deliveries made in 1959–60. First of two prototype J 32Bs flown on 7 January 1957; deliveries began in July 1958. Total Saab 32 production, 450, of which about 280 were A 32As. Last production delivery May 1960.
Notes: Design of a new attack aircraft to meet the special requirements of the Swedish Air Force began in 1946, a jet-powered type being specified, to replace the Saab 18B twin piston-engined types. Detail design and construction of a mock-up was authorised on 20 December 1948 and led to the appearance and production in substantial quantity of the Saab 32 Lansen. Three versions were built: the A 32A attack aircraft, the SK 32C two-seat trainer and the J 32B fighter. The J 32B is no longer in service but A 32As have been operational with four attack wings—F6, F7, F15 and F17—since 1956 and were only slowly being superseded by the AJ 37 Viggen in 1973 (see page 40). Also in service until full introduction of the reconnaissance version of the Viggen are two squadrons of the S 32C with F11 at Nyköping. F7 was working up on the Viggen in 1973, and F15 and F6 will commence conversion during 1974.

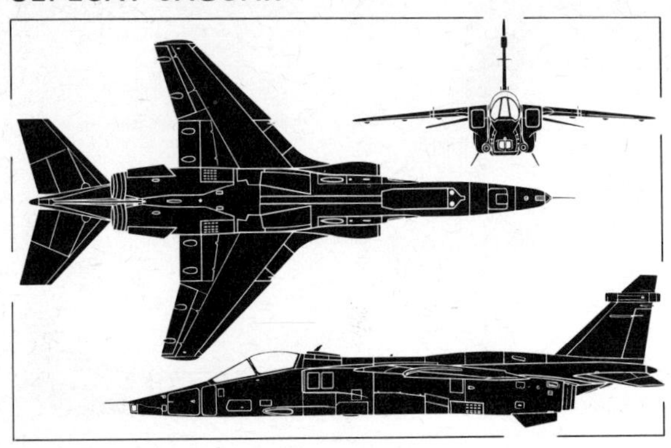

The photograph (above) and general arrangement silhouette (left) both depict the Jaguar GR Mk 1 of the RAF with nose laser range-finder and tail-mounted avionics pack

Country of Origin: France/United Kingdom.
Type: Tactical strike fighter and advanced trainer.
Power Plant: Two 4,620 lb st (2 100 kgp) dry and 7,140 lb st (3 240 kgp) with reheat Rolls-Royce Turboméca RT.172 Adour 102 turbofans.
Performance: (GR Mk 1 at typical weight) Max speed, 820 mph (1 320 km/h) or Mach 1·1 at 1,000 ft (305 m), 1,057 mph (1 700 km/h) or Mach 1·6 at 32,810 ft (10 000 m); cruise with max ordnance, 430 mph (690 km/h) or Mach 0·65 at 39,370 ft (12 000 m); range with external fuel for lo-lo-lo mission profile, 450 mls (724 km), for hi-lo-hi mission profile, 710 mls (1 142 km); ferry range, 2,270 mls (3 650 km).
Weights: Normal take-off, 23,000 lb (10 430 kg); max take-off, 32,600 lb (14 790 kg).
Dimensions: Span, 28 ft 6 in (8,69 m); length (A and GR Mk 1), 50 ft 11 in (15,52 m); length (E and T Mk 2), 53 ft 10 in (16,42 m); height, 16 ft 0½ in (4,89 m); wing area, 260·3 sq ft (24,18 m²).
Accommodation: (Jaguar A and GR Mk 1) Pilot only; (Jaguar E and T Mk 2) Two in tandem, full dual control.
Armament: Fixed forward firing armament comprising (Jaguar E and A) two 30-mm DEFA cannon, (GR Mk 1) two 30-mm Aden cannon, (T Mk 2) one 30-mm Aden cannon. One fuselage and four wing hardpoints carry max ordnance load of 10,000 lb (4 536 kg).
Status: Eight prototypes built, first flight dates being: Jaguar E, 8 September 1968 and 11 February 1969; Jaguar A, 29 March and 27 May 1969; Jaguar M, 14 November 1969; Jaguar S (GR Mk 1) 12 October 1969 and 12 June 1970; Jaguar B (T Mk 2) 30 August 1971. First production Jaguar E flown 2 November 1971; first production Jaguar A flown 20 April 1972; first GR Mk 1 on 11 October 1972, first T Mk 2 on 22 March 1973. Production deliveries to *l'Armée de l'Air* started in May 1972; first deliveries to RAF in June 1973. Production programme (planned), 200 Jaguar E and A for *l'Armée de l'Air*; 165 GR Mk 1 and 35 T Mk 2 for RAF.
Notes: Nine RAF squadrons to fly Jaguars include Nos 6, 54 and 41 in No 38 Group, Strike Command, and six in RAF Germany. The first *l'Armée de l'Air* unit is the 7e *Escadre* at St. Dizier. Early production aircraft have Adour 101 engines without part-throttle reheat. The RAF's Jaguar GR Mk 1 differs from the French single-seat model in having a nose-mounted laser rangefinder and a tail-mounted avionics pack.

SUD-AVIATION SO-4050 VAUTOUR

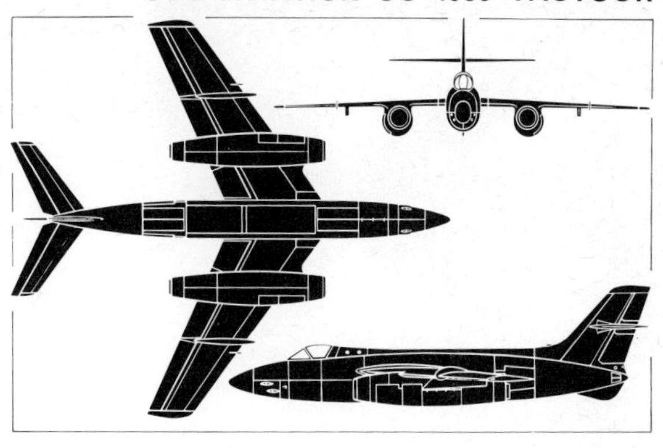

The photograph (above) depicts a Vautour IIN of the 30e Escadre of the Armée de l'Air and the general arrangement silhouette (right) illustrates the Vautour IIA

Country of Origin: France.
Type: (Vautour IIA) Ground attack and (IIN) all-weather fighter aircraft.
Power Plant: Max speed 721 mph (1 160 km/h) at sea level, initial rate of climb, 11,800 ft/min (60 m/sec); time to reach 32,810 ft (10 000 m), 5 min, service ceiling, 44,290 ft (13 500 m); normal range, 1,550 mls (2 500 km); max range with drop tanks, 3,730 mls (6 000 km).
Weights: Empty, about 22,000 lb (10 000 kg); take-off weight, according to mission, 30,865–39,680 lb (14 000–18 000 kg).
Dimensions: Span, 49 ft 6½ in (15,10 m); length, 51 ft 1½ in (15,57 m); height, 14 ft 9 in (4,50 m); wing area, 474 sq ft (45 m²).
Accommodation: (Vautour IIA) Pilot only (IIN) pilot and observer.
Armament: Four 30-mm DEFA cannon and (IIA) one ASM under fuselage plus four underwing pylons for rocket pods, bombs or ASMs; (IIN) provision in fuselage for retractable rocket packs plus four underwing pylons for rocket pods or AAMs.
Status: Prototype SO-4050-001 Vantour N first flown on 16 October 1952; -002 Vautour A flown on 16 December 1963; -003 Vautour B flown on 5 December 1954; first of six pre-production aircraft (SO-4050-04 Vautour 8 flown on 17 March 1955); first production Vautour II flown 30 April 1956. Production totals, Vautour IIA, 30; Vautour IIB, 40 and Vautour IIN, 70 (plus prototypes and pre-production).
Notes: Design of the Vautour began in June 1951, the object being to produce a single-seat twin-jet attack aircraft armed with air-to-air ground missiles. This version became the Vautour IIA in its production guise, two-seat variants also being designed for tactical bombing for *l'Armée de l'Air* but only the Vautour IIB and IIN reached operational status in France, the Vautour IIAs being transferred to Israel. One squadron of the latter still operated with the latter in 1973, serving in long-range attack rôle, and Israel has also received from France a few Vautour IINs which have been modified for electronic reconnaissance and counter-measures duties. In France, two *escadrons* of the 30e *Escadre de Chasse Tous-Temps* still operated Vautour IINs from their base at Reims in 1973, but the type was earmarked for early retirement from both the French and the Israeli forces, and is scheduled to give place in the former to the Mirage F1 during 1974–75, by which time it will have completed some 20 years of service with the *Armée de l'Air.*

SUKHOI SU-7BM

The photograph (above) and general arrangement silhouette (left) illustrate the standard Sukhoi Su-7BM with twin ventral tanks and rocket pods fitted

Country of Origin: USSR.
NATO Code Name: *Fitter-A.*
Type: Ground attack fighter.
Power Plant: One 14,200 lb (6 440 kg) dry and 22,046 lb (10 000 kg) with reheat Lyulka AL-7F-1 turbojet.
Performance: Max speed (clean), 720 mph (1 160 km/h) or Mach 0·95 at 1,000 ft (305 m), 1,056 mph (1 700 km/h) or Mach 1·6 at 39,370 ft (12 000 m); max speed in high-drag configuration (e.g., two rocket pods and two 132 Imp gal/600 l drop tanks), 790 mph (1 270 km/h) or Mach 1·2 at 39,370 ft (12 000 m); max speed in max dry thrust and high-drag configuration, 530 mph (853 km/h) at 1,000 ft (305 m); combat radius (hi-lo-hi mission profile), 285 mls (460 km).
Weights: Empty, 19,000 lb (8 620 kg); normal loaded, 26,455 lb (12 000 kg); max take-off, 29,600 lb (13 425 kg).
Dimensions: Span, 29 ft 3½ in (8,93 m); length (including probe), 57 ft 0 in (17,37 m); height, 15 ft 5 in (4,70 m).
Accommodation: Pilot only.
Armament: Two 30-mm NR-30 cannon with 70 rpg and external loads such as two 550-lb (250-kg) bombs and two UV-16-57 pods each containing 16 57-mm rockets distributed between four stores stations (two under wings and two under fuselage).
Status: Prototypes allegedly flown in 1955 with production deliveries of the initial service version to the Soviet Air Forces commencing 1958–59.
Notes: Currently the standard ground attack fighter of the Soviet Air Forces and serving with the air forces of Czechoslovakia, Egypt, East Germany, Hungary, India and Poland, the Su-7 has been progressively improved since first introduced into service, the Su-7BM featuring an uprated engine, a revised brake chute housing, repositioned pilot head, the provision of 30-mm cannon in place of 23-mm weapons and other modifications. The variant supplied to India is designated Su-7BMK and a tandem two-seat conversion training variant is designated Su-7U (*Moujik*). A variable-geometry version of the Su-7 (*Fitter-B*) was demonstrated publicly in July 1967. This is essentially similar to the standard Su-7BM apart from pivoting outer wing panels, the fixed portion being increased in thickness to accommodate the wing sweep actuation mechanism. The variable-geometry version has entered production and service as the Su-17 (*Fitter-C*).

VOUGHT A-7 CORSAIR II

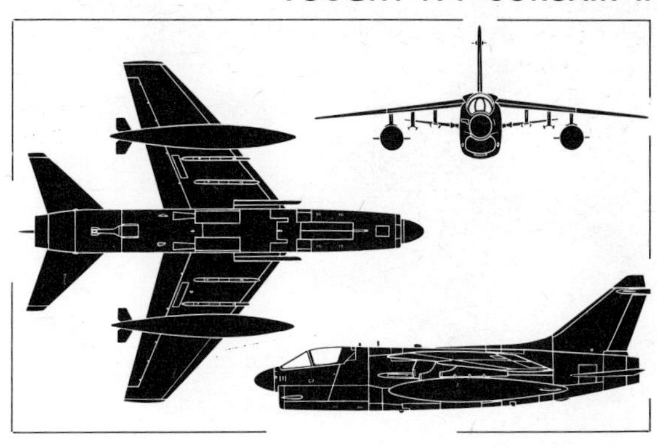

The photograph above depicts two A-7Ds of the USAF's 57th Fighter Weapons Wing. The general arrangement silhouette illustrates the A-7E with long-range ferry tanks

Country of Origin: USA.
Type: Tactical attack aircraft.
Power Plant: One (A-7A) 11,350 lb st (5 150 kgp) Pratt & Whitney TF30-P-6 or (A-7B and A-7C) 12,200 lb st (5 534 kgp) TF30-P-8 or (A-7D) 14,250 lb st Allison (Rolls-Royce RB.168 Spey) TF41-A-1 or (A-7E and YA-7H) 15,000 lb st (6 800 kgp) TF41-A-2 turbofan.
Performance: (A-7D) Max speed (clean), 698 mph (1 123 km/h); tactical radius with internal fuel and 3,600 lb (1 623 kg) external weapon load, 700 mls (1 127 km); ferry range, over 4,000 mls (6 437 km).
Weights: Empty, 19,490 lb (8 840 kg); max take-off, 42,000 lb (19 050 kg).
Dimensions: Span, 38 ft $8\frac{1}{2}$ in (11,79 m); length, 46 ft $1\frac{1}{2}$ in (14,05 m); length (YA-7H), 48 ft $5\frac{1}{2}$ in (14,84 m); height, 16 ft 0 in (4,88 m); wing area, 375 sq ft (34,84 m²).
Accommodation: Pilot only or (YA-7H) two pilots in tandem.
Armament: One 20-mm M61-A-1 multi-barrel gun in fuselage. Two external weapon stations on fuselage and six underwing pylons carry a maximum combined load of 15,000 lb (6 804 kg) including bombs, ASMs, AAMs, rockets, gun pods or fuel tanks.
Status: First of seven development A-7As flown on 27 September 1965; deliveries to operational units began October 1966; first A-7B flown on 6 February 1968; first A-7C flown on 25 November 1968; first A-7D (TF30 engine) flown on 5 April 1968; first A-7D (TF41 engine) flown on 26 September 1968; YA-7H prototype flown on 29 August 1972. Production totals, A-7A, 199; A-7B, 196; A-7D (planned procurement) 387; A-7C, 67; A-7E (planned procurement) 494.
Notes: The A-7A entered service with VA-147 in February 1967 and was followed in due course by the A-7B with uprated engine. After the USAF had adopted a version of the Corsair II with a licence-built Rolls-Royce Spey engine, as the A-7D, the USN also adopted this engine, plus updated avionics, in the A-7E. Production delays with the new engine resulted in the first 67 A-7Es receiving the TF30-P-8 powerplant and being redesignated A-7C. The single YA-7H two-seat trainer was converted from the 68th A-7E (ie, the first with TF41 engine). No procurement of the A-7H had been funded up to the end of 1973. During 1973, studies were being undertaken for an advanced attack single-seater based on the YA-7H.

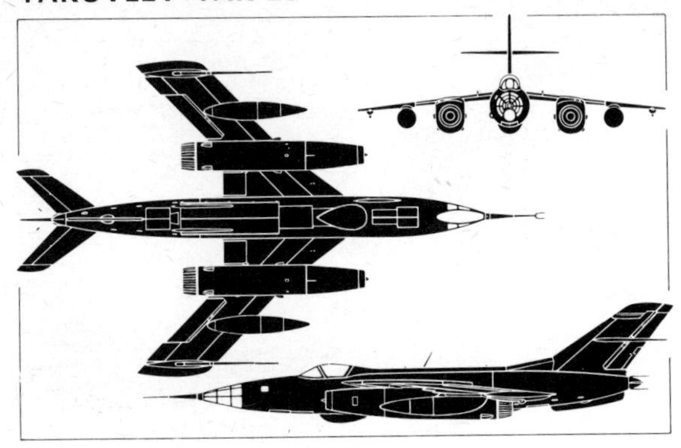

The photograph (above) and the general arrangement silhouette (left) both illustrate the standard Yak-28 tactical strike and reconnaissance aircraft

Country of Origin: USSR.
NATO Code Name: *Brewer.*
Type: Tactical strike and reconnaissance aircraft.
Power Plant: Two 10,140 lb (4 600 kg) dry and 13,670 lb (6 200 kg) with reheat Tumansky R-11 turbojets.
Performance: (Estimated) Max speed (clean), 760 mph (1 225 km/h) or Mach 1·15 at 39,370 ft (12 000 m), 725 mph (1 167 km/h) or Mach 0·95 at sea level; tactical radius, 230 mls (370 km/h) at 630 mph (1 015 km/h) at sea level, 575 mls (925 km) at 570 mph (917 km/h) at 36,090 ft (11 000 m); initial climb, 28,000 ft/min (142,2 m/sec); service ceiling, 55,000 ft (16 765 m).
Weights: (Estimated) Normal loaded, 35,000 lb (15 876 kg).
Dimensions: (Estimated) Span, 41 ft 0 in (12,5 m); length, 70 ft 0 in (21,34 m); height, 13 ft 0 in (3,96 m).
Accommodation: Pilot and observer.
Armament: One 30-mm NR-30 cannon in starboard side of forward fuselage. Internal weapons bay believed to house two 1,100-lb (500-kg) bombs. Stores stations outboard of engine nacelles can accommodate ASMs, 550-lb (250-kg) or 1,100-lb (500-kg) bombs, or tandem rocket pods for 57-mm missiles.

Status: Flown in prototype form in 1960 with production deliveries to the Soviet Air Forces commencing 1963-64. Currently standard tactical strike-recce equipment.
Notes: Evolved in parallel with the Yak-28P all-weather interceptor as a successor to the Ilyushin Il-28, the Yak-28 is currently serving in three versions: the standard tactical strike model, the Yak-28R tactical reconnaissance model with weapons bay deleted, this space being occupied by fuel tankage and non-optical reconnaissance sensors and ECM equipment, and the Yak-28U (*Maestro*) advanced operational conversion trainer which retains the weapons bay but has the glazed nose deleted and a second cockpit for the pupil inserted ahead and below the standard cockpit. The initial production version of the Yak-28 had a shorter (by approximately 2 ft 6 in/76 cm) fuselage, shorter engine air intakes and shorter wingtip probes, but later production aircraft feature similar fuselage length, engine nacelles and wingtip probes to those of the Yak-28P. The weapons bay is positioned well aft of the CG and is presumably intended to accommodate quite modest loads, and the aircraft is usually seen flying with flush-fitting underwing fuel tanks. The Yak-28 has now largely supplanted the Il-28 in V-VS service.

BOEING B-52 STRATOFORTRESS

The photograph above depicts a B-52D with 3,000 US gal (11 355 1) auxiliary tanks and the general arrangement silhouette (right) illustrates the B-52H

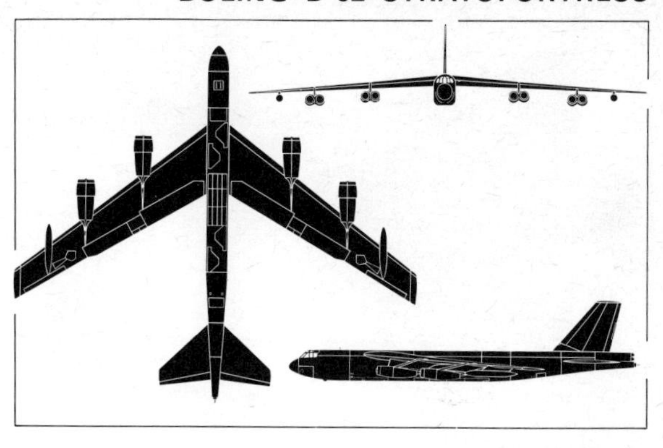

Country of Origin: USA.
Type: Long-range strategic bomber.
Power Plant: Eight (B-52C & D) 10,900 lb st (4 944 kgp) dry or 12,500 lb st (5 670 kgp) with water injection Pratt & Whitney J-57-P-29W turbojets or (B-52E, F, G) 11,200 lb st (5 080 kgp) dry or 13,750 lb st (6 242 kgp) with water injection J57-P-43W turbojets or (B-52H) 17,000 lb st (7 718 kgp) TF33-P-3 turbofans.
Performance: (B-52H) Max speed, 650 mph (1 040 km/h) at 50,000 ft (15 250 m); typical cruising speed, 565 mph (909 km/h) at 36,000 ft (10 973 m); service ceiling, approx 55,000 ft (16 764 m); unrefuelled range, 12,500 mls (20 117 km).
Weights: (B-52H) Max take-off, 488,000 lb (221 350 kg).
Dimensions: Span, 185 ft 0 in (56,42 m); length, 157 ft 7 in (48,03 m); height, 40 ft 8 in (12,39 m); wing area, 4,000 sq ft (371,6 m^2).
Accommodation: Crew of six comprising two pilots, gunner and ECM operator on flight deck upper level and navigator and bombardier at lower level.
Armament: Internal weapons bay with provision for up to 60,000 lb (27 215 kg) of conventional or nuclear bombs. Underwing provision (B-52C, D, E and F) for up to 12 750-lb (340-kg) bombs on each of two pylons or (B-52G and H) one AGM-28 Hound Dog stand-off ASM each side, plus ALE-25 diversionary rocket pods. Tail armament of four 0·50-in (12,7 mm) guns in MD-9 remotely controlled turret or (B-52H only) one 20-mm ASG-21 multi-barrel cannon.
Status: First and second prototypes (XB-52 and YB-52) first flown on 2 October 1952 and 15 April 1952; B-52A flown on 5 August 1954; B-52B on 25 January 1955; B-52C on 9 March 1956; B-52D on 4 June 1956; B-52E on 3 October 1957; B-52F on 6 May 1958; B-52G on 26 September 1958 and B-52H on 6 March 1961. Production totals: B-52A, 3; RB-52B, 27; B-52B, 23; B-52C, 35; B-52D, 170; B-52E, 100; B-52F, 89; B-52G, 193; B-52H, 102, total 743, completed October 1962.
Notes: In 1972 work began to adapt 96 B-52Gs and Hs in the SAC inventory to allow them each to carry 20 Boeing AGM-69A SRAM ASNs—eight in the bomb-bay and six beneath each wing. In another programme in 1973, more than 200 B-52Gs and Hs were being fitted with forward-looking IR sensors (FLIR) in twin under-nose radomes.

DASSAULT MIRAGE IVA

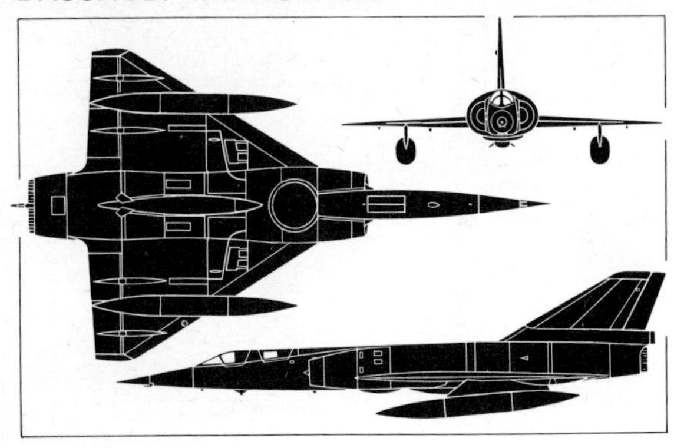

The photograph (above) and the general arrangement silhouette (left) depict the Mirage IVA which currently equips three Escadres of the Armée de l'Air

Country of Origin: France.
Type: Medium-range strategic bomber.
Power Plant: Two 10,360 lb st (4 700 kgp) dry and 15,435 lb (6 700 kgp) with reheat SNECMA Atar 9K turbojets.
Performance: Max speed, 1,454 mph (2 340 km/h) at 40,000 ft (13 125 m); max stabilised speed, 1,222 mph (1 966 km/h) at 60,000 ft (19 685 m); time to climb to 36,090 ft (11 000 m), 4 min 15 sec; service ceiling, 65,600 ft (20 000 m); tactical radius (supersonic to target, return at 595 mph/958 km/h at 40,000 ft (13 125 m) 770 mls (1 240 km); max range with drop tanks, 2,485 mls (4 000 km).
Weights: Empty, 31,967 lb (14 500 kg); normal loaded, 69,666 lb (31 600 kg); max take-off, 73,800 lb (33 475 kg).
Dimensions: Span, 38 ft $10\frac{1}{2}$ in (11,85 m); length, 77 ft 1 in (23,49 m); height, 17 ft $8\frac{1}{2}$ in (5,40 m); wing area, 839·6 sq ft (78,00 m^2).
Accommodation: Pilot and navigator in tandem.
Armament: (Strategic mission) One 60 KT free-falling bomb, semi-recessed in fuselage. (Tactical strike mission) up to 16 1,000-lb (454-kg) bombs externally under wings (six each side) and fuselage (four), or four AS.37 Martel anti-radar missiles.
Status: Prototype first flown on 17 June 1959; three pre-production models flown on 12 October 1961, 1 June 1962 and 23 January 1963 respectively. First production Mirage IVA flown on 7 December 1963. Production totals, one prototype, three pre-production, 62 full production, all for *l' Armée de l'Air*; production completed 1967.
Notes: The Mirage IVA was developed to provide France with a nuclear deterrent force, capable of striking at targets in the Soviet Bloc territory with French-produced nuclear bombs. In this rôle, it achieved full operational status in March 1968 when the 62nd and last example was delivered to complete the *"Force de Frappe"*. The aircraft are distributed between three *Escadres de Bombardement*, the 91e et Mont-de-Marsan, the 93e at Istres and the 94e at Avord. Each *Escadre* has three four-aircraft *Escadrons*, one at the HQ base and two detached to other bases. Supporting the Mirage IVAs are 12 C-135 tankers, divided into three squadrons, one at each *Escadre* base. Four Mirage IVAs also operate in the strategic reconnaissance rôle, with a special equipment pod under the fuselage. The Mirage IVA is scheduled to remain in service until 1980 when France's deterrent becomes wholly missile-based.

ENGLISH ELECTRIC (BAC) CANBERRA

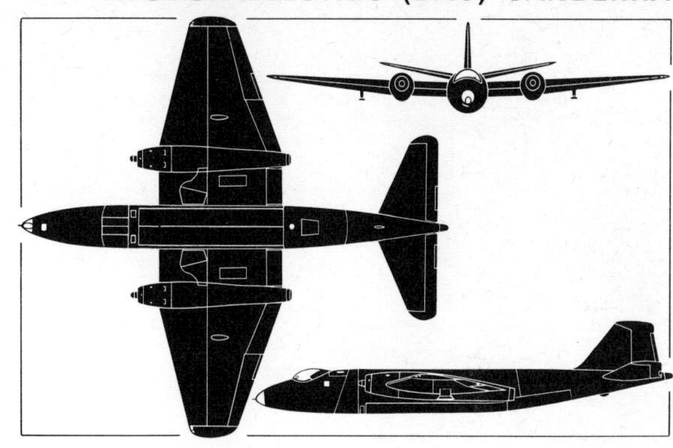

The photograph above depicts a Canberra B Mk 62 of the Argentine Air Force carrying Matra rocket launchers and the general arrangement silhouette illustrates the basic B Mk 6

Country of Origin: United Kingdom.
Type: Tactical bomber, reconnaissance, trainer and target tug.
Power Plant: Two (B Mk 6) 7,400 lb st (3 357 kgp) Rolls-Royce Avon 109 or (PR Mk 9) 10,050 lb st (4 560 kgp) Avon 206.
Performance: (B(I) Mk 6) Max speed 518 mph (834 km/h) at sea level and 580 mph (933 km/h) at 35,000 ft (10 668 m); typical cruising speed, 402 mph (647 km/h) at 20,000 ft (6 096 m); initial rate of climb, 3,400 ft/min (17,2 m/sec); max operational altitude, 48,000 ft (14 630 m); max range, interdictor rôle, 800 mls (1 287 km); max ferry range, 3,630 mls (5 842 km).
Weights: (B(I) Mk 6) Normal take-off (clean), 51,448 lb (23 336 kg); normal take-off (with tip tanks), 55,590 lb (25 215 kg); max take-off, 56,250 lb (25 514 kg).
Dimensions: Span, 63 ft 11½ in (19,50 m); span over tip tanks, 65 ft 6 in (19,96 m); span (PR Mk 9), 67 ft 10 in (20,67 m); length, 65 ft 6 in (19,96 m); length (PR Mk 9), 66 ft 8 in (20,32 m); height, 15 ft 8 in (4,77 m); wing area, 960 sq ft (89,19 m²); wing area (PR Mk 9), 1,045 sq ft (97,08 m²).
Accommodation: Pilot and navigator; (and bomb-aimer).
Armament: Internal bomb bay with capacity for 6,000 lb (2 722 kg) or (interdictor versions) 3,000 lb (1 360 kg) of bombs plus four 20-mm cannon in a ventral pack. Provision under wings for two pylons carrying up to 1,000 lb (454 kg) of ordnance each.
Status: Prototype first flown 13 May 1949; first B Mk 2 flown on 23 April 1950; PR Mk 3 on 19 March 1950; T Mk 4 on 6 June 1952; B Mk 6 on 26 January 1954; PR Mk 7 on 28 October 1953; B (I) Mk 8 on 23 July 1954; PR Mk 9 on 8 July 1955. Production totals: Prototypes, 4; B Mk 2, 418; PR Mk 3, 36; T Mk 4, 58; B Mk 5, 1; B Mk 6, 99; B (I) Mk 6, 22; PR Mk 7, 74; B(I) Mk 8, 57; PR Mk 9, 23; B(I) Mk 12, 16; T Mk 13, 1; PR Mk 57, 10; B(I) Mk 58, 71. Other marks by conversion. Total UK production 902, plus 56 B Mk 20 and T Mk 21 built in Australia and 403 B-57 variants built in USA.
Notes: Variants continuing in use in UK are T Mk 4, PR Mk 9, E Mk 15 (radio calibration), T Mk 17, TT Mk 18, T Mk 19 and T Mk 22. Overseas users include Argentina (B Mk 62, T Mk 64); Ethiopia (B Mk 52); Ecuador (B Mk 6); India (PR Mk 57, B(I) Mk 58, B Mk 66, T Mk 67); Peru (B Mk 72, T Mk 74, B(I) Mk 78); Rhodesia (B Mk 2, T Mk 4); South Africa (T Mk 4, B(I) Mk 6) and Venezuela (B Mk 82, PR Mk 83, T Mk 84 and B(I) Mk 88).

GENERAL DYNAMICS (CONVAIR) FB-111A

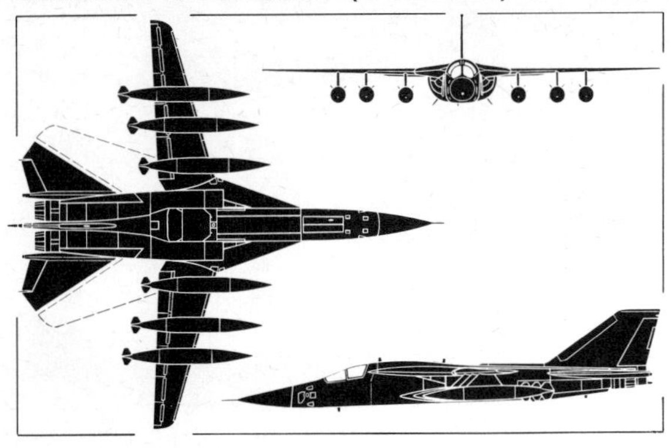

Both the photograph (above) and the general arrangement silhouette (left) illustrate the FB-111A which currently equips two wings of the USAF Strategic Air Command

Country of Origin: USA.
Type: Supersonic strategic bomber.
Power Plant: Two 12,500 lb st (5 670 kgp) dry and 20,350 lb st (9 230 kgp) with reheat Pratt & Whitney TF30-P-7 turbofans.
Performance: Max speed, 1,450 mph (2 334 km/h) at 36,000 ft (10 973 m), equivalent to Mach 2·5; max speed at sea level, about 910 mph (1 470 km/h), equivalent to Mach 1·2; service ceiling, over 60,000 ft (18 288 m); range, about 2,500 mls (4 023 km) without flight refuelling.
Weights: Empty, approximately 47,500 lb (21 545 kg); normal loaded approximately 100,000 lb (45 360 kg).
Dimensions: Span (spread), 70 ft 0 in (21,34 m); span (fully swept-back), 33 ft 11 in (10,34 m); height, 17 ft 1½ in (5,22 m).
Accommodation: Pilot and navigator side-by-side in pressurised escape capsule.
Armament: Internal weapon bay can accommodate two 750 lb (340-kg) bombs or other weapons of similar weight. Eight underwing pylons can carry up to 48 750-lb (340-kg) bombs or a maximum load of 37,500 lb (17 000 kg) of assorted weapons or six Boeing AGM-69A SRAM missiles (subject to limitation on external weapon loads when wings are fully swept).
Status: Prototype (F-111A conversion) first flown on 30 July 1967, first production FB-111A flown on 13 July 1968. Deliveries began October 1969; production of 76 completed 1971.
Notes: The FB-111A was developed from the F-111A fighter (see page 19) to provide Strategic Aerospace Command with a supersonic bomber to replace the B-58 Hustler and early versions of the long-range subsonic B-52 Stratofortress (see page 63). A programme to acquire 210 examples was announced by Department of Defense in December 1965 but this requirement was later cut to 76, to equip two Wings with two squadrons each—the 50th Bomb Wing at Pease AFB, NH and the 380th Strategic Aerospace Wing at Plattsburgh, NY. The FB-111A differs from the F-111 in having the larger span wing originally developed for the US Navy F-111B but not used on USAF versions of the fighter. It also has beefed-up structure and undercarriage to permit the higher operating weights when carrying a full fuel and bomb load, plus uprated engines and revised avionics for its strategic role. Like the F-111, the FB-111A has terrain-following radar and can fly at supersonic speed at about 200 ft (60 m) above the ground on automatic control.

HAWKER SIDDELEY VULCAN

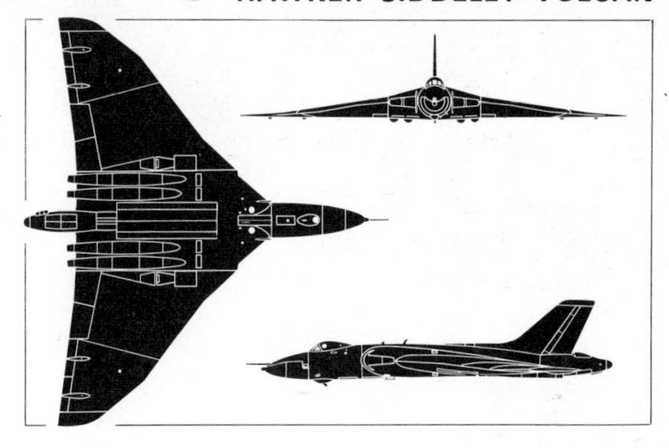

The photograph (above) and the general arrangement silhouette (right) depict the Vulcan B Mk 2, the former illustrating an aircraft assigned to the Near East Air Force (RAF)

Country of Origin: United Kingdom.
Type: Long-range tactical bomber.
Power Plant: Four 20,000 lb st (9 072 kgp) Rolls-Royce (Bristol) Olympus 301 or 17,000 lb st (7 710 kg) Olympus 201 turbojets.
Performance: (B Mk 2) Max speed, 645 mph (1 038 km/h) at 40,000 ft (12 192 m) equivalent to Mach 0·98; max cruising speed, 627 mph (1 010 km/h) at 55,000 ft (16 750 m), equivalent to Mach 0·95; service ceiling, 65,000 ft (19 812 m); high-level tactical radius, 2,300 mls (3 700 km); tactical radius with low-level attack on target, 1,725 mls (2 780 km); combat radius with one flight-refuelling, 2,875 mls (4 630 km).
Weights: (B Mk 2) Maximum take-off, over 180,000 lb (81 646 kg).
Dimensions: (B Mk 2) Span, 111 ft 0 in (33,83 m); length, 99 ft 11 in (30,45 m); height, 27 ft 2 in (8,28 m); wing area, 3,964 sq ft (368,3 m²).
Accommodation: Flight crew of five comprising two pilots side-by-side and (facing aft behind pilots) navigator, air electronics officer and radar operator.
Armament: Internal bomb-bay can accommodate tactical or nuclear weapons, up to 21 1,000-lb (454-kg) HE bombs or similar loads.
Status: Prototype (Avro 698) first flown on 30 August 1952 with Avon engines; second prototype flown on 3 September 1953 with Olympus engines. First production B Mk 1 flown on 4 February 1955, total 45 built (Mk 1 and 1A), complete by April 1959. Prototype B Mk 2 first flown on 31 August 1957; first production B Mk 2 flown on 30 August 1958. Deliveries began July 1960, production of approximately 50 complete by end 1964.
Notes: The Vulcan was developed to become one of the V-force bombers with which the RAF maintained Britain's strategic nuclear deterrent until this task was transferred to the Royal Navy's Polaris-equipped submarines. Subsequently, two squadrons retained Vulcan 2s equipped with Blue Steel stand-off bombs for low-level tactical strike, but RAF Strike Command's three Vulcan squadrons (Nos 617, 44 and 101) are now in service for overland strike duties, at high or low level, using conventional HE free-fall or retarded bombs. Two other Vulcan squadrons, Nos 9 and 35, are normally based in Cyprus. No 27 Squadron was reformed on Vulcans in 1973 to replace the strategic reconnaissance squadron of Victor SR Mk 2s.

ILYUSHIN IL-28

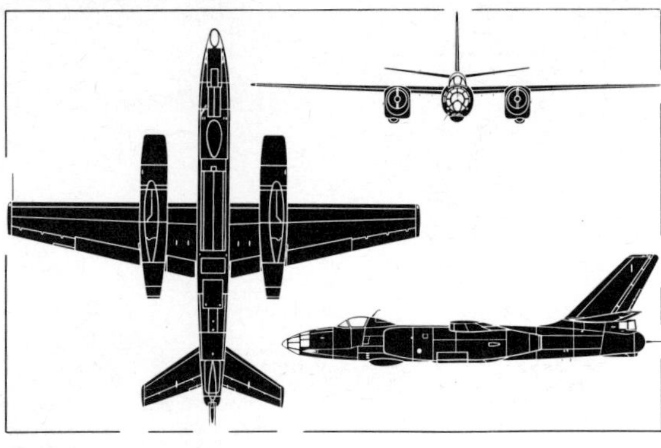

The photograph above illustrates an Il-28R with tandem radomes of the Rumanian Air Force, the general arrangement silhouette depicting the standard Il-28 tactical bomber

Country of Origin: USSR.
NATO Code Name: *Beagle* (and *Mascot*).
Type: Light tactical bomber and patrol aircraft (and trainer).
Power Plant: Two 5,952 lb st (2 700 kgp) Klimov VK-1 turbojets.
Performance: Max speed, 559 mph (900 km/h) at 14,765 ft (4 500 m) and 500 mph (805 km/h) at 39,370 ft (12 000 m); typical cruising speed, 472 mph (760 km/h) at 3,280 ft (1 000 m); initial rate of climb, 2,953 ft/min (15 m/sec); service ceiling, 40,355 ft (12 300 m); range with max fuel, 1,355 mls (2 180 km) at 32,810 ft (10 000 m).
Weights: Empty equipped, 28,417 lb (12 890 kg); normal take-off, 40,565 lb (18 400 kg); max loaded, 46,297 lb (21 000 kg).
Dimensions: Span (excluding tip tanks), 70 ft 4¾ in (21,45 m); length, 57 ft 10¾ in (17,65 m); wing area, 654·44 sq ft (60,8 m²).
Accommodation: Crew of three comprising pilot, navigator/bombardier and radio-operator/tail gunner.
Armament: Internal stowage for normal bomb load of 2,205 lb (1 000 kg) with provision (Il-28T) for two small torpedoes to be carried. Two fixed forward firing 23-mm NR-23 cannon in fuselage nose and two 23-mm NR-23 cannon in Il-K6 tail turret.

Status: Prototype first flown on 8 August 1948 followed by two more prototypes. First production aircraft flown early 1950 with first deliveries to operational squadrons later in 1950. Several thousand built, including Il-28U and Il-28T versions and licence production in China.
Notes: The Ilyushin Il-28 was developed in competition with a design by the Tupolev bureau, the latter being adopted for service with the Soviet Naval Air Force (AV-MF) as the Tu-14T but eventually being superseded by the Ilyushin design, which meanwhile had been adopted for large-scale service by the Soviet Air Forces (V-VS). In addition to the basic Il-28 tactical bomber, the V-VS used the Il-28U (Nato code name *Mascot*) which featured a second, raised cockpit for the instructor behind the standard cockpit, and several versions of the Il-28R tactical reconnaissance version, some examples of which featured tandem ventral radomes. Only in limited frontline service with the V-VS in 1973, the Il-28 still serves with several other air forces, more than 20 of which received examples from the Soviet Union, and the AV-MF still used the Il-28T torpedo-bomber in some numbers. The Il-28 remains standard light bomber equipment with China.

TUPOLEV TU-16

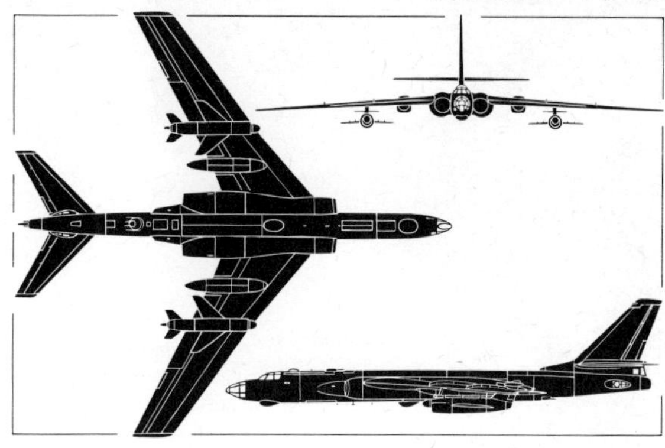

The photograph above depicts the basic Tu-16 (Badger-A) bomber and the general arrangement silhouette (right) illustrates the Kennel missile-carrying version (Badger-B)

Country of Origin: USSR.
NATO Code Name: *Badger.*
Type: Long-range strategic bomber, flight refuelling tanker and maritime reconnaissance aircraft.
Power Plant: Two 20,950 lb st (9 500 kgp) Mikulin AM-3M turbojets.
Performance (Estimated): Max speed, 587 mph (945 km/h) at 35,000 ft (10 700 m); service ceiling, 42,650 ft (13 000 m); range with max bomb-load, 3,000 mls (4 800 km); range with 6,600 lb (3 000 kg) bomb-load, 3,975 mls (6 400 km).
Weights: Normal take-off, about 150,000 lb (68 000 kg).
Dimensions: Span, 110 ft 0 in (33,5 m); length, 120 ft 0 in (36,5 m); height, 35 ft 6 in (10,8 m); wing area, about 1,820 sq ft (169 m²).
Accommodation: Flight crew of five or six including two pilots, navigator, radio operator and gunners.
Armament: Six 23-mm cannon, in pairs in dorsal, ventral and tail turrets. Seventh 23-mm gun fixed to fire forwards, in starboard side of front fuselage, in *Badger -A, -B, -E, -F* and *-G.* Internal weapon bay with ability to carry up to 19,800 lb (9 000 kg) load. Two wing strong points to carry (*Badger-B*) *Kennel* or (*Badger-G*) *Kelt* anti-shipping missiles. Provision under fuselage (*Badger-C*) to carry *Kipper* stand-off bomb.
Status: Prototype first flown about 1952; first public appearance 1954. Production in Soviet Union complete, with up to 2,000 built, but continuing in China.
Notes: The Tu-16 is the production derivative of the Tu-88 prototype, developed as the first Soviet long-range jet bomber. The initial *Badger-A* version carried only free-fall bombs but those remaining in service have now been modified to *Badger-E* for photo-reconnaissance with cameras in the bomb-bay, or *Badger-F* with special electronic gear in underwing pods. *Badger-B* carried two *Kennel* turbojet-powered missiles under the wings for anti-shipping use and was operated by the Soviet Navy, but has now been superseded by *Badger-G,* which is similar with provision to carry two rocket-powered *Kelt* missiles. *Badger-C* is a Soviet Air Force version with additional target-acquisition radar in the nose and a *Kipper* stand-off bomb under the fuselage. For maritime reconnaissance, *Badger-D* carries extensive electronic sensors and no missiles. A substantial number remained in the Soviet strategic bomber force in 1973.

TUPOLEV TU-22

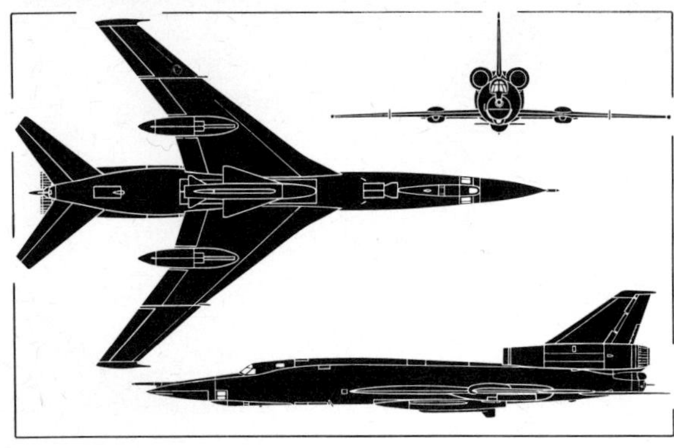

The photograph above illustrates the standard Tu-22 (Blinder-A) and the general arrangement silhouette (left) depicts the missile-carrying version (Blinder-B)

Country of Origin: USSR.
NATO Code Name: *Blinder.*
Type: Long-range medium bomber and strike reconnaissance.
Power Plant: Two (approx) 27,000 lb st (12 250 kgp) turbojets with reheat.
Performance (Estimated): Max speed, without external stores, 990 mph (1 590 km/h) or Mach 1·5 at 39,370 ft (12 000 m), 720 mph (1 160 km/h) or Mach 0·95 at 1,000 ft (305 m); normal cruise, 595 mph (960 km/h) or Mach 0·9 at 39,370 ft (12 000 m); tactical radius on standard fuel for high altitude mission, 700 mls (1 125 km); service ceiling, 60,000 ft (18 290 m).
Weights (Estimated): Max take-off, 185,000 lb (84 000 kg).
Dimensions (Estimated): Span, 91 ft 0 in (27,74 m); length, 133 ft 0 in (40,50 m); height, 17 ft 0 in (5,18 m); wing area, 2,030 sq ft (188,6 m²).
Accommodation: Crew of three in tandem, with visual aiming and navigation station forward of pilot and radar operator behind.
Armament: Free-fall weapons housed in bay in centre fuselage or (*Blinder-B*) one Kitchen ASM carried semi-recessed under centre fuselage. Remotely-controlled 23-mm cannon in tail barbette for rear defence.
Status: First public appearance in 1961, in the Aviation Day fly-past over Moscow (ten examples). Believed to have reached operational status with the Soviet Air Force in 1965.
Notes: The Tu-22 is the successor to the subsonic Tu-16 in Soviet medium-bomber formations and with shore-based maritime strike elements of the Soviet Naval Air Arm. The basic version is identified by the Air Standards Co-ordinating Committee of NATO as *Blinder-A* while the version that carries a stand-off missile is *Blinder-B*. Also in service is a training version which has a raised rear cockpit with dual controls, for an instructor behind the standard pilot cockpit, replacing the radar operator's position. A maritime reconnaissance version of the aircraft, with cameras in the weapon bay and additional electronic equipment, is identified as *Blinder-C*. Recent production models of the Tu-22 have displayed a number of modifications including enlarged air intakes, nacelles and exhaust orifices, suggesting that more powerful engines have now been installed. The flight refuelling probe has also been extended, by comparison with the earlier *Blinder-A* standard, and a version equipped for electronic countermeasures duties has been reported.

TUPOLEV "BACKFIRE"

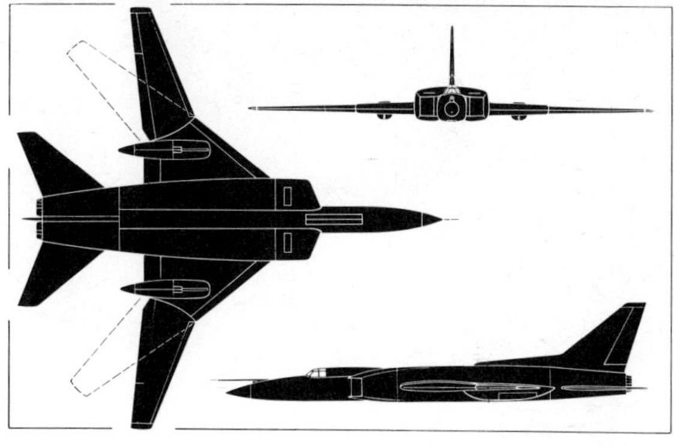

The model photograph above and the general arrangement silhouette (right) depicting the variable-geometry Backfire bomber should be considered as provisional

Country of Origin: USSR.
NATO Code Name: *Backfire.*
Type: Strategic bomber.
Power Plant: Two turbofans with afterburners, probably of Kuznetsov origin and with a probable augmented thrust of about 40,000 lb st (18 144 kgp) each.
Performance (Estimated): Max speed, about Mach 2·2 to 2·5; operating ceiling, about 50,000 ft (15 240 m); range, over 4,500 mls (7 242 km) at subsonic speed.
Weights: Normal loaded, about 275,000 lb (124 740 kg).
Dimensions: Not known.
Accommodation: Crew of two or three.
Armament: Primary attack armament comprises one ASM-6 solid-propellant stand-off missile with a range of about 460 mls (740 km).
Status: First reported 1969; prototype observed July 1970.
Notes: Development of this variable-geometry strategic bomber, given the name *Backfire* by the Air Standards Co-ordinating Committee, is believed to have started in the Soviet Union in the mid-sixties, the first reported sighting of a prototype taking place in July 1970, at an airfield near the Tupolev works at Kazan in Central Asia. This sighting, and most other known details of the *Backfire*, are the result of reconnaissance satellite monitorings of Soviet activity. By 1973, it was thought that about a dozen *Backfires* were in flight test and introduction into service was expected in 1975 or 1976. The *Backfire* is the first Soviet strategic bomber combining supersonic capability with sufficient range to hit targets in Continental USA, using a stand-off missile launched some distance from the borders, or with small nuclear missiles delivered in a low-altitude supersonic dash. The design origin of *Backfire* is not known with certainty, although some reports ascribe it to the Tupolev bureau because of the location of the prototype. Other bureaux with the technical qualifications to develop such an aircraft include those of Ilyushin and Myasishchev. The engines are most probably related to the NK-144 units used in the Tu-144 supersonic transport (see *Civil* volume). Comparable from some aspects with the USAF's Rockwell (North American) B-1 variable-geometry strategic bomber which was in the prototype assembly phase late 1973 and which will match *Backfire's* assumed performance in key areas. The fairings on the substantial fixed portion of the wing are believed to accommodate the wing pivot mechanism.

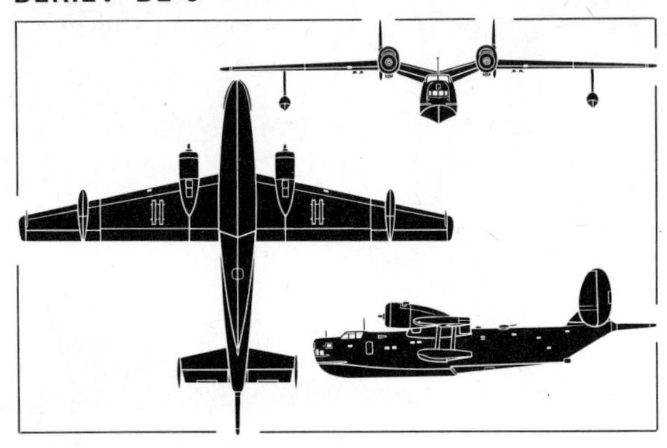

The photograph (above) depicts the Be-6 in its original maritime patrol form with tail turret and the general arrangement silhouette (left) shows the MAD tail boom fitted

Country of Origin: USSR.
NATO Code Name: *Madge.*
Type: General purpose flying-boat.
Power Plant: Two 2,000 hp Shvetsov ASh-73TK radial piston-engines.
Performance: Maximum speed, 258 mph (415 km/h) at 7,875 ft (2 400 m); cruising speed, 235 mph (377 km/h) at sea level, 248 mph (400 km/h) at 8,200 ft (2 500 m); range, 3,000 mls (4 900 km).
Weights: Empty, 41,500 lb (18 825 kg); loaded 51,600 lb (23 400 kg).
Dimensions: Span, 108 ft 3 in (33,0 m); length, (without MAD) 77 ft 4 in (23,57 m); length, (with MAD) 84 ft 0 in (25,60 m); height, 25 ft 1 in (7,64 m); wing area, 1,290 sq ft (120 m²).
Accommodation: Crew of eight, including nose and dorsal gunners.
Armament: One 23-mm NS-23 cannon in nose; two similar guns in tail turret retroactively replaced on most aircraft by a magnetic anomaly detection stinger) and two in dorsal remotely-controlled barbette.

Status: First flown in 1949. Entered service early 'fifties. Several hundred built; production ended early 'sixties.
Notes: Following the decision to concentrate all Soviet seaplane development in the design bureau headed by Georgi M Beriev at Taganrog, design work began shortly after the end of World War II on a new maritime patrol and reconnaissance 'boat, and prototypes had been built and tested by 1947. Initially the design was identified as the LL-143 but following its acceptance for production it received the designation Be-6. The production version, flown for the first time in 1949, embodied some re-design of the forward hull, with provision aft of the hull step for a retractable radome housing a search scanner, and more powerful engines were fitted. At a later stage, the tail turret was removed and an extended fairing for MAD equipment was added. The Be-6 became the most widely-used flying-boat of the *Morskaya Aviatsiya* by the late 'fifties, but by 1965 was beginning to be phased out as the turboprop-powered Be-12 entered service (see page 73). No longer in front line service for maritime reconnaissance or patrol, the Be-6 now remains in use for general duties, including transport of personnel and stores and submarine co-operation tasks, particularly in the Arctic.

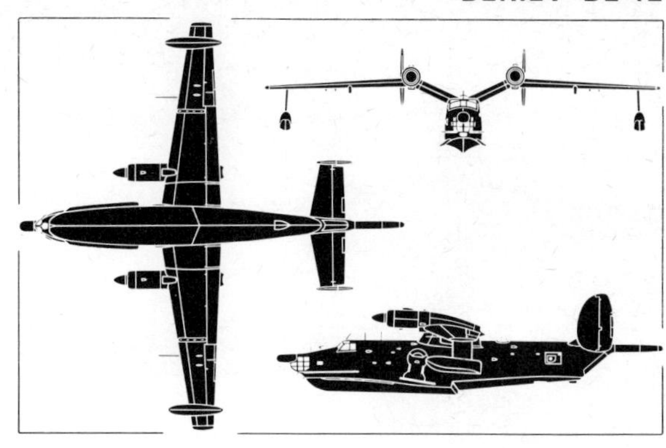

Both the photograph (above) and general arrangement silhouette (right) depict the standard production Be-12, the largest amphibian flying boat in military service

Country of Origin: USSR.
NATO Code Name: *Mail.*
Type: Maritime reconnaissance amphibian.
Power Plant: Two 4,000 shp Ivchenko AI-20D turboprops.
Performance (Estimated): Max speed 379 mph (610 km/h) at 10,000 ft (3 050 m); typical patrol speed, 199 mph (320 km/h) at 985 ft (300 m); initial rate of climb, 3,000 ft/min (15,2 m/sec); service ceiling, 37,000 ft (4 025 km); max range, 2,485 mls (4 000 km).
Weights (Estimated): Max take-off, 65,035 lb (29 500 kg).
Dimensions (Approximate): Span, 97 ft 6 in (29,70 m); length, 101 ft 11¼ in (32,9 m); height, 22 ft 11½ in (7,0 m); wing area, 1,030 sq ft (95,69 m²).
Accommodation: Crew of five–six, including two pilots, navigator, radar operator and MAD operator.
Armament: Offensive armament of homing torpedoes, depth bombs, mines or rockets carried externally on wing pylons. Internal stowage for sono-buoys, flares, etc, in a bay aft of the hull's single step, and this may also provide stowage for additional depth bombs or mines.

Status: Prototype reported to have flown in 1960. Believed to have entered service with the Soviet Navy during 1965–66.
Notes: The Be-12 was the largest amphibian flying-boat in military service in the early 'seventies, although the Shin Meiwa US-1 under development in Japan (see page 85) will be larger. Named Tchaika (Gull) in the Soviet Union, the Be-12 was first seen in public at a Soviet Aviation Day fly-past over Moscow in 1961, and was apparently developed to replace the piston-engined Be-6 in service with *Morskaya Aviatsiya* (Soviet Navy Aviation). It is known to have been in service by 1967, and probably earlier, and is now standard equipment in units of the Soviet Northern and Black Sea fleets. Some examples have also been observed in the Mediterranean, possibly operating from Syrian bases. In three series of special flights, in 1964, 1968 and 1970, Be-12s set up a number of international records, subsequently confirmed by the FAI, and these indicated that the aircraft could reach an altitude of 39,977 ft (12 155 m) without any payload and could lift a payload of 22,266 lb (10 100 kg) to a height of 6,560 ft (2 000 m) and that it could average a speed of 343 mph (551 km/h) over 500-km and 1 000-km closed circuits, although with only a small payload.

The photograph (above) and general arrangement silhouette (left) depict the standard Alizé, the former depicting an aircraft of the Indian Navy's No 310 Squadron

Country of Origin: France.
Type: Carrier-borne anti-submarine attack aircraft.
Power Plant: One 2,100 eshp Rolls-Royce Dart R.Da. 21 turboprop.
Performance: Max speed, 285 mph (460 km/h) at sea level, 292 mph (470 km/h) at 10,000 ft (3 048 m); normal patrol speed, 144 mph (232 km/h) at 1,500 ft (457 m); initial rate of climb, 1,380 ft/min (7,0 m/sec); service ceiling, 20,500 ft (6 248 m); endurance (search configuration at 1,500 ft (457 m)) 5 hr 12 min; ferry range, 1,785 mls (2 870 km).
Weights: Empty equipped, 12,566 lb (5 700 kg); max take-off, 18,100 lb (8 200 kg).
Dimensions: Span, 51 ft 2 in (15,6 m); length, 45 ft 6 in (13,86 m); height, 16 ft 5 in (5,0 m); wing area, 387·5 sq ft (36 m²).
Accommodation: Pilot and radar operator side-by-side in cockpit, with second radar operator in sideways-facing seat to rear.
Armament: Internal weapon bay can accommodate one torpedo or three depth bombs of 353-lb (160-kg) weight or similar loads. Underwing provision for two 353-lb (160-kg) or 385-lb (175-kg) depth bombs and six 5-in (12,7-cm) rockets or two ASMs.

Status: First of two Br 960 prototypes flown on 3 August 1951; aerodynamic prototype of Br 1050 flown on 26 March 1955; prototype Alizé flown on 6 October 1956. Delivery of production aircraft began March 1959. Production totals, two prototypes, three pre-production aircraft, 87 production aircraft; deliveries completed 1962.
Notes: The Alizé was developed in 1954 from the Br 960 Vultur, which had an Armstrong Siddeley Mamba turboprop in the nose and a Hispano-Suiza Nene turbojet in the rear fuselage. This dual power plant configuration was abandoned for the Alizé, a large retractable radome taking the space that the turbojet had occupied. *Aéronvale* acquired a production batch of 75, and three squadrons are equipped with the type—Flotilles 4F and 9F to serve aboard the aircraft carriers *Foch* and *Clémenceau,* and 6F to train crews for these units. In addition, 12 Alizés were acquired by the Indian Navy in 1961 and are used (with two more purchased from *Aéronavale* in 1967) to equip No 310 Squadron, "The Cobras", at INS *Garuda,* one four-plane flight being normally deployed aboard the carrier INS *Vikrant.* No fixed-wing successor for the Alizé is forseen by the *Aéronavale,* after its phase-out in the mid 'seventies.

BREGUET ATLANTIC

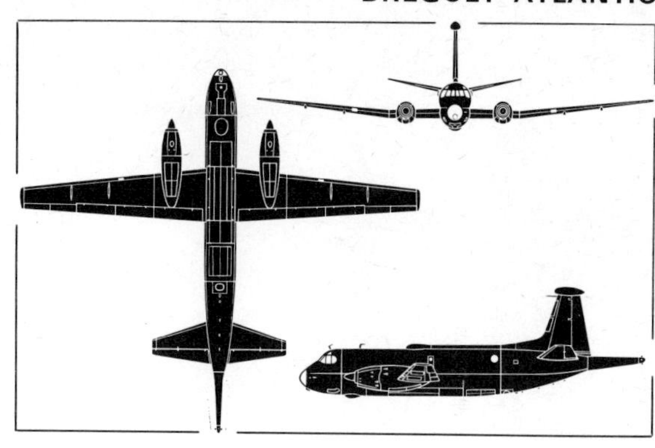

Both photograph (above) and general arrangement silhouette (right) depict the standard Atlantic maritime patrol aircraft, the former illustrating an Aéronavale example

Country of Origin: France.
Type: Long-range maritime patrol aircraft.
Power Plant: Two 6,105 ehp Rolls-Royce (Snecma-built) Type R Ty 20 Mk 21 turboprops.
Performance: Max speed, 409 mph (658 km/h) at high altitudes; cruising speeds, 363 mph (584 km/h) at 19,685 ft (6 000 m), 342 mph (550 km/h) at 26,250 ft (8 000 m); cruising speed for maximum endurance, 195 mph (320 km/h); initial rate of climb, 2,450 ft/min (12,44 m/sec); service ceiling, 32,800 ft (10 000 m); patrol endurance at range of 620 mls (1 000 km) from base, 12 hrs; max endurance 18 hrs; max range, with 10 per cent reserve, 4,950 mls (7 970 km).
Weights: Empty, 52,900 lb (24 000 kg); max take-off, 95,900 lb (43 500 kg).
Dimensions: Span, 119 ft 1¼ in (36,30 m); length, 104 ft 1½ in (31,75 m); height, 37 ft 1¾ in (11,33 m); wing area, 1,295·3 sq ft (120,34 m²).
Accommodation: Crew of 12 comprising two pilots, three observers and, in the central tactical compartment, a tactical co-ordinator, navigator, two sono-buoy operators, radio operator, radar operator and ECM/MAD operator.
Armament: Internal weapons bay can accommodate four homing torpedoes or nine acoustic torpedoes or depth charges, mines, etc. Four underwing pylons can each carry one AS.12 or similar ASM.
Status: First of two prototypes flown on 21 October 1961; first of two pre-production aircraft (with lengthened, production-standard fuselage), flown on 25 February 1963; first production model flown on 19 July 1965. Deliveries began in December 1965. Production totals, 40 for France, 20 for Germany, 9 for Netherlands and 18 for Italy; production completed 1973.
Notes: The Atlantic was designed by Breguet company, design and construction being shared between companies in France, Germany, Belgium and the Netherlands; when an Italian order for the aircraft was placed in 1968, several Italian companies were also brought into the manufacturing consortium. The Atlantic equips three *Aéronavale Flotilles*, Nos 21F, 22F and 24F, two *Staffeln* of MFG 3 of the West German *Marineflieger* (including five aircraft modified for ECM duties), No 321 Squadron of the Dutch *Marine Luchtvaartdienst* and the 87° and 88° *Gruppi* of 41º *Stormo* of Italy's Marinavia.

CANADAIR CP-107 ARGUS

Anti-submarine/maritime reconnaissance

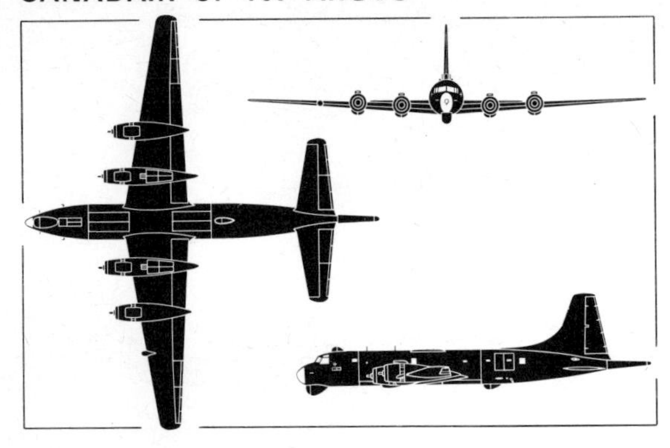

Both the photograph (above) and general arrangement silhouette (left) depict the Argus Mk 2 of the Canadian Armed Forces' Maritime Command

Country of Origin: Canada.
Type: Long-range maritime reconnaissance aircraft.
Power Plant: Four 3,400 hp (dry) and 3,700 hp with water injection Wright R-3350-EA-1 Turbo-Compound piston engines.
Performance: Max speed, 315 mph (507 km/h) at 20,000 ft (6 096 m) and 288 mph (463 km/h) at sea level; max cruising speed, 230 mph (370 km/h) at sea level; economical cruising speed, 223 mph (359 km/h) at 5,000 ft (1 524 m); typical patrol speed, 190 mph (306 km/h) at 1,000 ft (305 m); initial rate of climb, 1,700 ft/min (8,6 m/sec); service ceiling, over 20,000 ft (6 100 km); max range, 5,900 mls (9 495 km); patrol endurance, 12 hrs at 830 mls (1 336 km) from base, 8 hrs at 1,210 mls (1 947 km) and 4 hrs at 1,590 mls (2 558 km); cruising endurance, 24 hrs.
Weights: Empty, 81,000 lb (36 740 kg); normal loaded, 148,000 lb (67 130 kg).
Dimensions: Span, 142 ft 3½ in (43,37 m); length, 128 ft 3 in (39,09 m); height, 36 ft 8½ in (11,19 m); wing area, 2,075 sq ft (192,76 m²).
Accommodation: Crew of 15, comprising three pilots, three navigators, two flight engineers and seven electronics operators.
Armament: Internal weapons bays (two in tandem) accommodate up to 8,000 lb (3 630 kg) of varied stores including homing torpedoes, depth bombs or mines. Provision for one pylon under each outer wing with a capacity of 3,800 lb (1 725 kg) each.
Status: Prototype first flown on 29 March 1957. Production deliveries began May 1958. Production totals, prototype, 1; Mk 1, 12, Mk 2, 20. Production completed July 1960.
Notes: The Canadair CL-28 was designed to meet an RCAF requirement for a replacement for the Avro Lancaster MR Mk 10, still in service in 1953, and was based on the wing and tail unit of the Bristol Britannia, with a new fuselage. Piston engines were substituted for the Proteus turboprops, to obtain the necessary long endurance, and a new fuselage was designed, with weapons bays fore and aft of the centre-section structure, search radar in a nose radome and MAD gear in a long rail "stinger". The Mk 2 version of the Argus differed from the initial production batch in having a smaller chin radome and improved equipment. Four squadrons of the Maritime Command of the RCAF—later the Canadian Armed Forces—were originally equipped with the Argus, these being Nos 404, 405, 407 and 415, and these continued to operate in 1973.

GRUMMAN S-2 TRACKER (AND C-1 TRADER)

The photograph above depicts an S-2E Tracker of the Royal Australian Navy's No 816 Squadron. The general arrangement silhouette (right) illustrates the S-2D

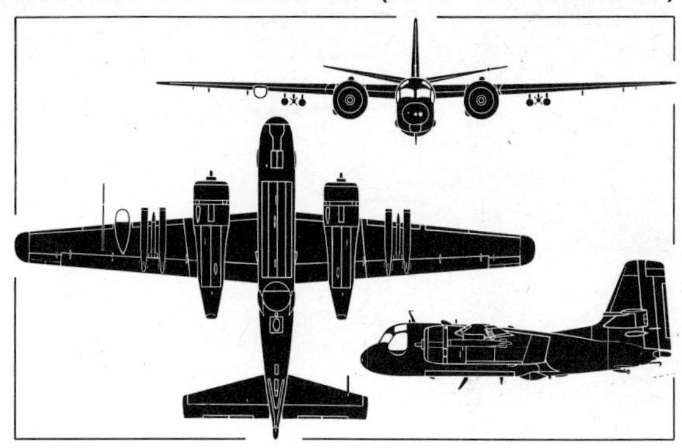

Country of Origin: USA.
Type: Carrier or land-based anti-submarine aircraft (and COD transport).
Power Plant: Two (XS2F-1) 1,425 hp Wright R-1820-76 or (S-2A, -2B, -2C) 1,525 hp R-1820-82 or (S-2D, -2E, -2F, -2G) 1,525 hp R-1820-82WA piston radials.
Performance: (S-2E) Max speed, 265 mph (426 km/h); cruising speed, 150 mph (241 km/h) at 1,500 ft (450 m); initial rate of climb, 1,390 ft/min (7,0 m/sec); service ceiling, 21,000 ft (6 400 m); range, 1,300 mls (2 095 km); endurance with max fuel, 10 per cent reserves, 9 hrs.
Weights: Empty, 18,750 lb (8 505 kg); max take-off, 29,150 lb (13 222 kg).
Dimensions: Span (S-2A, -2B, -2C), 69 ft 8 in (21,23 m), (S-2D, -2E, -2F, -2G), 72 ft 7 in (22,13 m); length (S-2A, -2B, -2C), 42 ft 3 in (12,88 m), (S-2D, -2E, -2F, -2G), 43 ft 6 in (13,26 m); height (S-2A, -2B, -2C), 16 ft 3½ in (4,96 m), (S-2D, -2E, -2F, -2G), 16 ft 7½ in (5,06 m); wing area (S-2A, -2B, -2C), 485 sq ft (45,06 m²), (S-2D, -2E, -2F, -2G), 496 sq ft (46,08 m²).
Accommodation: Crew of four comprising pilot, co-pilot/navigator/radio operator, radar operator and MAD operator.
Armament: Internal weapons bay accommodates two homing torpedoes, two Mk 101 depth bombs, four 385 lb (175 kg) depth charges or similar loads. Provision under wings for six pylons to carry torpedoes (ferry only), 5-in (12,7 cm) rockets or 250-lb (114-kg) bombs.
Status: First of two prototypes (XS2F-1) flown on 3 July 1953; first S2F-1 (S-2A) flown on 30 April 1953; first S2F-2 (S-2C) flown on 12 July 1954; first S2F-3 (S-2D) flown on 25 July 1962; first S2F-3S (S-2E) flown in September 1960; first S-2G flown in 1972. Production totals: XS2F-1, 2; S-2A, 655; S-2C, 60; S-2D, 119; S-2E, 245, grand total 1,181. In addition, 100 built by de Havilland Canada as CS2F-1. Production completed 1968.
Notes: Production versions for US Navy and export were S-2A (originally S2F-1) with smaller wing and fuselage; S-2C with enlarged bomb-bay and S-2D/S-2E with increased dimensions, more fuel and updated equipment. Conversions were S-2B and S-2F, S-2As with equipment updates, and S-2G, modified S-2Es introduced in 1972 for operation until 1976. Based on the Tracker, the C-1A (originally TF-1) had a new 9-seat fuselage; 87 were built.

HAWKER SIDDELEY NIMROD

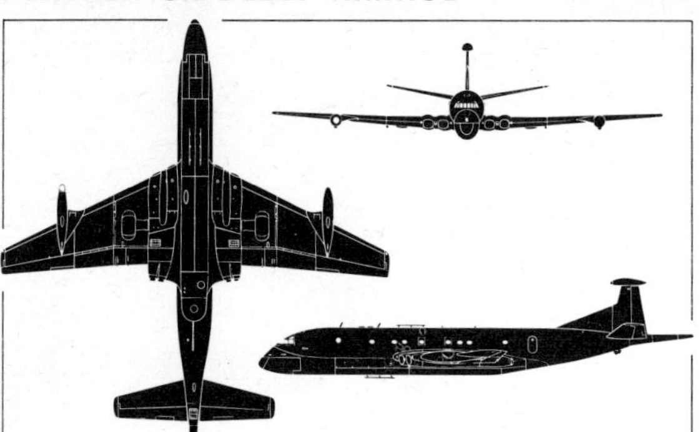

Both the photograph (above) and the general arrangement silhouette (left) illustrate the Nimrod MR Mk 1 which currently equips five squadrons of RAF Strike Command

Country of Origin: United Kingdom.
Type: Long-range maritime patrol aircraft.
Power Plant: Four Rolls-Royce Spey RB.168 Mk 250 turbofans of approx 11,500 lb st (5 217 kgp) each.
Performance: Max speed, 575 mph (926 km/h); max transit speed, 547 mph (880 km/h); economical transit speed, 490 mph (787 km/h); typical ferry range, 5,180–5,755 mls (8 340–9 265 km); typical endurance, 12 hrs.
Weights: Normal take-off, 175,500 lb (79 605 kg); max overload, 192,000 lb (87 090 kg).
Dimensions: Span, 114 ft 10 in (35,00 m); length, 126 ft 9 in (38,63 m); height, 29 ft 8½ in (9,01 m); wing area, 2,121 sq ft (197,05 m²).
Accommodation: Normal flight crew of three (two pilots and a flight engineer) plus a nine-man team in the tactical compartment including two navigators and seven sensor operators. Provision for conversion as emergency trooper carrying 45 passengers in the rear pressure cabin.
Armament: Internal weapons bay, approximately 50-ft (15-m) long, can accommodate homing torpedoes, mines, depth charges, bombs etc. Underwing provision for two pylons each to carry two ASMs such as the Aérospatiale AS.12.
Status: Prototypes (converted Comet 4C airframes) first flown on 23 May and 31 July, 1967, respectively. First production MR Mk 1 flown on 28 June 1968, deliveries of 46 ordered by RAF began on 2 October 1969. Three R Mk 1 delivered to RAF 1972.
Notes: Designed as a replacement for the long-serving Shackletons of the former Coastal Command, the Nimrod was derived from the Comet commercial airliner, using substantially the same wing and tail unit, with the same basic fuselage cross section but with the addition of an unpressurised pannier under the fuselage to accommodate the weapons bay and operational equipment. The fuselage was shortened by 6 ft 6 in (1,98 m) and Spey turbofans substituted for the original Avons. The total Nimrod force in Strike Command comprises five squadrons (Nos 42, 120, 201, 203 and 206) of which No 203 is based in Malta. In addition three Nimrods for electronic reconnaissance duties, and designated R Mk 1s, serve with No 51 Squadron. All Nimrod MR Mk 1s are scheduled to undergo an update programme to introduce new avionics and sensors, when they will become MR Mk 2s.

ILYUSHIN IL-38

Both the photograph (above) and the general arrangement silhouette (right) depict the Il-38, now a standard Soviet long-range maritime patrol aircraft

Country of Origin: USSR.
NATO Code Name: *May.*
Type: Long-range maritime patrol aircraft.
Power Plant: Four 4,250 ehp Ivchenko AI-20M turboprops.
Performance (Estimated): Max continuous cruise, 400 mph (645 km/h) at 15,000 ft (4 570 m); normal cruise, 370 mph (595 km/h) at 2,000 ft (610 m); max range, 4,500 mls (7 240 km); loiter endurance, 12 hrs at 2,000 ft (610 m).
Weights (Estimated): Empty equipped, 80,000 lb (36 287 kg); max take-off, 140,000 lb (63 500 kg).
Dimensions: Span, 122 ft 9 in (37,40 m); length, 131 ft 0 in (39,92 m); height, 33 ft 4 in (10,17 m); wing area, 1,507 sq ft (140,0 m²).
Armament: Internal weapons bay for depth bombs, homing torpedoes, etc. Wing hardpoints for external ordnance loads.
Accommodation: Normal crew believed to consist of 12 members, of which four are on flight deck and remainder in tactical compartment in cabin, to operate sensors and indicators and coordinate data flow to other aircraft and ships.
Status: The Il-38 reportedly flew in prototype form during 1967–68, and entered service with the Soviet naval air arm early in 1970. Examples were observed in service with Egyptian Air Force insignia during 1972 but it is believed that these were Soviet Navy aircraft temporarily deployed to Egyptian bases.
Notes: The Il-38 was evolved from the Il-18 commercial transport (see *Civil* volume) in a similar manner to the development of the Lockheed Orion (page 82) from the Electra transport. Apart from some strengthening, the wings, tail assembly and undercarriage are similar to those of the Il-18, these components having been married to an entirely new fuselage housing anti-submarine warfare systems, a tactical operations compartment and weapons bays. The fuselage ends in a "sting" fairing containing MAD (Magnetic Anomaly Detection) equipment. By comparison with the Il-18, the wing is positioned farther forward on the fuselage, to compensate for the heavier weight of equipment in the forward fuselage. Il-38s have been observed operating in the Mediterranean area and over the seas around the Soviet Union, apparently in a reconnaissance rôle monitoring the sea forces of Western powers. This type is apparently replacing converted Tu-16 and Tu-20 bombers currently serving with Soviet naval long-range maritime patrol units.

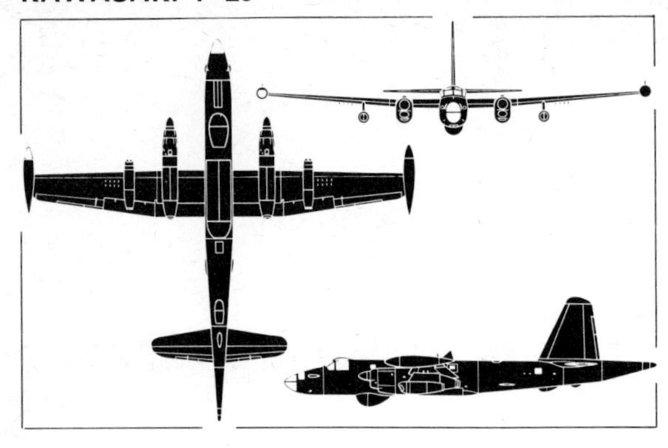

The P-2J, illustrated by the photograph above and the general arrangement silhouette (left), is a Japanese turboprop-powered derivative of the Lockheed P-2H Neptune

Country of Origin: Japan.
Type: Maritime patrol and anti-submarine patrol aircraft.
Power Plant: Two 2,850 ehp General Electric T64-IHI-10 turboprops and two 3,085 lb st (1 400 kgp) Ishikawajima J3-IHI-7C turbojets.
Performance: Max permissible speed (dive) 403 mph (649 km/h); max cruising speed, 250 mph (402 km/h); economical cruising speed, 230 mph (370 km/h); initial rate of climb, 1,800 ft/min (9,2 m/sec); service ceiling, 30,000 ft (9 150 m); range with max fuel, 2,765 mls (4 450 km).
Weights: Empty, 42,500 lb (19 277 kg); max take-off weight, 75,000 lb (34 019 kg).
Dimensions: Span (over tip-tanks), 97 ft 8½ in (29,78 m); length, 95 ft 10¾ in (29,23 m); height, 29 ft 3½ in (8,93 m); wing area, 1,000 sq ft (92,9 m²).
Accommodation: Crew of 12, comprising two pilots on flight deck and ten-man tactical team in fuselage.
Armament: Internal stowage for up to 8,000 lb (3 629 kg) of weapons which can include homing torpedoes, mines, bombs, etc. Provision for up to 16 5-in (12,7 cm) rockets under wings.

Status: Prototype (converted P-2H) first flown on 21 July 1966. First production P-2J flown on 8 August 1969. Deliveries to JMSDF began in October 1969 with first squadron equipped by February 1971. Production of 46 scheduled for completion by March 1974 with options on second batch of 45 included in third national defence programme.
Notes: After the Kawasaki company had built 46 P-2H Neptunes under licence for the JMSDF (see page 81) it undertook development of the P-2J to provide a successor with improved operational capability. Using basically the same wing and tail unit as the Neptune, the P-2J has a lengthened fuselage which allows it to accommodate a more modern search and control system, comparable with that in the Lockheed Orion. Fuel capacity is increased and a change of power plant, to Japanese-built versions of the T64 turboprop, results in a major reduction in empty weight with no significant loss of performance. Like the final production series of Neptunes, the P-2J has underwing jet engines to boost the combat speed, these being J3s of indigenous design and construction. Current plans call for completion of production in 1976, but in 1973 consideration was being given to extended production.

LOCKHEED P-2 NEPTUNE

The SP-2H Neptune, illustrated by the photograph above and the general arrangement silhouette (right), is now restricted to Naval Reserve units in US service

Country of Origin: USA.
Type: Maritime reconnaissance and patrol bomber.
Power Plant: Two 3,500 hp Wright R-3350-32W radial engines and (P-2E, P-2G, P-2H only) two 3,400 lb st (1 540 kgp) Westinghouse J-34-WE-36 turbojets.
Performance: (P-2H) Max speed, 403 mph (648 km/h); max speed (piston engine only), 356 mph (573 km/h) at 10,000 ft (3 050 m), 305 mph (490 km/h) at 8,500 ft (2 590 m) and 262 mph (422 km/h) at sea level; patrol speed, 173–207 mph (278–333 km/h) at 1,000 ft (305 m); service ceiling, 22,000 ft (6 700 m); normal range, 2,200 mls (3 540 km); ferry range, 3,685 mls (5 930 km).
Weights: (P-2H) Empty, 49,935 lb (22 650 kg); max loaded, 79,895 lb (36 240 kg).
Dimensions: Span, 103 ft 10 in (31,65 m); length, 91 ft 8 in (27,94 m); height, 29 ft 4 in (8,94 m); wing area, 1,000 sq ft (92,9 m²).
Accommodation: Crew of seven comprising pilot, co-pilot, navigator-bombardier, radar/MAD operator, radio-operator, sonobuoy operator and ordnanceman.

Armament: (P-2H) Optional provision for dorsal turret containing two 0·50-in (12,7 mm) machine guns. Internal weapons bay has provision for up to 8,000-lb (3 629-kg) load which can include bombs up to 2,000 lb (907 kg) each, two 2,165-lb (982-kg) torpedoes or two 2,000-lb (907-kg) sea mines or twelve 325-lb (147-kg) depth charges. Underwing provision for up to 16 5-in (12,7 cm) rockets.
Status: Prototype (XP2V-1) first flown on 17 May 1945; XP2V-2 flown on 7 January 1947; P2V-3 flown on 6 August 1948; P2V-4 flown on 14 November 1949; P2V-5 flown on 29 December 1950; P2V-6 flown on 16 October 1952; P2V-7 flown on 26 April 1954. Production total for US Navy, 838, delivered 1946–62, including XP2V-1, 2; P2V-1, 15; P2V-2, 60; P2V-3, 83; P2V-4, 52; P2V-5, 424; P2V-6, 83; P2V-7, 119, plus production for export.
Notes: Design of the Neptune began in 1941 and it enjoyed a 15-year production life, over 1,000 being produced. Original designations P2V-1 to P2V-7 were changed to P-2D to P-2H in 1962. Front-line use by US Navy ended in 1970 but the Neptune still served with Naval Reserve units in 1973 and with overseas forces, including Argentina, Brazil, Chile, France, and Japan.

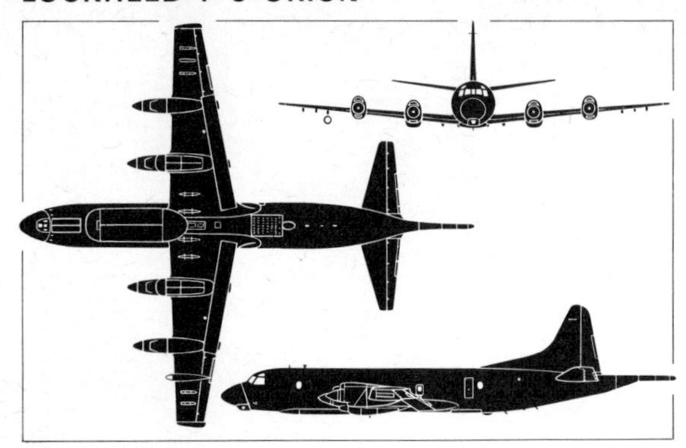

The photograph above illustrates a P-3B of No 333 Squadron of the Royal Norwegian Air Force and the general arrangement silhouette (left) depicts the P-3C version of the Orion

Country of Origin: USA.
Type: Long-range maritime patrol aircraft.
Power Plant: Four (P-3A) 4,500 eshp Allison T56-A-IOW or (P-3B, C, D, E, F) 4,910 eshp T56-A-14 turboprops.
Performance: (P-3B, 3C) Max speed, 473 mph (761 km/h) at 15,000 ft (4 570 m); economical cruise speed, 378 mph (608 km/h) at 25,000 ft (7 620 m); typical patrol speed, 237 mph (381 km/h) at 1,500 ft (450 m); initial rate of climb, 1,950 ft/min (9,9 m/sec); service ceiling, 28,300 ft (8 625 m); mission radius with 3-hr search, 1,550 mls (2 494 km); max range, 4,766 mls (7 670 km).
Weights: (P-3B, 3C) Empty, 61,491 lb (27 890 kg); max warload, 20,000 lb (9 071 kg); normal take-off, 135,000 lb (61 235 kg); max loaded, 142,000 lb (64 410 kg).
Dimensions: Span, 99 ft 8 in (30,37 m); length, 116 ft 10 in (35,61 m); height, 33 ft 8½ in (10,29 m); wing area, 1,300 sq ft (120,77 m²).
Accommodation: Flight crew of five and five-man tactical team in cabin. Provision for up to 50 combat troops to be carried in emergency trooping rôle.
Armament: Weapons bay in fuselage can house one 2,000 lb (908 kg) mine, three 1,000 lb (454 kg) mines, two Mk 101 nuclear depth bombs with four Mk 43, 44 or 46 torpedoes or eight Mk 54 depth bombs or similar loads. Total of ten underwing pylons which can carry mines or rockets, or torpedoes for ferry purposes only. Max underwing load is 6,000 lb (2 721 kg).
Status: Aerodynamic prototype (modified Electra) first flown on 19 August 1958; full prototype (YP3V-1/YP-3A) flown on 25 November 1959; first P-3A flown on 15 April 1961; first P-3C flown on 18 September 1968. Totals of 157 P-3As and 124 P-3Bs delivered to US Navy, being followed by 202 P-3Cs. Ten P-3Bs supplied to RAAF; five to RNZAF and five to R Norwegian AF. Three P-3As delivered Spain in 1973, and four P-3Fs ordered by Iran for 1974 delivery.
Notes: Deliveries of P-3A to US Navy began in August 1962, with later batches introducing Deltic system of improved avionics. Some converted to WP-3A for weather reconnaissance. P-3B introduced new engines and P-3C has A-NEW control system based on a digital computer. Two EP-3Bs and 10 P-3Bs became EP-3Es with special reconnaissance equipment and the single RP-3D, with a 17-man crew, is a modified P-3C to map the Earth's magnetic field.

LOCKHEED S-3A VIKING

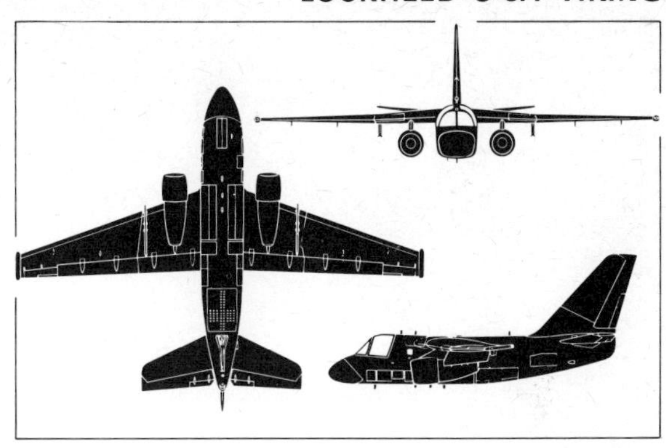

The S-3A Viking, illustrated by the photograph (above) and the general arrangement silhouette (right) is scheduled to enter US Navy service during 1974

Country of Origin: USA.
Type: Carrier-borne anti-submarine aircraft.
Power Plant: Two 9,275 lb st (4 207 kgp) General Electric TF34-GE-2 turbofans.
Performance (Estimated): Max speed, 507 mph (816 km/h); max cruising speed, 403 mph (649 km/h); typical loiter speed, 184 mph (257 km/h) at sea level; initial rate of climb, over 4,200 ft/min (21,3 m/sec); service ceiling, over 35,000 ft (10 670 m); combat range, 2,300 mls (3 705 km); ferry range, 3,450 mls (5 560 km).
Weights (Estimated): Empty, 26,600 lb (12 065 kg); normal loaded, 42,500 lb (19 277 kg); max take-off, 47,000 lb (21 320 kg).
Dimensions: Span, 68 ft 8 in (20,93 m); length, 53 ft 4 in (16,26 m); height, 22 ft 9 in (6,93 m); wing area, 598 sq ft (55,56 m²).
Accommodation: Crew of four comprising two pilots side-by-side on flight deck and tactical co-ordinator and sensor operator in cabin.
Armament: Internal weapons bays accommodate four torpedoes, bombs, depth charges or mines. Additional weapon stowage on two wing pylons, for bombs, depth charges, rockets, etc.
Status: First of eight development and evaluation S-3As flown on 21 January 1972, with completion of this batch by April 1973. Production orders for 93 up to FY74 budget against planned procurement of 186 for 12 squadrons. Production aircraft deliveries in second half of 1973 with Fleet Introduction in February 1974.
Notes: Development of an anti-submarine aircraft to replace the Grumman S-2 Tracker (page 77) began in 1967 when the US Navy issued a Request for Proposals, with the contract award to Lockheed announced in August 1969. Teamed with Lockheed–California in the submission was LTV, which has responsibility for design and manufacture of the wing, engine pods, tail unit and landing gear. US Navy planning provides for acquisition of 186 Vikings. A key design feature of the S-3A was a high speed capability for transit between carrier and search area, and contributing to this objective is the use of retractable MAD and FLIR radomes and a retractable in-flight refuelling probe. Drop tanks can be carried on the wing pylons to extend the ferry range, and the fifth aircraft in the development batch was completed as a flight refuelling tanker for trials. Several design variants of the Viking have been proposed, including ASW command and control, utility transport and various electronic countermeasures versions.

MYASISHCHEV M-4

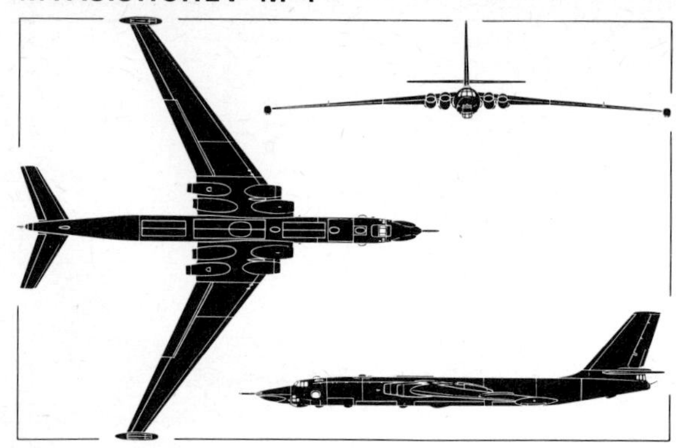

The photograph above illustrates the maritime patrol version (Bison-B) of the original M-4 strategic bomber and the silhouette (left) depicts the more powerful model (Bison-C)

Country of Origin: USSR.
NATO Code Name: *Bison.*
Type: Long-range maritime reconnaissance aircraft and flight refuelling tanker.
Power Plant: Four (*Bison-A*) 19,180 lb st (8 700 kgp) Mikulin AM-3D turbojets or (*Bison-C*) 28,660 lb st (13 000 kgp) Type D-15 turbojets.
Performance: Max speed, 560 mph (901 km/h) at 36,000 ft (10 973 m); typical cruising speed, 520 mph (835 km/h) at 40,000 ft (12 192 m); operational ceiling, 45,000 ft (13 700 m); range, 7,000 mls (11 250 km) with 10,000 lb (4 500 kg) bomb load.
Weights (Estimated): Max take-off, 350,000 lb (158 760 kg).
Dimensions (Estimated): Span, 170 ft 0 in (51,82 m); length, 162 ft 0 in (49,38 m).
Accommodation: Flight crew of five or six.
Armament: Two 23-mm guns each in (*Bison-A*) four and (*Bison-B* and *-C*) two remotely-controlled turrets above and below fuselage and in tail turret. Large internal weapon-bay for nuclear or conventional bombs.

Status: First public appearance (over Moscow) May 1954; in service as a bomber since 1955, for maritime reconnaissance since 1964 and as a flight-refuelling tanker.
Notes: The big Myasishchev bomber assigned the name *Bison* by the Air Standards Co-ordinating Committee was the Soviet Union's first long-range strategic jet bomber, being a contemporary of the Boeing B-52 and playing a similar rôle in the Cold War during the late 'fifties and early 'sixties. The basic bomber version was *Bison-A*, but this was adapted in 1964 as *Bison-B* for the maritime reconnaissance rôle, with a new nose radome, a flight refuelling probe, and various special items of electronic equipment, indicated by an assortment of radomes and "blisters" under the fuselage. Two of the four fuselage gun barbettes were deleted on this version. Also used for maritime reconnaissance, *Bison-C* is similar but has a redesigned nose of better streamline shape, with the flight refuelling probe centred. One such version called Type 201-M set up a series of payload-to-height records in 1959 and was powered by the D-15 engines as indicated above. About 50 Bisons were reported in service as flight refuelling tankers in 1973, in addition to those used for maritime duties.

SHIN MEIWA SS-2 (PS-1)

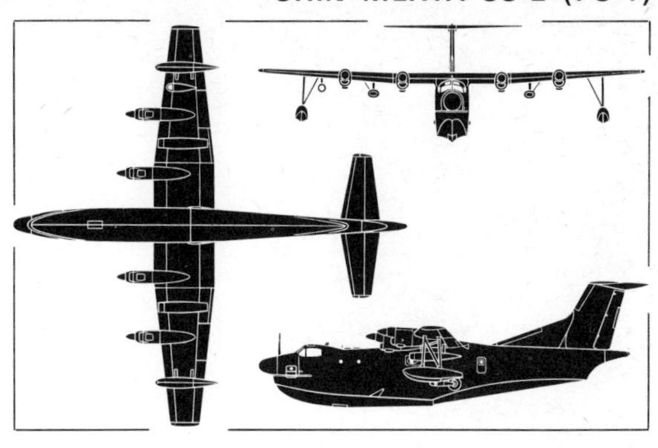

The PS-1 maritime patrol flying boat, illustrated by the photograph (above) and the general arrangement silhouette (right), entered service with the Maritime Self-Defence Force in 1973

Country of Origin: Japan.
Type: Long-range maritime patrol flying boat.
Power Plant: Four 3,060 ehp Ishikawajima-built General Electric T64-IHI-10 turboprops.
Performance: Max speed, 340 mph (547 km/h) at 5,000 ft (1 525 m); normal cruise, 265 mph (426 km/h) at 5,000 ft (1 525 m); initial rate of climb, 2,264 ft/min (6,9 m/sec); service ceiling 29,530 ft (9 000 m); normal range 1,347 mls (2 168 km); ferry range 2,948 mls (4 744 km); max endurance, 15 hrs.
Weights: Empty equipped, 58,000 lb (26 300 kg); normal take-off, 79,366 lb (36 000 kg); max take-off, 99,208 lb (45 000 kg).
Dimensions: Span, 108 ft $8\frac{3}{4}$ in (33,14 m); length, 109 ft 11 in (33,50 m); height, 31 ft $10\frac{1}{2}$ in (9,71 m); wing area, 1,462 sq ft (135,8 m^2).
Accommodation: Flight crew of three, comprising two pilots and a flight engineer. Tactical team of seven, comprising co-ordinator, two sonar operators, MAD operator, radar operator, radio operator and navigator, in upper deck compartment.
Armament: Weapons compartment in upper deck, aft of tactical compartment, accommodates four 330-lb (150-kg) anti-submarine bombs. One weapons pod beneath each wing between the engine nacelles, each containing two homing torpedoes. Pylon beneath each wing tip can carry three 5-in (12,7-cm) rockets.
Status: Two prototypes (PX-S) first flown, respectively, on 5 October 1967 and 14 June 1968 and delivered to JMSDF on 31 July and 30 November 1968. Two pre-production models (PS-1) delivered in 1972. Initial production deliveries in 1973 with procurement of 14 planned in 1972–76 defence programme.
Notes: The PX-S was developed in Japan to meet a requirement of the Maritime Self-Defence Force for a new anti-submarine flying-boat. The two prototypes were tested by the 415th Flight Test Squadron at Iwakuni, after which the JMSDF adopted the designation PS-1 for the production model, which is identified by Shin Meiwa as the SS-2. The initial unit formed to operate the PS-1 was the 31st *Kokuh-gun* (Air Group) also at Iwakuni, which had the two prototypes, two pre-production models and the first five production aircraft on strength by March 1973. Design of an amphibian version, the SS-2A, was begun in June 1970, and three examples of this version were ordered by the JMSDF as US-1s, the first US-1-equipped rescue unit being planned for 1976.

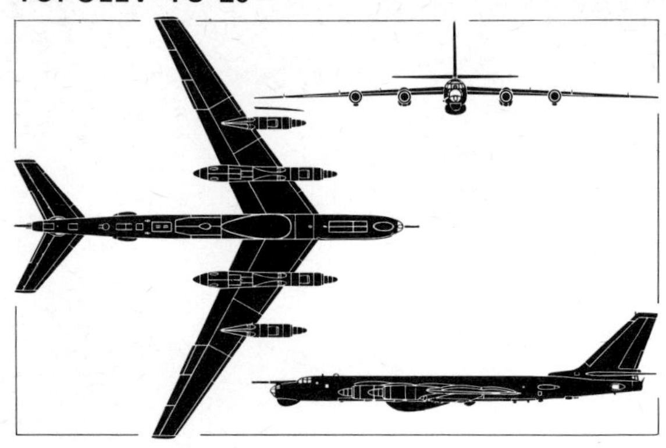

The photograph (above) and general arrangement silhouette (left) both illustrate the so-called Bear-D, the version of the Tu-20 featuring a large ventral radome

Country of Origin: USSR.
NATO Code Name: *Bear.*
Type: (Bear-A and -E) Long-range strategic bomber and (Bear-B, -C, -D and -F) maritime reconnaissance and patrol.
Power Plant: Four 14,795 ehp Kuznetsov NK-12M turboprops.
Performance: Max speed over target, 500 mph (805 km/h) at 41,000 ft (12 500 m); max range, 7,800 mls (12 550 km) with 25,000 lb (11 340 kg) bomb load.
Weights (Estimated): Max take-off, 340,000 lb (154 220 kg).
Dimensions: Span, 159 ft 0 in (48,50 m); length, 155 ft 10 in (47,50 m); wing area (estimated) 3,150 sq ft (292,6 m²).
Accommodation: Crew of five or six for bomber rôle, including a tail gunner in separate, isolated compartment; main cockpit area connected with rest and relief-crew area in centre fuselage by pressurised crawl-way above bomb-bay. Maritime reconnaissance versions have additional observers/equipment operators in this amidships compartment.
Armament: Maximum bomb load, about 25,000 lb (11 340 kg) for long-range missions. Defensive armament of six 23-mm cannon, two each in remotely-controlled dorsal and ventral barbettes and two in tail barbette.
Status: Flight testing (Tu-95 prototype) began late summer of 1954; public debut (seven examples) over Moscow on 3 July 1955. Deliveries (*Bear-A*) began second half of 1956. First sightings of *Bear-B,* June 1961, of *Bear-C,* September 1964 and *Bear-D* mid-1967. Operational force of about 100 in 1973.
Notes: The Tu-95 prototypes were the final product of a long evolution process started when the Soviet Union put a pirated copy of the Boeing B-29 into production as the Tu-4. Identified as the Tu-20 in its operational form, *Bear* is the largest turboprop aircraft to achieve a front-line combat rôle, and *Bear-A,* the basic strategic bomber, has been followed by a number of variants with special equipment for maritime reconnaissance. These are known by the NATO code-names *Bear-B* (with a wide under-nose radome and provision to carry *Kangaroo* ASM); *Bear-C* (with additional avionics for maritime rôle); *Bear-D* (with a large ventral radome and equipped as an anti-shipping missile control aircraft); and *Bear-F* (with enlarged rear fairings behind the inboard engines). *Bear-E* is similar to *Bear-A* but has a number of camera windows in the weapon bay.

AVRO SHACKLETON

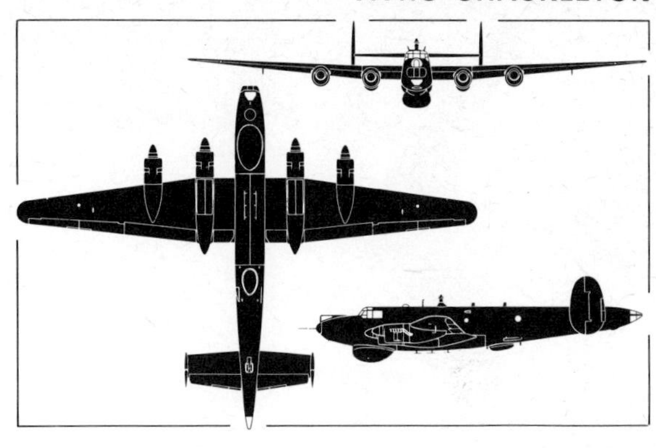

The photograph (above) and the general arrangement silhouette (right) both illustrate the Shackleton AEW Mk 2 currently operated by the RAF's No 8 Squadron

Country of Origin: United Kingdom.
Type: Airborne early warning aircraft.
Power Plant: Four 2,450 hp Rolls-Royce Griffon 57A 12-cylinder liquid-cooled engines, plus (Mk 3 only) two 2,500 lb st (1 134 kgp) Rolls-Royce Bristol Viper 203 turbojets.
Performance: (MR Mk 3) Max speed, 302 mph (486 km/h) at 12,000 ft (3 658 m); max cruising speed, 253 mph (407 km/h) at 10,000 ft (3 050 m); long-range cruising speed, 200 mph (322 km/h); initial rate of climb, 850 ft/min (4,3 m/sec); service ceiling, 19,200 ft (5 852 m); range, 3,660–4,215 mls (5 890–6 780 km).
Weights: (MR Mk 3) Empty, 57,800 lb (26 218 kg); normal loaded, 85,000 lb (38 555 kg); max take-off, 100,000 lb (45 360 kg).
Dimensions: (MR Mk 3) Span, 119 ft 10 in (36,53 m); length, 92 ft 6 in (28,19 m); height, 23 ft 4 in (7,11 m); wing area, 1,458 sq ft (135,45 m²).
Accommodation: Flight crew of three plus eight in tactical compartment in fuselage.
Armament: Two 20-mm cannon in nose.
Status: First of three prototypes flown on 9 March 1949. Deliveries of MR Mk 1 began on 28 September 1950; deliveries of MR Mk 2 began in October 1952. First MR Mk 3 flown on 2 September 1971; deliveries began in January 1972. Production quantities: Prototypes, 3; MR Mk 1, 77; MR Mk 2, 69 (12 converted to AEW Mk 2); MR Mk 3, 42 (including 8 for SAAF).
Notes: The Shackleton was derived indirectly from the famous Lancaster bomber by way of the Lincoln, having the latter's wing, tail unit and undercarriage with a new fuselage. Production passed through three variants, of which the Mk 1 and Mk 2 had tailwheel undercarriages and, respectively, short and long fuselages, and the Mk 3 had a nosewheel undercarriage and (through a later modification programme) jet boost engines in the inboard nacelles. After 20 years of service, these Shackletons have mostly been replaced by Nimrods (see page 78), but seven of the eight Mk 3s supplied to the South African Air Force are still in service with No. 35 Squadron based at DF Malan airfield, and the RAF's No. 8 Squadron has 12 AEW Mk 2s for airborne early warning. These are modified MR Mk 2s with APS-20 search radar under the nose, in place of the retractable radome farther aft, and during 1973 these were to receive additional weapons bay fuel tanks to raise endurance from 12 to 14 hours.

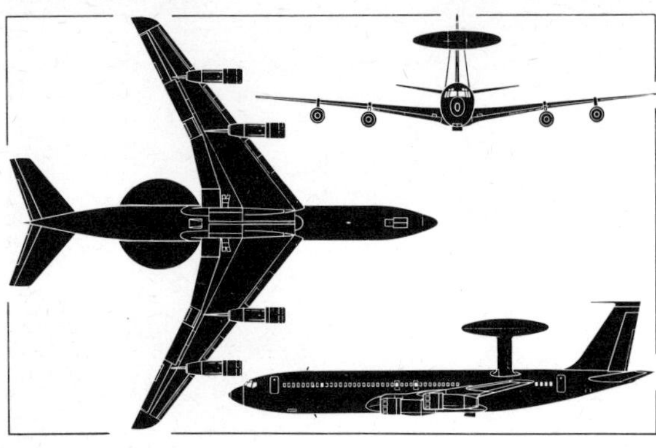

The photograph (above) and the general arrangement silhouette (left) both illustrate one of the two EC-137D prototypes for the E-3A airborne early warning and control system aircraft

Country of Origin: USA.
Type: Airborne warning and control system development aircraft.
Power Plant: Four 21,000 lb st (9 525 kgp) Pratt & Whitney TF33-P-7 turbofans.
Performance: No details of performance released for publication. Maximum and cruising speeds comparable with those of Boeing 707-320C on which the E-3A is based (see *Civil* volume). Mission requirement is for a 7-hr search at a distance of 1,150 mls (1 850 km) from base, cruising at 29,000 ft (8 840 m).
Weights: No details of weights released for publication. Weight of external rotodome is 12,500 lb (5 670 kg) and maximum take-off weight is approximately 330,000 lb (149 685 kg).
Dimensions: Span, 145 ft 9 in (44,42 m); length, 152 ft 11 in (46,61 m); height, 42 ft 5 in (12,93 m); wing area, 3,050 sq ft (283,4 m²).
Accommodation (EC-137D prototype): Flight crew of four, comprising aircraft commander, pilot, navigator and flight engineer; AWACS crew of 13, comprising battle commander, an eight-man air defence operations team and a four-man communication and electronics maintenance team.

Status: Two EC-137D prototypes first flown on 9 February 1972. Programme comprises one pre-production YE-3A and 39 production models, with deliveries to begin early in 1977. The EC-137Ds and YE-3A will be converted to production standard.
Notes: The programme to develop a new airborne warning and control system (AWACS) for the USAF, based on the Boeing 707 airframe, began in July 1970, the first step being to build two EC-137D prototypes. These were "minimum modifications" of the commercial 707s, fitted with large 30 ft (9,12 m) diameter rotodomes above the rear fuselage to accommodate the surveillance radar. Both prototypes made their first flights on the same day, being fitted with competitive radars by Westinghouse and Hughes, and tests of the radars were completed on 31 August 1972 leading to selection of the Westinghouse equipment. The AWACS system includes extensive sensors, communications and navigation equipment plus the largest-capacity airborne computer yet flown. The production E-3A is intended to meet the requirements of both Tactical Air Command and Aerospace Defence Command, with no change of equipment. Its capabilities include surveillance, detecting, tracking and weapons selection and direction.

DOUGLAS A-3 SKYWARRIOR

Both photograph (above) and general arrangement silhouette (right) illustrate the EKA-3B Skywarrior electronic countermeasures aircraft of the US Navy

Country of Origin: USA.
Type: Electronic countermeasures and flight refuelling tanker.
Power Plant: Two 10,500 lb st (4 763 kgp) dry and 12,400 lb st (5 625 kgp) with water injection Pratt & Whitney J57-P-10 turbojets.
Performance: (A-3B) Max speed, 610 mph (982 km/h) at 10,000 ft (3 050 m) and 560 mph (901 km/h) at 36,000 ft (10 973 m) ; service ceiling, 41,000 ft (12 497 m) ; tactical radius, 1,050 mls (1 690 km) ; max range, 2,900 mls (4 667 km).
Weights: (A-3B) Empty, 39,409 lb (17 875 kg) ; normal loaded, 73,000 lb (33 112 kg) ; max take-off, 82,000 lb (37 195 kg).
Dimensions: Span, 72 ft 6 in (22,10 m) ; length, 76 ft 4 in (23,27 m) ; height, 22 ft $9\frac{1}{2}$ in (6,94 m) ; height (EA-38) 23 ft 6 in (7,16 m) ; wing area, 812 sq ft (75,44 m^2).
Accommodation: (EA-3B) Flight crew of three and four electronics operators in pressurized fuselage.
Armament: Bomber versions had internal provision for up to 12,000 lb (5 443 kg) and two 20-mm cannons in remotely-controlled tail barbette. Recce and tanker versions usually unarmed.
Status: Prototype (XA3D-1) first flown on 28 October 1952; YA3D-1 first flown on 16 September 1953; YRA-3B first flown on 22 July 1958; EA-3B first flown on 10 December 1958; TA-3B first flown on 29 August 1959. Production quantities: A-3A, 50; A-3B, 164; EA-3B, 24; RA-3B, 30; TA-3B, 12. Other variants by conversion. Production completed 1961.
Notes: Development of the Skywarrior was started in parallel with work on the super-carriers of the USS *Forrestal* type that were evolved soon after World War II, and at the time of first flight the XA3D-1 prototype was the largest carrier-based aircraft in the world. The A-3A and A-3B bomber versions, the latter with more powerful engines and provision for flight refuelling, equipped US Navy bomber squadrons until the mid-sixties but were retired after the Navy relinquished its strategic bombing rôle and the Skywarrior squadrons re-equipped with RA-5Cs. Many A-3Bs were then converted to KA-3B refuelling tankers, becoming the only carrier-based tankers in the Fleet inventory, with a capacity of 1,300 US gal (4 925 l) and the addition of ECM equipment then produced the EKA-3B version. The EKA-3Bs were extensively used by the Navy throughout the Vietnam campaign, two aircraft being attached to each operational carrier; a total of 30 was converted.

DOUGLAS B-66 DESTROYER

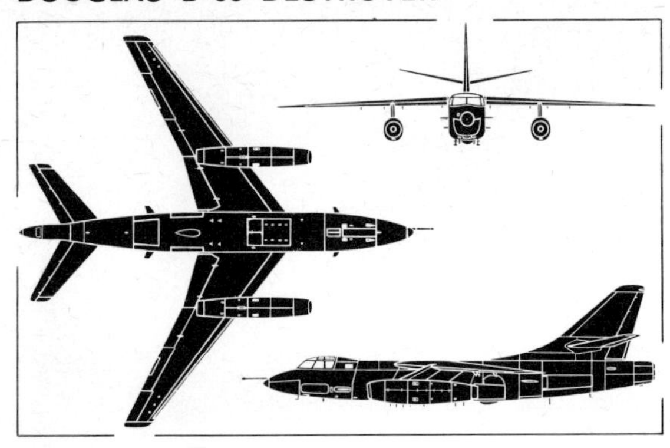

The photograph above depicts an EB-66C and the general arrangement silhouette (left) illustrates the EB-66E, both of which are electronic jamming and countermeasures aircraft

Country of Origin: USA.
Type: Electronic countermeasures aircraft.
Power Plant: (EB-66E) Two 10,000 lb st (4 536 kgp) Allison J71-A-13 turbojets.
Performance: Max speed, 594 mph (956 km/h) at 36,000 ft (10 973 m); range, more than 1,500 mls (2 414 km).
Weights: (B-668) Empty, 42,369 lb (19 218 kg); max take-off, 83,000 lb (37 650 kg).
Dimensions: Span, 72 ft 6 in (22,10 m); length 75 ft 2 in (22,91 m); height, 23 ft 7 in (7,19 m); wing area, 780 sq ft (72,46 m²).
Accommodation: (EB-66B) Flight crew of three and four radar operators. (EB-66E) Crew of three.
Armament: Original bomber versions had internal stowage for up to 15,000 lb (6 804 kg) of bombs and two 20-mm cannon in remotely-controlled tail barbette. Electronic countermeasures versions are unarmed.
Status: First of five RB-66A development aircraft flown on 28 June 1954; first production RB-66B flown on 28 June 1954; first B-66B flown on 4 January 1955; first RB-66C flown on 29 October 1955; first WB-66D delivered on 26 June 1957. Production completed June 1958, totals as follow: RB-66A, 5; RB-66B, 145; B-66B, 72; RB-66C, 36; WB-66D, 36. Other variants by conversion.
Notes: The Destroyer was evolved as a tactical light bomber and reconnaissance aircraft, serving in both these rôles with the USAF from early 1956 until the late 'sixties. Since these variants were phased out of service, the primary rôle of the Destroyer has been electronic countermeasures, for which purpose many of the original aircraft were converted. The first ECM model was the EB-66B (modified B-66B) with a four-man compartment in the bomb bay and radomes at the nose and tail (replacing the tail guns) and at the wing tips. This type entered service over Vietnam, operating from bases in Thailand, in 1968 and was supplemented by the EB-66C (converted RB-66C) and EB-66E (modified RB-66B). The electronic equipment in the EB-66E, primarily intended to jam enemy radars while USAF and USN aircraft went into attack, was almost entirely automatic and was operated by the aircraft's normal flight crew of three. In addition to its use in South East Asia, up to and beyond the end of the Vietnam war, the EB-66 was used by one Tactical Electronic Warfare Squadron based with the USAFE in Germany.

FAIREY GANNET 3

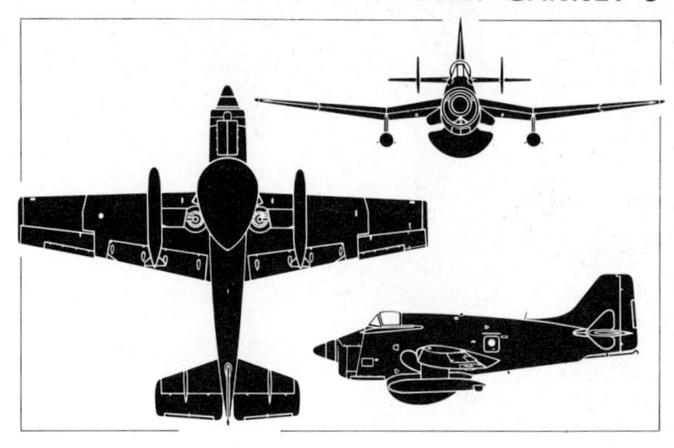

The photograph (above) and general arrangement silhouette (right) both depict the Gannet AEW Mk 3, the former illustrating an aircraft of the Royal Navy's No 849 Squadron

Country of Origin: United Kingdom.
Type: Airborne early warning aircraft.
Power Plant: One 3,875 ehp Rolls-Royce Bristol Double Mamba 102 turboprop.
Performance: Max speed, 250 mph (402 km/h) at 5,000 ft (1 524 m); service ceiling, 25,000 ft (7 625 m); range, about 800 mls (1 287 km); endurance, 5–6 hrs at 130–140 mph (209–225 km/h).
Weights: Estimated gross weight, 24,000 lb (10 886 kg).
Dimensions: Span, 54 ft 6 in (16,56 m); length, 44 ft 0 in (13,41 m); height, 16 ft 10 in (5,15 m); wing area, 490 sq ft (45,5 m²).
Accommodation: Pilot and two radar observers.
Status: Prototype Fairey GR.17/45 first flown on 19 September 1949. First production Gannet 1 flown on 9 July 1953. Aerodynamic prototype of Gannet AEW Mk 3 first flown on 20 August 1958 and first production example flown on 2 December 1958. Production totals: prototypes, 3; AS Mk 1, 181; T Mk 2, 38; AEW Mk 3, 44; AS Mk 4, 75; T Mk 5, 8. Export deliveries, included in these totals, comprised 40 Mk 1 and 2 to R Australian Navy; 15 Mk 4 and 1 Mk 5 to German Navy and 17 Mk 4 and one Mk 5 to Indonesian Air Force.
Notes: The Gannet was designed to a late World War II specification for a ship-board anti-submarine aircraft to take advantage of the then-new turboprop engine, and became standard equipment with Royal Navy and RAF front-line squadrons from 1955 to 1960. Alongside the AS and dual-control training versions, the Fairey company developed the AEW Mk 3 airborne early warning variant, with a redesigned fuselage, modified tail unit and strengthened undercarriage. The principal new feature was the installation of APS-20 search radar in a large radome replacing the weapons bay. The Gannet AEW Mk 3 was used to equip No 849 Squadron, RN, starting in 1960, and this unit provided detached flights to serve aboard all aircraft carriers operated by the RN up to 1973, when only HMS *Ark Royal* remained in service. With the run-down of the FAA, some Gannet 3s were scrapped and their radars were transferred for use in Shackleton AEW Mk 2s (see page 87). Five Gannet 4s were converted for use as shore–ship communications aircraft, with radar removed and designated COD Mk 4 (Carrier On-board Delivery) and these remained in service, based at RAF Lossiemouth, to support operations from HMS *Ark Royal*.

GRUMMAN E-2 HAWKEYE

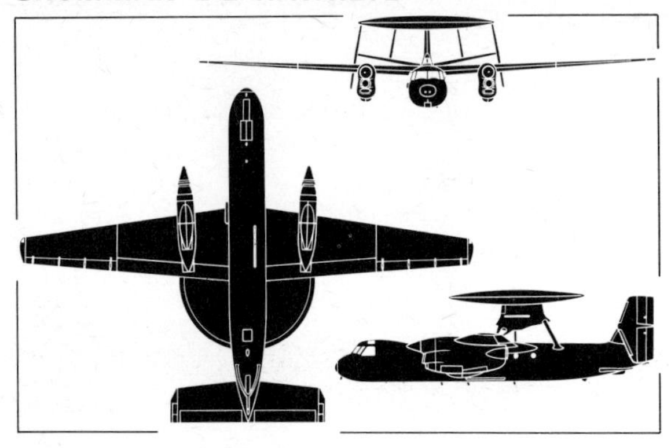

The photograph (above) and the general arrangement silhouette (left) both depict the E-2B version of the Hawkeye airborne early warning aircraft

Country of Origin: USA.
Type: (E-2) Airborne early warning aircraft and (C-2) transport.
Power Plant: Two 4,050 shp Allison T56-A-8/8B turboprops.
Performance: (E-2C) Max speed, over 363 mph (584 km/h); cruising speed, 310 mph (409 km/h); service ceiling, 28,100 ft (8 565 m); ferry range, 1,722 mls (2 771 km).
Performance: (C-2A) Max speed, 352 mph (567 km/h); typical cruising speed, 297 mph (478 km/h); initial rate of climb, 2,330 ft/min (11,8 m/sec); service ceiling, 28,800 ft (8 780 m); range, 1,650 mls (2 660 km).
Weights: (E-2C) Empty, 37,616 lb (17 062 kg); max internal fuel load, 12,400 lb (5 624 kg); max take-off, 51,490 lb (23 355 kg).
Weights: (C-2A) Empty, 31,154 lb (14 131 kg); max take-off, 54,830 lb (24 870 kg).
Dimensions: (E-2) Span 80 ft 7 in (24,56 m); length, 57 ft 7 in (17,55 m); height, 18 ft 4 in (5,59 m); wing area, 700 sq ft (65,03 m^2).
Dimensions: (C-2A) Span, 80 ft 7 in (24,56 m); length, 56 ft 8 in (17,27 m); height, 15 ft 11 in (4,85 m); wing area, 700 sq ft (65,03 m^2).

Accommodation: (E-2) Flight crew of two and team of three for ATDS (Airborne Tactical Data System) operation.
Accommodation: (C-2A) Flight crew of two, dual controls. Up to 39 troops or 20 stretchers and four attendants or cargo.
Status: Prototype (W2F-1) first flown 21 October 1960; first fully-equipped prototype flown on 19 April 1961; development aircraft with AN/APS-111 flown on 17 August 1965; prototype E-2B flown on 20 February 1969; prototype E-2C flown on 20 January 1971; first production E-2C flown in 1973. Production of E-2A totalled 59 (all later converted to E-2B), completed in 1967. First C-2A flown on 18 November 1964; production of 25 completed in 1971. Planned total E-2C procurement, 28.
Notes: The E-2A, B and C differ primarily in radar and avionics, the E-2A and E-2B having an AN/APS-96 long-range radar, while the E-2C has AN-APS/111, and the E-2B and C having Litton L-304 multi-purpose computer. Two US Navy units, VAW-11 and VAW-12, provide detached units of E-2s for service aboard operational aircraft carriers, and some TE-2As are in use for training. Derived from the E-2, the C-2A has a completely new fuselage to carry personnel and supplies between ships and shore stations.

GRUMMAN EA-6B PROWLER

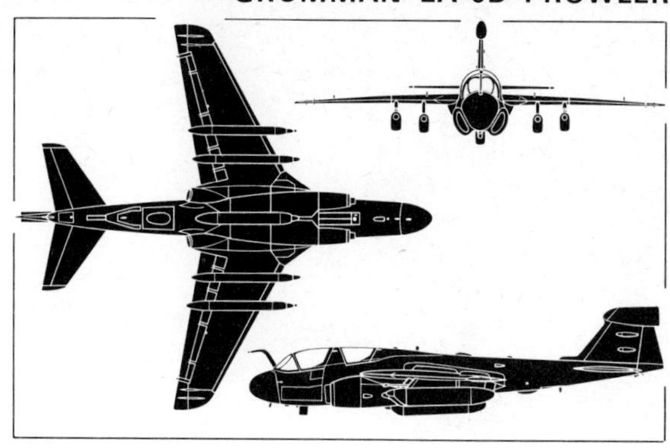

The photograph (above) and the general arrangement silhouette (right) both illustrate the EA-6B Prowler, the former depicting a Prowler of US Navy Squadron VAQ-129

Country of Origin: USA.

Type: Carrier-borne electronic warfare aircraft.

Power Plant: Two 9,300 lb st (4 218 kgp) Pratt & Whitney J52-P-8A turbojets.

Performance: Max speed, 599 mph (964 km/h) at sea level; average cruising speed, 466 mph (750 km/h) service ceiling, 38,000 ft (11 582 m); ferry range (with five 250 Imp gal / 1 136 l external tanks,) 2,475 mls (3 982 km).

Weights: Empty, 34,580 lb (15 686 kg); typical mission weight, 51,000 lb (23 133 kg); normal take-off weight, 58,500 lb (26 535 kg); max overload, 63,177 lb (28 656 kg).

Dimensions: Span, 53 ft 0 in (16,15 m); length, 59 ft 5 in (18,11 m); height, 16 ft 3 in (4,95 m); wing area, 529 sq ft (49,15 m²).

Accommodation: Pilot and electronic countermeasures operator side-by-side in forward cockpit; two operators side-by-side in rear cockpit.

Armament: None.

Status: Prototype EA-6B flown on 25 May 1968. Production deliveries for US Navy began January 1971 and operational deployment began mid-1972. Total procurement, 45 funded up to mid-1973.

Notes: The first electronic countermeasures version of the Grumman A-6 Intruder (see page 51) was the EA-6A, first flown in 1963 and originally developed under the designation A2F-1Q. This was a two-seater, like the A-6A, but had the bombing/navigation system equipment deleted to make room for more than 30 antenae required to detect, locate, classify, record and jam enemy transmissions. Production totalled 27 for US Marine Corps use. The EA-6B is a more extensive revision of the same basic design, but the front fuselage is lengthened by 40 in (1,02 m) to allow two extra seats to be inserted aft of the pilots, the additional crew being needed to handle extra electronics equipment. Like the EA-6A, the EA-6B has a large radome atop the fin, and it carries 8,000 lb (3 629 kg) of avionics internally and a 950-lb (431-kg) pod on fuselage centreline and four wing stations. The EA-6B can also be used in the ECM escort or penetration rôles, and its range and altitude capability are compatible with the A-6 Intruder. Because of its changed rôle and the engineering changes to the airframe, the EA-6B was renamed Prowler, and procurement began in FY 1969 for US Navy squadrons. An expanded capability (EXCAP) version entered service in 1973.

GRUMMAN OV-1 MOHAWK

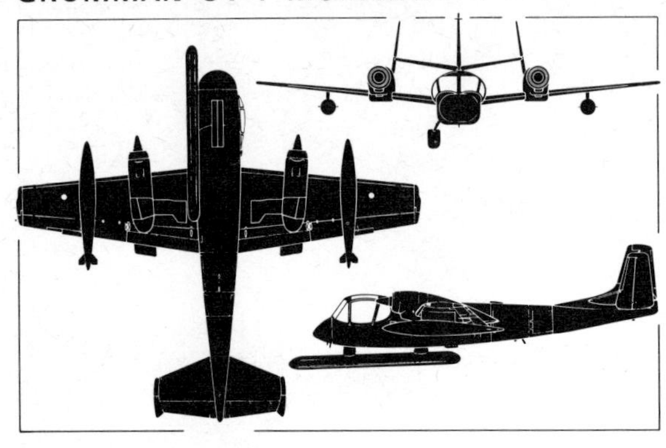

Both the photograph (above) and the general arrangement silhouette (left) depict the final production version of the Mohawk, the OV-1D with side-looking airborne radar

Country of Origin: USA.
Type: STOL observation aircraft.
Power Plant: Two 1,150 shp Lycoming (OV-1A, -1B, -1C) T53-L-7 or L-15 or (OV-1D) 1,400 shp T53-L-701 turboprops.
Performance: (OV-1A) Max level speed, 308 mph (496 km/h) at 5,000 ft (1 520 m); max cruising speed, 304 mph (489 km/h); economical cruising speed, 207 (334 km/h); initial rate of climb, 2,950 ft/min (15,0 m/sec); max range, 1,410 mls (2 270 km).
Performance: (OV-1D) Max level speed, 305 mph (491 km/h) at 10,000 ft (3 050 m); economical cruising speed, 207 mph (334 km/h); initial rate of climb, 3,618 ft/min (18,3 m/sec); service ceiling, 25,000 ft (7 620 m); max range, 1,011 mls (1 627 km).
Weights: (OV-1A) Empty equipped, 9,937 lb (4 507 kg); normal take-off, 12,672 lb (5 748 kg); maximum loaded, 15,031 lb (6 818 kg).
Weights: (OV-1D) Empty equipped, 12,054 lb (5 467 kg); normal take-off, 15,544 lb (7 051 kg); max loaded, 18,109 lb (8 214 kg).
Dimensions: Span (OV-1A, -1C), 42 ft 0 in (12,80 m); span (OV-1B, -1D), 48 ft 0 in (14,63 m); length, 41 ft 0 in (12,50 m); height, 12 ft 8 in (3,86 m); wing area (OV-1A, -1C), 330 sq ft (30,65 m^2); wing area (OV-1B, -1D), 360 sq ft (33,45 m^2).
Accommodation: Crew of two side-by-side, optional dual control.
Status: Prototype (YOV-1A) first flown 14 April 1959. First deliveries (OV-1A) 1961. Production completed December 1970. Production quantities: YOV-1A, 9; OV-1A, 64; OV-1B, 101; OV-1C, 133; OV-1D, 37.
Notes: As the Grumman G-134, the Mohawk design won joint Army and Marine Corps backing for development as a battlefield surveillance aircraft, with STOL capability and able to operate from unprepared fields. The Marine Corps version, designated OF-1, was dropped before first flight, and the programme continued only for the Army, the initial designation of AO-1 later being changed to OV-1. Nine test examples were completed in 1959 and the first production orders were placed. Three versions were built in parallel, designated according to rôle equipment carried—OV-1A with cameras, OV-1B with side-looking airborne radar and OV-1C with cameras and infra-red sensors. Final production version was the OV-1D, with an improved camera installation plus provision for carrying SLAR or IR equipment and many earlier models have been converted to OV-1D standard.

LOCKHEED EC-121 WARNING STAR

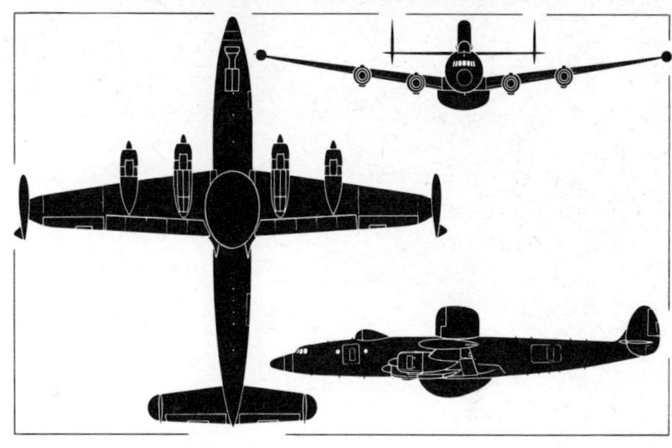

The photograph above depicts an EC-121Q of the US Navy and the general arrangement silhouette illustrates an EC-121H airborne early warning aircraft of the USAF

Country of Origin: USA.
Type: Electronic reconnaissance and transport aircraft.
Power Plant: Four 3,250 hp Wright R-3350-91 18-cylinder radial piston engines.
Performance (C-121G): Max speed, 368 mph (592 km/h) at 20,000 ft (6 096 m); typical cruising speed, 259 mph (417 km/h) at 10,000 ft (3 048 m); initial rate of climb, 1,100 ft/min (5,6 m/sec); service ceiling, 22,300 ft (6 797 m); range with 31,640 lb (14 350 kg), 2,085 mls (3 355 km); ferry range, 4,330 mls (6 968 km).
Performance (RC-121D): Max speed, 321 mph (517 km/h) at 20,000 ft (6 096 m); initial rate of climb, 845 ft/min (4,3 m/sec); service ceiling, 20,600 ft (6 279 m); range, 4,600 mls (7 400 km).
Weights (C-121G): Empty equipped, 75,133 lb (34 080 kg); normal take-off, 133,000 lb (60 327 kg); max overload, 145,000 lb (65 770 kg).
Weights (RC-121D): Empty, 80,611 lb (36 564 kg); normal take-off, 143,600 lb (65 135 kg).
Dimensions: Span, 123 ft 0 in (37,50 m); length, 116 ft 2 in (35,41 m); height, 24 ft 8 in (7,52 m); wing area, 1,650 sq ft (153,3 m²).

Accommodation (C-121G): Flight crew of five, plus provision for full relief crew and up to 72 troops or 47 stretchers plus two attendants.
Accommodation (EC-121K): Crew of nine comprising plane commander, two pilots, navigator, two flight engineers, CIC officer, CIC assistant and radioman.
Status: Prototype (C-69) first flown on 9 January 1943. Production for USAF/USN comprised: C-69, 21; C-69C, 1; PC-121A, 9; VC-121B, 1; C-121C, 33; RC-121C, 10; RC-121D, 72; VC-121E, 1; YC-121F, 2; C-121G, 32; R7V-1, 18; R7V-2, 2; WV-1, 2; WV-2, 136; WV-3, 8. Other military-use Constellations were built as commercial models.
Notes: USAF Air Defence Command relies upon EC-121Hs (modified RC-121Ds) for AEW and command control duties, linked with NORAD SAGE installation and other specially-equipped versions designated EC-121K, P, Q, R, S and T operated in SE Asia and elsewhere. For maritime patrol duties the Indian Air Force uses eight ex-Air India L-1049Gs which currently equip the service's No 6 Sqdn, these aircraft having been fitted with the necessary avionics for the search rôle, and some L-749As are used for search and rescue by *l'Armée de l'Air*.

LOCKHEED SR-71

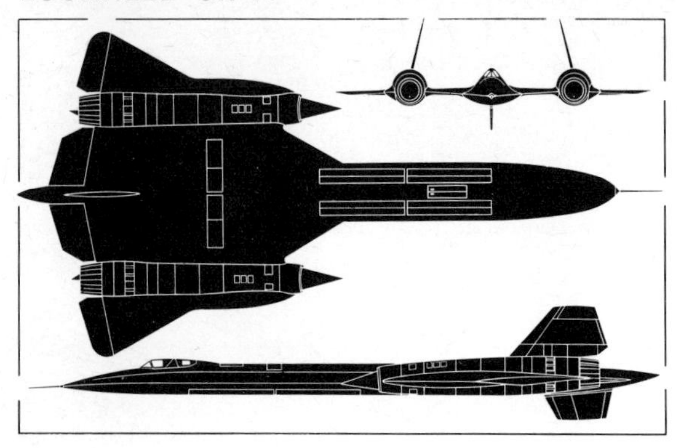

Both the photograph (above) and the general arrangement silhouette (left) illustrate the SR-71 long-range strategic reconnaissance aircraft

Country of Origin: USA.
Type: Long-range strategic reconnaissance aircraft.
Power Plant: Two 32,500 lb st (14 740 kgp) with afterburning Pratt & Whitney J58 (JT11D-20B) turbojets.
Performance (Estimated): Max speed, over Mach 3, equivalent to approximately 2,000 mph (3 218 km/h) at high altitudes; operating ceiling, above 80,000 ft (24 384 m); range, about 3,000 mls (4 828 km) at Mach 3 at 79,000 ft (24 080 m); max endurance, 1½ hrs at Mach 3.
Weights (Estimated): Max take-off, 170,000 lb (77 110 kg).
Dimensions: Span, 55 ft 7 in (16,95 m); length, 107 ft 5 in (37,74 m); height, 18 ft 6 in (5,64 m).
Accommodation: Crew of two in tandem, comprising pilot and reconnaissance systems officer.
Status: First of three A-11 (YF-12A) prototypes flown on 26 April 1962; first SR-71A flown on 22 December 1964. Production batch of about 24 built for USAF, completed 1968.
Notes: Design of the Lockheed A-11 began in 1959, to produce an aircraft of such high performance that it could penetrate enemy airspace without fear of being intercepted. Responsibility for the design and prototype construction was entrusted to the well-known Lockheed "Skunk Works", the secret experimental department run by C L Johnson and the first flight was made from a secret base in Nevada known as "Kelly Johnson's ranch". The initial trio of A-11 prototypes were evaluated as air defence fighters armed with eight AIM-47A AAMs and designated YF-12A and YF-12C, but the only operational rôle for which the design was accepted was strategic reconnaissance. Delivery of production SR-71As equipped for this rôle began in January 1966, the aircraft being used to equip a unit of the 9th Strategic Reconnaissance Wing of SAC at Beale AFB, where they soon became known as Black Birds because of their all-black finish. For training purposes, the Wing had two SR-71Bs with a raised rear cockpit containing a second set of controls, and one of these was later replaced by a similarly modified SR-71C. The SR-71 uses special fuel, designated JP-7, and a special version of the KC-135 refuelling tanker therefore had to be assigned to the SR-71 force, being designated KC-135Q. The SR-71's equipment carried internally ranges from multiple-sensor high-performance systems for interdiction reconnaissance to strategic systems capable of extreme-altitude specialised surveillance.

LOCKHEED U-2

The photograph above depicts a two-seat U-2D and the general arrangement silhouette (right) illustrates the single-seat U-2B with wing pinion tanks

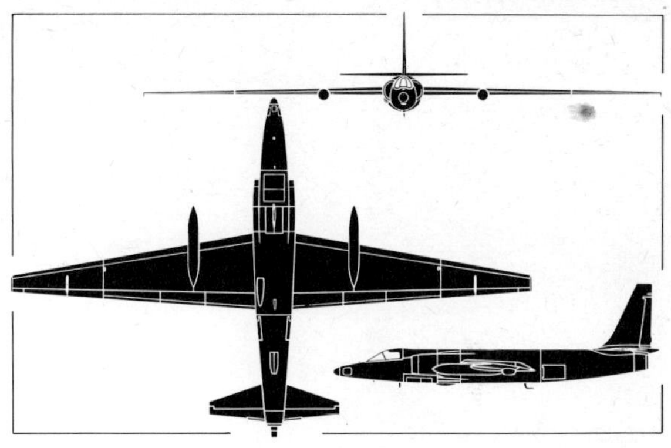

Country of Origin: USA.
Type: High altitude and special purpose reconnaissance aircraft.
Power Plant: One (U-2A) 11,200 lb st (5 080 kgp) J57-P-13A or (U-2B, C, D) 17,000 lb st (7 711 kgp) Pratt & Whitney J75-P-13 turbojet.
Performance: Max speed, 528 mph (850 km/h) at 40,000 ft (12 192 m); cruising speed, 460 mph (740 km/h); service ceiling, approx 80,000 ft (24 384 m); range, up to 4,000 mls (6 440 km); max endurance, over 8 hrs.
Weights: Normal take-off, 15,850 lb (7 190 kg); max overload, 17,270 lb (7 835 kg); with slipper tanks on wings.
Dimensions: Span, 80 ft 0 in (24,38 m); length, 49 ft 7 in (15,11 m); height, 13 ft 0 in (3,96 m); wing area, about 565 sq ft (52,5 m²).
Accommodation: Pilot only or (U-2D only) two pilots.
Status: Prototype first flown on 1 August 1955; production completed by 1958. Production totals, 48 U-2A/B/C and five U-2D.
Notes: The U-2 gained international publicity in 1960 when an example operated on behalf of the US Government was shot down on a reconnaissance flight over the Soviet Union. Its development, with this type of mission the primary objective, had begun at the end of 1954 at the Lockheed "Skunk Works" directed by C L Johnson, the first flight being made by Tony Le Vier after only eight months work. Single-seat models are designated U-2A, -2B or -2C according to equipment and power plant, some also having been re-designated WU-2A for operation by SAC on fall-out sampling missions, later taken over by RB-57Fs (see page 98). The two-seat U-2Ds were used for training and other special tasks. Although overflights were stopped after the incident on May Day 1960, U-2s have continued to be used for photo and electronic survey flights around the Soviet Union, over China, North Vietnam and elsewhere, some being operated by the Chinese Nationalist Air Force from Taiwan. In 1973, a major U-2 activity was in connection with monitoring Chinese mainland communications, from specially-equipped U-2s flying outside Chinese airspace. Also in 1973, Lockheed conducted flight tests of a U-2 specially equipped with US Navy ocean surveillance sensors, and others were in use with NASA. The latter has used the U-2 to support general Earth resources programmes, flying sustained missions at 65,000 ft (19 810 m) over ecological test areas, and to conduct observations in astronomy, high-altitude atmospheric physics and geophysics.

MARTIN/GENERAL DYNAMICS RB-57F CANBERRA

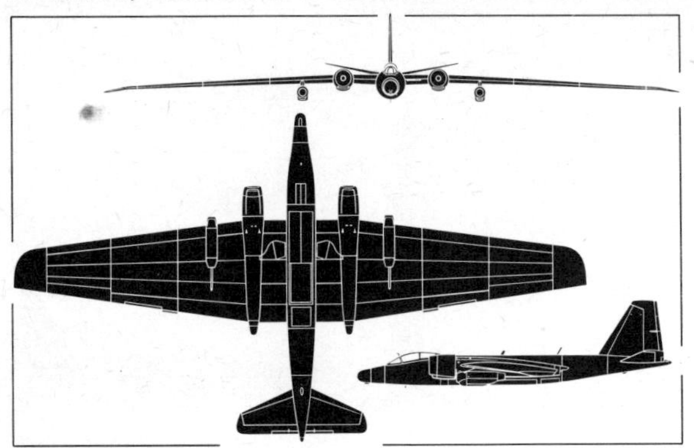

Both the photograph (above) and the general arrangement silhouette (left) illustrate the RB-57F high-altitude reconnaissance version of the Canberra

Country of Origin: USA.
Type: High altitude reconnaissance.
Power Plant: Two 18,000 lb st (8 165 kgp) Pratt & Whitney TF33-P-11A turbofans and two 3,300 lb st (1 500 kgp) Pratt & Whitney J60-P-9 turbojets.
Performance: Typical reconnaissance mission speed, 460 mph (740 km/h) at 60,000 ft (18 288 m); max speed, over 500 mph (805 km/h); normal time to climb to 60,000 ft (18 288 m), 35–40 min; service ceiling, about 70,000 ft (21 336 m); range, about 3,000 mls (4 830 km); endurance, over 6 hrs at 60,000 ft (18 288 m).
Weights: Gross, 63,000 lb (28 576 kg).
Dimensions: Span, 122 ft 5 in (37,32 m); length, 69 ft 0 in (21,03 m); height, 19 ft 0 in (5,79 m); wing area, 2,000 sq ft (185,8 m²).
Accommodation: Pilot and navigator/observer, in tandem.
Status: First production B-57A flown on 20 July 1953; first B-57B flown on 28 June 1954; first B-57C flown on 30 December 1954. Production totals, B-57A, 8; RB-57A, 67; B-57B, 202; B-57C, 38; RB-57D, 20; B-57E, 68.
Notes: The Martin B-57 was a licence-built version of the British Canberra (page 65), used by the USAF as a tactical night intruder, light bomber, reconnaissance bomber and target tug. One squadron of B-57Gs (converted B-57Bs) remains in service in 1973 with the ANG, carrying LLTV, laser rangefinder and forward-looking IR equipment. To provide a special high-altitude reconnaissance version of the B-57, General Dynamics undertook a major redesign of the airframe, producing the RB-57F in 1964. Modifications included a completely new three-spar wing of almost double the original span, revised control system, new engines plus underwing boost engines, nose radar and special electronic equipment. A total of 21 was converted by GD from B-57Bs and RB-57Ds and since 1964 the type has been operated by the 58th Weather Reconnaissance Squadron, USAF, at Kirtland AFB, New Mexico, for specialised high altitude duties, including collection of air samples following nuclear detonations in China and elsewhere. In performing these duties, RB-57Fs are detached to bases around the world, and have operated from Panama, Argentina, Alaska and Japan. The underwing auxiliary engines are not always carried and the mounting points, plus two others outboard, can be used to carry special equipment or long-range tanks.

NORTH AMERICAN ROCKWELL A-5 VIGILANTE

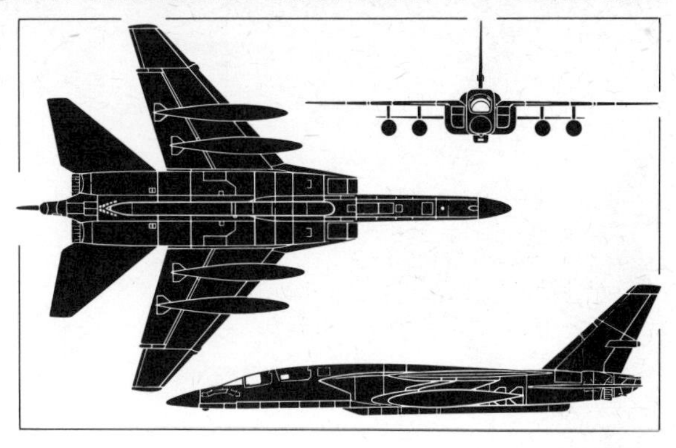

The photograph (above) and the general arrangement silhouette (right) illustrate the RA-5C Vigilante shipboard long-range reconnaissance aircraft of the US Navy

Country of Origin: USA.
Type: Carrier-based reconnaissance aircraft.
Power Plant: Two 11,870 lb st (5 395 kgp) dry and 17,860 lb st (8 118 kgp) with reheat General Electric J79-GE-10 turbojets.
Performance: Max speed, about Mach 2·1, 1,385 mph (2 230 km/h) at 40,000 ft (12 192 m); max low-level cruise, 633 mph (1 019 km/h); cruising speed for long-range 560 mph (901 km/h) at 40,000 ft (12 192 m); service ceiling, 64,000 ft (19 507 m); tactical radius, 1,000 mls (1 610 km); range, about 3,000 mls (4 828 km).
Weights: Normal take-off, 66,818 lb (30 310 kg).
Dimensions: Span, 53 ft 0 in (16,15 m); length, 75 ft 10 in (23,11 m); height, 19 ft 5 in (5,92 m); wing area, 769 sq ft (71,44 m^2).
Accommodation: Pilot and navigator/observer in tandem.
Armament: Provision for tactical nuclear or conventional bombs, rockets or other stores on four underwing pylons.
Status: Prototype (YA3J-1) first flown on 31 August 1958. Production deliveries began mid-1961. First YA-5C (limited) flown on 29 April 1962; first A-5C flown on 30 June 1962; first A-5C Phase II flown in March 1969, production completed 1971. Production quantities: YA3J-1, 2; A3J-1/A-5A, 57; YA-5C (limited), 6; A3J-3P/RA-5C, 91, all for US Navy.
Notes: The Vigilante was designed to operate from the US Navy's largest attack carriers with the ability to carry a nuclear or conventional bomb load over a substantial range, and the need for a rocket motor to boost take-off performance was originally foreseen, but later abandoned. A unique feature of the initial production version (originally A3J-1, later A-5A) was the linear bomb bay between the two engines, the bombs being ejected through an orifice at the rear. This version lacked the "saddle-back" fuel tank which was introduced in the projected A-5B, but production of the latter was cancelled when the US Navy relinquished its strategic nuclear bombing rôle. Production switched to the RA-5C version, carrying cameras and side-looking radar in and beneath the fuselage, and all the A-5As were converted to this configuration in addition to new production in two batches, the first of which had 10,900 lb st (4 944 kgp) dry or 17,000 lb st (7 711 kgp) with reheat J79-GE-8 engines. The RA-5C entered service aboard the USS *Ranger* in 1964 and remained a major type in the US Navy inventory in 1973.

TUPOLEV "MOSS"

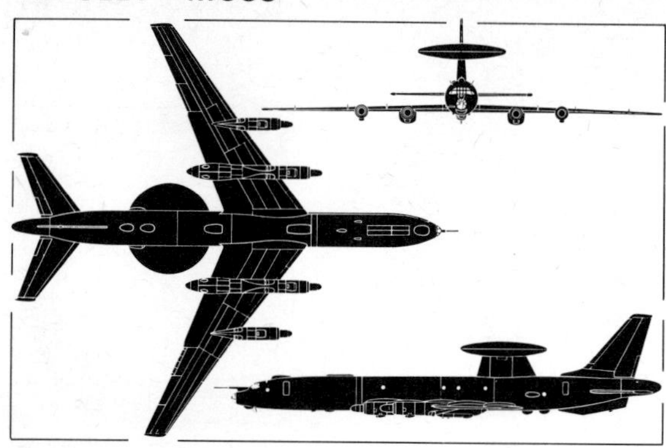

The photograph (above) and the general arrangement silhouette (left) illustrate the airborne early warning and control system aircraft that is known to NATO as "Moss"

Country of Origin: USSR.
NATO Code Name: *Moss.*
Type: Airborne early warning and control system aircraft.
Power Plant: Four 14,795 ehp Kuznetsov NK-12 MV turboprops.
Performance (Estimated): Max continuous cruise, 460 mph (740 km/h) at 25,000 ft (7 620 m); max unrefuelled range, 4,000+ mls (440+ km); service ceiling 39,000 ft (11 890 m).
Weights (Estimated): Normal max take-off, 360,000 lb (163 290 kg).
Dimensions: Span, 168 ft 0 in (51,20 m); length, about 188 ft 0 in (57,30 m); height, 51 ft 0 in (15,50 m); wing area, 3,349 sq ft (311,1 m²).
Accommodation: Flight crew of four and systems operation crew of about 10–12, plus provision for relief and maintenance crews.
Status: The AWACS aircraft assigned the reporting name *Moss* by the Air Standards Co-ordinating Committee became known to Western intelligence agencies in the mid 'sixties, and first appeared in service with the Soviet Air Forces in 1970.
Notes: Essentially an adaptation of the Tu-114 commercial transport, and apparently retaining the wings, tail surfaces, power plant and undercarriage of the earlier aircraft, the Tupolev AWACS type is primarily intended to locate low-flying intruders and to vector interceptors towards them. The dominating feature of the aircraft—like that of comparable AWACS types operated by the USAF, such as the Boeing E-3A (page 88) and by the USN such as the E-2 Hawkeye (page 92)—is the pylon-mounted saucer-shaped early warning radar scanner, in its housing with a diameter of approximately 37·5 ft (12,000 m). In converting the Tu-114 design for the AWACS rôle, provision was made for in-flight refuelling (from Tu-16 *Badger* or M-4 *Bison* tankers), the probe being on the forward fuselage just ahead of the windscreen. To avoid major structural and system redesign in the front fuselage, the fuel line is carried from the probe along the starboard side of the fuselage to the wing leading edge, and thus into the integral wing tanks. Various sensor-housing blisters on the fuselage and at the tailplane tips are also indicative of the special rôle of *Moss,* examples of which are frequently intercepted and photographed over the seas around the UK, USA and Japan and over the Mediterranean. Apart from its fighter control task it may be assumed that *Moss* will also be employed in enabling strike aircraft to elude interceptors.

YAKOVLEV YAK-27R

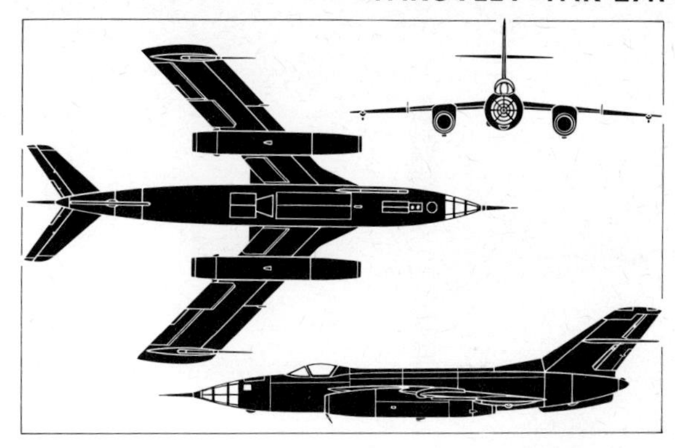

The Yak-27R tactical reconnaissance aircraft, illustrated by the photograph (above) and the general arrangement silhouette (right) has seen some 14 years service with the V-VS

Country of Origin: USSR.
NATO Code Name: *Mangrove.*
Type: Tactical reconnaissance aircraft.
Power Plant: Two 5,730 lb (2 600 kg) dry and 7,275 lb (3 300 kg) with reheat Tumansky RD-9 turbojets.
Performance: (Estimated) Max speed (clean), 686 mph (1 104 km/h) or Mach 0·95 at 36,090 ft (11 000 m); low-altitude cruise, 570 mph (917 km/h) or Mach 0·75; tactical radius (clean), 200 mls (320 km) at 1,000 ft (305 m), 500 mls (805 km) at 36,090 ft (11 000 m); initial climb, 18,000 ft/min (91,4 m/sec); service ceiling, 50,000 ft (15 250 m).
Weights: (Estimated) Normal loaded, 22,000 lb (9 980 kg); max take-off, 25,000 lb (11 340 kg).
Dimensions: (Estimated) Span, 36 ft 1 in (11,00 m); length, 55 ft 0 in (16,76 m); height, 12 ft 6 in (3,80 m); wing area, 323 sq ft (30,0 m^2).
Accommodation: Pilot and observer.
Armament: Single 30-mm NR-30 cannon in starboard side of forward fuselage.
Status: Prototypes flown 1956–57 with initial deliveries to the Soviet Air Forces in 1959. Currently operational in numbers.
Notes: Progressive development of the Yak-25R tactical reconnaissance variant of the Yak-25 all-weather interceptor resulted in the Yak-27R optimised for the Tac-R rôle but possessing secondary tactical strike capability. The forward fuselage of the Yak-27R is essentially similar to that introduced by the Yak-25R and apart from some structural strengthening the airframe is similar to that of the basic Yak-25, changes including a sharply-swept (63 deg) extension of the leading-edge wing root, inset outriggers and the adoption of an afterburning version of the RD-9 turbojet. Developed in parallel was an all-weather fighter version, the Yak-27P (*Flashlight-C*) with similar aerodynamic refinements and afterburners to those of the Yak-27R but tandem crew accommodation similar to that of the basic Yak-25, this combination being coupled with a sharply pointed nose radome to lessen drag and rain erosion. Production of the Yak-27P appears to have been restricted, possibly owing to the relatively marginal increase in performance that it offered over the Yak-25F. In addition to cameras, the Yak-27R includes sensors in radio and infra-red regions in its internal equipment and can carry underwing sensor pods.

YAKOVLEV "MANDRAKE"

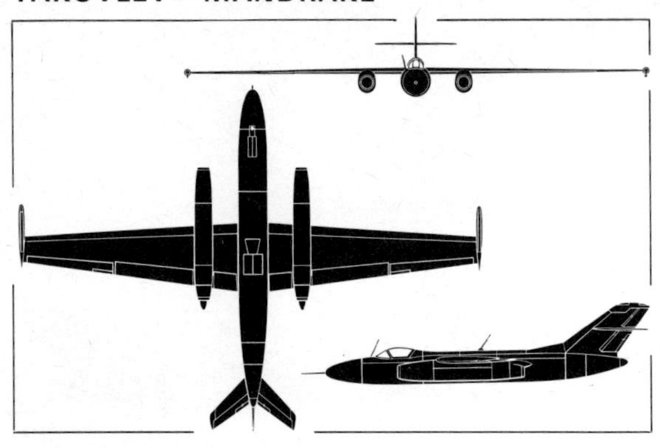

The Mandrake, illustrated by the photograph (above) and the general arrangement drawing (left), is a high-altitude reconnaissance derivative of the original Yak-25

Country of Origin: USSR.
NATO Code Name: *Mandrake.*
Type: High-altitude strategic reconnaissance aircraft.
Power Plant: Two 5,730 lb (2 600 kg) Tumansky RD-9 turbojets.
Performance: (Estimated) Max speed (clean), 470 mph (756 km/h) at 45,930 ft (14 000 m), 300 mph (483 km/h) at sea level; range cruise, 400 mph (644 km/h); normal operational ceiling, 65,000 ft (19 800 m); maximum ceiling, 70,000 ft (21 335 m); operational radius, 1,200 mls (1 930 km); maximum range, 2,500 mls (4 000 km).
Weights: (Estimated) Normal loaded, 21,000 lb (9 525 kg).
Dimensions: (Estimated) Span, 71 ft 0 in (21,64 m); length, 51 ft 0 in (15,54 m).
Accommodation: Pilot only.
Armament: Nil.
Status: Prototypes flown in the mid 'fifties and small numbers subsequently built for the Soviet Air Forces, some examples remaining in service late 1973.
Notes: The Yakovlev high-altitude strategic reconnaissance aircraft dubbed *Mandrake* by NATO (the official designation remaining obscure) was evolved by marrying the centre and aft fuselage, tail surfaces, power plants and zero-track tricycle undercarriage of the Yak-25 to a new forward fuselage and long-span high aspect ratio wing, thus evolving an aircraft in much the same category as the Lockheed U-2 and capable of operating at what were considered in the late 'fifties and early 'sixties to be extreme altitudes. In view of the relatively limited development effort involved the *Mandrake* may be considered a reasonably successful high-altitude performer, probably commencing cruise at approximately 56,000 ft (17 070 m) and drifting up to 65,000 ft (19 800 m) as fuel is burned off. It may be assumed that the *Mandrake* features a "wet wing" and it is known that some examples have afterburning RD-9 engines. Equipment is likely to include a long-focus camera for horizon-to-horizon photography and electronic sensing devices to pick up and record radio and radar transmissions. The *Mandrake* is not believed to have been used extensively in the European theatre although, while its cruise altitude would not guarantee immunity in deep reconnaissance missions, it should be suitable for probing defences. However, it has been employed for clandestine reconnaissance flights over China, India and Pakistan in recent years.

AERITALIA G.222

The G.222 general-purpose military transport, illustrated by the photograph (above) and the general arrangement silhouette (right) is entering production for the Italian Air Force

Country of Origin: Italy.
Type: General-purpose military transport.
Power Plant: Two 3,400 shp General Electric T64-P4D turbo-props.
Performance (estimated): Max level speed, 329 mph (530 km/h) at 15,000 ft (4 575 m); cruising speed, 273 mph (440 km/h) at 14,750 ft (4 500 m); initial rate of climb, 1,890 ft/min (9,6 m/sec); service ceiling, 25,000 ft (7 620 m); mission range, 2,019 mls (3 250 km) with 11,025 lb (5 000 kg) load; range with max fuel, 3,262 mls (5 250 km).
Weights: Empty, 29,321 lb (13 300 kg); max payload, 20,943 lb (9 500 kg); normal take-off, 54,000 lb (24 500 kg); max overload, 57,320 lb (26 000 kg); max landing, 56,217 lb (25 500 kg).
Dimensions: Wing span, 94 ft 6 in (28,80 m); length, 74 ft 5½ in (22,70 m); height, 32 ft 1¾ in (9,80 m); wing area, 969 sq ft (90,0 m²).
Accommodation: Flight crew of three (two pilots and radio operator/flight engineer) and seats for up to 44 equipped troops or 32 paratroops, or 36 stretchers plus eight seats or assorted military stores and supplies up to max payload limit.

Status: First and second prototypes flown on 18 July 1970 and 22 July 1971 respectively. Forty-seven ordered by Italian Air Force.
Notes: Ordered by the Italian Air Force as a replacement for Fairchild C-119s, the G.222 originated with the Fiat company, which was awarded a research project contract in 1963 to cover preliminary work on a V/STOL military transport. The basic design was to use a combination of cruise turboprops and lift jets for V/STOL capability, but only the alternative conventional configuration, without the lift jets, was accepted for development, Fiat building two prototypes and a static test specimen at their Turin plant. These prototypes have 2,970 shp CT64-820 engines and are unpressurised; the production model, with uprated engines, has provision for pressurisation. Manufacture of the G.222 is distributed throughout the Italian aerospace industry with final assembly at a new plant set up by the state-owned Aeritalia in the South of Italy. Current plans call for the re-equipment of two Italian Air Force squadrons currently operating the Fairchild C-119 with the G.222 during the mid 'seventies, and a number of variants of the basic design for both military and commercial tasks are currently projected, including airborne early warning and ECM rôles.

ANTONOV AN-12

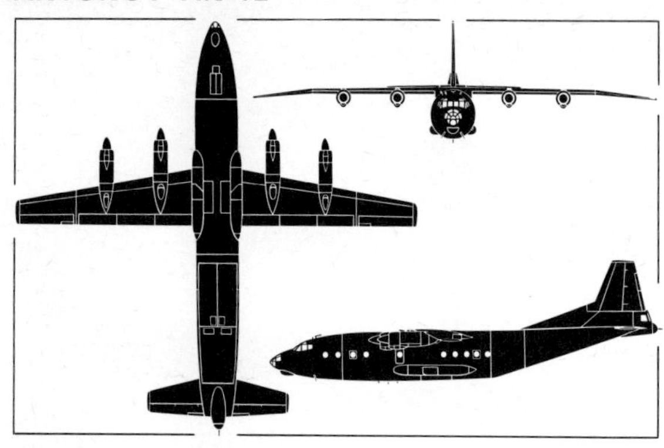

Both photograph (above) and general arrangement silhouette (left) illustrate the An-12 medium range transport, the former depicting an An-12 operated by the Polish Air Force

Country of Origin: USSR.
NATO Code Name: *Cub.*
Type: Medium range freighter and troop transport.
Power Plant: Four 4,000 ehp Ivchenko AI-20K turboprops.
Performance: Max speed 444 mph (715 km/h); max cruise, 373 mph (600 km/h); normal cruise, 342 mph (550 km/h) at 25,000 ft (7 500 m); initial rate of climb, 1,970 ft/min (10 m/sec); service ceiling, 33,500 ft (10 200 m); range 2,110 mls (3 400 km) with payload of 22,050 lb (10 000 kg) and 1-hr fuel reserve.
Weights: Max payload, 44,090 lb (20,000 kg); normal take-off, 119,050 lb (54 000 kg); max take-off, 134,480 lb (61 000 kg).
Dimensions: Span, 124 ft 8 in (38,0 m); length, 108 ft 3 in (33,0 m); height, 32 ft 3 in (9,83 m); wing area, 1,292 sq ft (120 m^2).
Accommodation: Crew of five (two pilots, radio operator, navigator and gunner) and provision for 14 passengers in addition to a full load of freight which may include wheeled or tracked vehicles, or surface-to-air and surface-to-surface missiles with support equipment.
Armament: Two 23-mm cannon in rear turret.

Status: Developed from An-10 commercial transport (see *Civil* volume) and first observed in 1959. Over 900 reported built. Production complete.
Notes: The An-12 was evolved from the commercial An-10A to meet Soviet Air Force requirements for a general purpose tactical transport carrying paratroops or miscellaneous Army loads. It differs from the commercial counterpart primarily in having a redesigned rear fuselage with an integral loading ramp—which may be lowered in flight for air-drop operations—a defensive tail gun position and revised tail surfaces. In addition to the Soviet Air Force, several foreign air forces have An-12s in service, including those of India (which received 16), Algeria, Egypt, Iraq and Poland. A civil version, the An-12B, has been produced—usually with the tail turret faired over—as a specialised freighter and is operated by Aeroflot and the airlines of some other Communist Bloc countries. Some An-12Bs operated by Aeroflot's Polar Aviation service in the Arctic and Antarctic operate on skis, being the largest aircraft so fitted, the skis being equipped with braking devices and warming equipment. The An-12 is currently the most widely-used of Soviet Air Force transports.

ANTONOV AN-22 ANTEI

The An-22, illustrated by the photograph above and the general arrangement silhouette (right) is currently the largest transport serving with the Soviet Air Forces

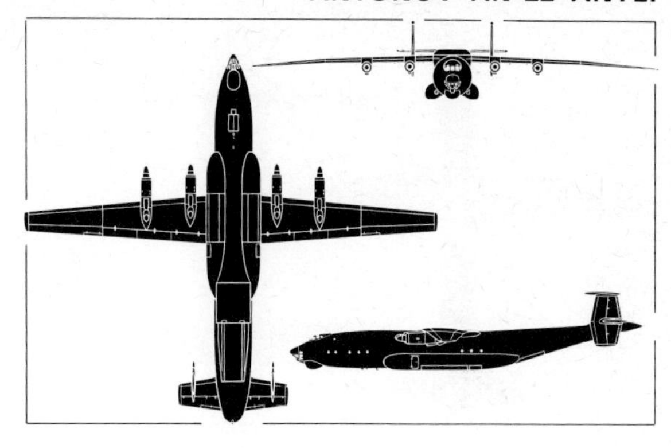

Country of Origin: USSR.
NATO Code Name: *Cock.*
Type: Long-range troop and supply transport.
Power Plant: Four 15,000 shp Kuznetsov NK-12MA turboprops.
Performance: Max speed, 460 mph (740 km/h); max cruise, 422 mph (679 km/h); range with max payload, 3,107 mls (5 000 km) at 404 mph (650 km/h); range with max fuel, 6,800 mls (10 950 km); cruising altitude, 26,250–32,800 ft (8 000–10 000 m).
Weights: Empty equipped, 251,327 lb (114 000 kg); max payload, 176,350 lb (80 000 kg); max fuel load, 94,800 lb (43 000 kg); max take-off, 551,156 lb (250 000 kg).
Dimensions: Span, 211 ft 3½ in (64,40 m); length, 189 ft 8 in (57,80 m); height, 41 ft 1 in (12,53 m); wing area, 3,713·6 sq ft (345 m²).
Accommodation: Crew of five-six (including two pilots, engineer, radio operator and navigator). Up to 29 passengers in addition to full freight load, which can include three tracked carriers for single Frog or twin Ganef surface-to-air missiles, self-propelled guns, etc.
Status: First of five prototypes flown on 27 February 1965. Production deliveries began in Spring of 1967. In production (1973) for both military and commercial use.
Notes: First appearing outside the Soviet Union in June 1965, at the Paris *Salon*, the An-22 was at that time the world's largest aeroplane and although now surpassed by the Lockheed C-5 Galaxy (page 129) it is still the largest *turboprop-powered* aircraft. It was developed primarily as a military transport for use by the Soviet Air Forces, with which it was in service by mid-1967. Current service versions differ from the initial batch in having a reconfigured nose section embodying two radars, whereas the navigation radar was initially carried under the starboard landing gear fairing. Of five prototypes, two were used by Aeroflot on experimental freight services in 1967, and a commercial production version was subsequently put into service by Aeroflot, primarily to undertake special supply operations in the Soviet Union's more remote and underdeveloped areas. Similar in general configuration to its much smaller predecessor, the An-12 (see opposite page), the An-22, known as the *Antei* (Antheus), is capable of taking-off in fully loaded condition within 1,420 yards (1 300 m) and landing within 875 yards (800 m). The present status of a commercial airbus version accommodating 300–350 passengers is uncertain.

ANTONOV AN-26

The An-26 freight, illustrated by the general arrangement silhouette (left) is a derivative of the An-24, an example of which in Somali service is depicted by the photo above

Country of Origin: USSR.
NATO Code Name: *Curl.*
Type: Short/medium-range military (and commercial) freighter.
Power Plant: Two 2,820 eshp Ivchenko AI-24T turboprops and (in starboard nacelle) one 1,984 lb st (900 kgp) Tumansky RU-19-300 turbojet.
Performance: Max speed 335 mph (540 km/h) at 19,685 ft (6 000 m); long-range cruise, 273 mph (440 km/h) at 22,965 ft (7 000 m); service ceiling, 24,935 ft (7 600 m); range with max payload, 807 mls (1 300 km); range with max fuel, 1,553 mls (2 500 km) with 3,307 lb (1 500 kg) payload and 1,280 lb (580 kg) fuel reserve.
Weights: Empty equipped, 37,258 lb (16 914 kg); max payload, 11,023 lb (5 000 kg); max take-off, 52,911 lb (24 000 kg).
Dimensions: Span, 95 ft 10 in (29,20 m); length, 78 ft 1 in (23,80 m); height, 28 ft 2 in (8,57 m); wing area, 780 sq ft (72,46 m^2).
Accommodation: Flight crew of four (two pilots, radio operator and navigator) and optional loadmaster. Optional seating for 38 troops on fold-up seats along each side of cabin. Direct rear-loading for freight or a variety of military vehicles, including the GAZ-69 and the UAZ-469, and provision for air-dropping over rear ramp. Provision for electrically- or manually-operated conveyor.
Status: First publicly displayed in May 1969 and believed in production for military and commercial use from 1969 onwards.
Notes: Derived from the An-24RT (see *Civil* Volume) the An-26 originated primarily to meet Soviet Air Force needs for a versatile tactical transport with rough-field capability. Its principal new feature was a redesigned rear fuselage of "beavertail" configuration, incorporating a unique two-position door, which can be lowered to form a conventional ramp for the loading of vehicles, or can be swung down and forward beneath the fuselage to permit a clear exit for paratroops and air-dropped supplies. Twin main wheels help spread the load for operations from unprepared landing sites and the jet booster in the starboard nacelle enhances the take-off performance, especially in hot or high conditions. A commercial version of the An-26 is reported to be in production for special freight-carrying duties and for use as an ambulance, and a few examples of the standard An-24 are operated as military transports in the Soviet Bloc nations.

BOEING C-97 STRATOFREIGHTER

The photograph above depicts a KC-97L of the Illinois Air Guard, one of nine ANG groups flying this type, and the general arrangement silhouette (right) also illustrates the KC-97L

Country of Origin: USA.
Type: Medium/long range troop and supply transport and flight-refuelling tanker.
Power Plant: Four 3,500 hp Pratt & Whitney R-4360-59 piston radial engines and (KC-97L only) two 5,200 lb st (2 360 kgp) General Electric J47-GE-23 turbojets.
Performance (KC-97G): Max speed, 375 mph (604 km/h); cruising speed, 300 mph (483 km/h); time to climb to 20,000 ft (6 096 m), 50 min; service ceiling, 35,000 ft (10 670 m); range, 4,300 mls (6 920 mm).
Weights: Empty, 82,500 lb (37 420 kg); max take-off, 175,000 lb (79 380 kg).
Dimensions: Span, 141 ft 3 in (43,05 m); length, 110 ft 4 in (33,63 m); height, 38 ft 3 in (11,66 m); wing area, 1,720 sq ft (159,7 m²).
Accommodation: Flight crew of five (two pilots, flight engineer, navigator-radio operator and boom operator). In troop carrying rôle, provision for up to 96 combat troops or 69 stretchers plus medical attendants.
Status: Prototype XC-97 flown on 15 November 1944 and pre-production YC-97 flown on 11 March 1947, entering service in October 1947. First KC-97A refuelling tanker flown in 1950. Total of 888 C-97 variants of all types built; final delivery on 16 November 1956.
Notes: The XC-97 (Boeing Model 367) originated as a Boeing proposal for a transport based on the wing of the B-29 Superfortress, and the first contract was placed by the USAAF, for three prototypes, in January 1943. Sub-variants from A to D were troop, passenger or casualty transports but major production of the design was concerned with the tanker variants with Boeing's unique Flying Boom system. Primarily for SAC, Boeing built 60 KC-97E, 159 KC-97F and 592 KC-97G, these remaining in service until superseded by KC-135s (see page 135). The KC-97L variant was fitted with underwing jet pods for tanker service with TAC. Nine Air National Guard groups were assigned to fly KC-97Ls in 1973 and other C-97 variants were in service as freighters. The Spanish Air Force has three KC-97Ls (as TK-1s) to refuel its F-4s and some KC-97Gs have been supplied to Israel to join locally-produced tanker conversions of surplus civil Stratocruisers. The Israeli air arm also employs the Stratocruiser as a swing-tail freighter.

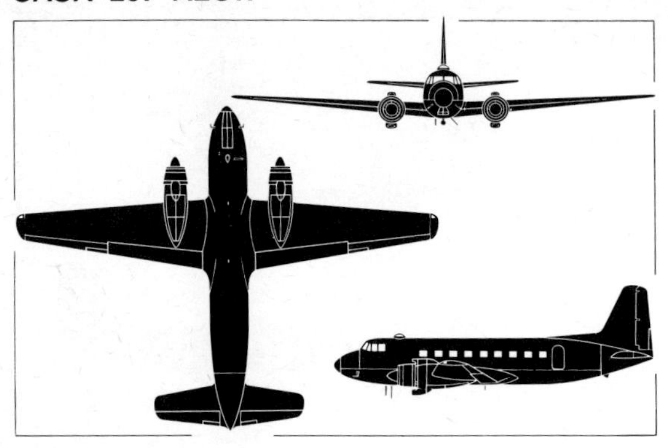

The photograph above illustrates a C-207-C of the Spanish 351 Escuadron and the general arrangement silhouette (left) depicts the C-207-A personnel transport

Country of Origin: Spain.
Type: Short/medium-range general purpose transport.
Power Plant: Two 2,040 hp Bristol Hercules 730 piston-radial engines.
Performance: Max level speed, 261 mph (420 km/h) at 4,920 ft (1 500 m); cruising speed at 64 per cent power, 222 mph (357 km/h) at 10,000 ft (3 050 m); initial rate of climb, 1,080 ft/min (5,5 m/sec); service ceiling, 27,887 ft (8 500 m); range with 6,600 lb (3 000 kg) payload, 1,460 mls (2 350 km).
Weights: (C-207-A) Empty equipped, 23,370 lb (10 600 kg); max payload, 6,806 lb (3 087 kg); max take-off (C-207-A), 35,275 lb (16 000 kg); (C-207-C), 36,376 lb (16 500 kg).
Dimensions: Span, 91 ft 2½ in (27,80 m); length 68 ft 5 in (20,85 m); height 25 ft 5 in (7,75 m); gross wing area, 923 sq ft (85,8 m²).
Accommodation: Flight crew of four and (C-207-A) up to 38 passengers or (C-207-C) up to 7,385 lb (3 350 kg) of freight or 37 troops.
Status: First flown 28 September 1955. Production of 10 C-207-A and 10 C-207-C completed in 1968.

Notes: The largest aircraft of original Spanish design built to date, the Azor was an attempt to produce a 30–38 seat airliner which could be used by the Spanish airlines Iberia and Aviaco, primarily on domestic and regional routes. However, the type found no commercial customers after prototype testing had been completed and production orders were placed only by the Spanish *Ministerio del Aire* on behalf of the Air Force. An initial batch of 10 C-207-As was built in personnel transport configuration and a second batch of 10 was then ordered for use as freighters. These were designated C-207-Cs and had re-stressed fuselages with double cargo-loading doors and provision for carrying up to 37 paratroops. The Azors serve with the Spanish Air Force nomenclature T-7A and T-7B, and are operated primarily by 351 *Escuadron del Mando de Transporte* at Getafe, Madrid. The two XT-7 prototypes remained in service in 1973 with the 406 *Escuadron de Experimentacion en Vuelo* at Torrejon. CASA (Construcciones Aeronáuticas SA) in 1973 was projecting a STOL transport, the CASA 401, which, powered by four turboprops, was intended to replace both the C-207 Azor and some of the Douglas C-54s currently operated by the Spanish Air Force from the late 'seventies.

CASA 212 AVIOCAR

The Aviocar, illustrated by the photograph (above) and the general arrangement silhouette (right), will begin to enter service with the Spanish Air Force during 1974

Country of Origin: Spain.
Type: STOL utility transport.
Power Plant: Two 775 ehp Garrett-AiResearch TPE331-5-251C turboprops.
Performance: Max speed, 227 mph (365 km/h) at sea level, 249 mph (400 km/h) at 12,000 ft (3 660 m); high-speed cruise, 243 mph (391 km/h) at 12,000 ft (3 660 m); initial rate of climb, 1,716 ft/min (8,5 m/sec); service ceiling, 24,606 ft (7 500 m); range with max payload, 205 mls (330 km); range with max fuel, 1,198 mls (1 927 km/h).
Weights: Empty equipped, 8,045 lb (3 650 kg); max payload, 4,410 lb (2 000 kg); max take-off, 13,890 lb (6 300 kg).
Dimensions: Span, 62 ft 4 in (19,00 m); length, 49 ft 10½ in (15,20 m); height, 20 ft 8 in (6,30 m); wing area, 430·6 sq ft (40,0 m²).
Accommodation: Flight crew of two; provision for 15 troops and a jumpmaster in paratroop versions, or up to 4,410 lb (2 000 kg) of supplies, including light vehicles.
Status: Prototypes first flown on 26 March and 23 October 1971. First pre-production model flown on 17 November 1972. Orders placed for 12 pre-production and 32 production Aviocars for Spanish Air Force and 28 for the Portuguese Air Force.
Notes: Design of the CASA 212 began in 1965 in response to a Spanish Air Force requirement for a small tactical transport to replace the Junkers Ju 52/3m tri-motor transports still in service. Early design studies were based on the use of two Turboméca Astazou or Pratt & Whitney PT6A-23 turboprops but AiResearch TPE 331s were finally selected to power the two prototypes ordered by the *Ministerio del Aire* in September 1968. These prototypes both flew with -201 versions of the engines rated at 755 ehp, later being re-engined with the more powerful -251Cs selected for production models. The pre-production batch under construction in 1972/73 comprised two navigation trainers and six for photographic survey duties with the Spanish Air Force (which has given the Aviocar the nomenclature T12), and four as commercial demonstrators, in which guise the Aviocar can seat 18 passengers. For delivery starting in 1974, the Spanish Air Force also ordered 29 Aviocars as freight paratroop transports and three more navigation trainers. It is anticipated that the Aviocar will supplant both the Ju 52/3m and Douglas C-47 in Spanish service.

CASA 352L (JU 52/3M)

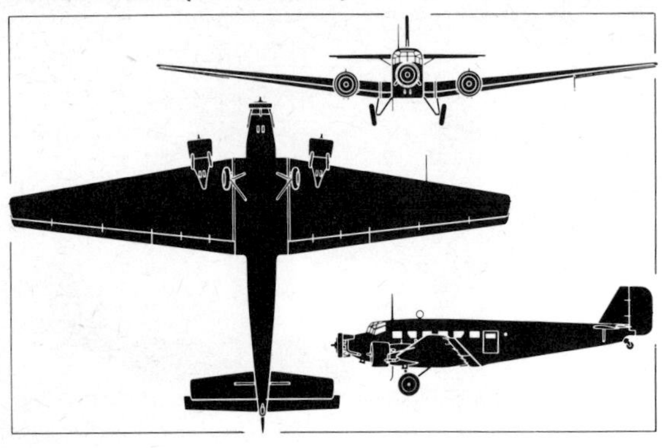

The photograph above depicts a CASA 352-L on the Spanish Air Force, the general arrangement silhouette illustrating the externally similar Ju 52/3m from which it derives

Country of Origin: Germany (and Spain).
Type: General purpose transport.
Power Plant: Three 725 hp BMW 132A or (CASA 352-L) 750 hp ENMASA Beta E-9C piston radials.
Performance (CASA 352-L): Max speed 157 mph (252 km/h); cruising speed, 145 mph (233 km/h); service ceiling, 23,300 ft (7 100 m); range, 615 mls (990 km).
Weights (CASA 352-L): Empty, 14,550 lb (6 600 kg); payload, 4,915 lb (2 230 kg); max take-off, 24,250 lb (11 000 kg).
Dimensions: Span, 95 ft 11½ in (29,25 m); length, 62 ft 0 in (18,90 m); height, 20 ft 0 in (6,10 m); wing area, 1,190 sq ft (110,6 m²).
Accommodation: Flight crew of three and accommodation for up to 18 troops.
Status: First Ju 52/3m flew in Germany in April 1931; deliveries began (for commercial use) early in 1932. Production total in Germany and by licences up to 1944, 4,825. Production of CASA 352-L began in 1945; 170 delivered by 1952.
Notes: The famous *Tante Ju* (Auntie Ju), one of the most widely known and most utilised of any aircraft type used by the *Luftwaffe* in World War II, originated for the commercial market, when the Junkers company adapted its single-engined Ju 52 design to have three engines. In this triple-engined form, the Ju 52/3m was adopted by many airlines and civil operators and 1,600 examples were built up to 1939, including those ordered by the *Luftwaffe* and some foreign air forces. Other than a few civil examples believed to survive in South America and one flying example in the USA, the sole survivors of German origin in 1973 were three still serving with the Swiss Air Force. For the Spanish Air Force, CASA built 170 Ju 52/3ms between 1945 and 1952 as the CASA 352L with ENMASA Beta engines, and these for a time provided the primary transport force, carrying the Air Force nomenclature T 2B. No longer equipping front-line units, they were still being used in 1973 for miscellaneous duties including paratroop training with Nos 404 and 405 *Escuadrónes* at Alcantarilla. Others serve for general transport and liaison duties as *Aviones de Base* by Nos 401, 402 and 403 *Escuadrónes* at Cuatro Vientos, and No 511 *Escuadrón* at Getafe. The last Ju 52/3m alias CASA 352L transports serving with the Spanish Air Force were expected to be phased out during 1974–75, its successor being the CASA 212 Aviocar.

DE HAVILLAND CANADA DHC-5 BUFFALO

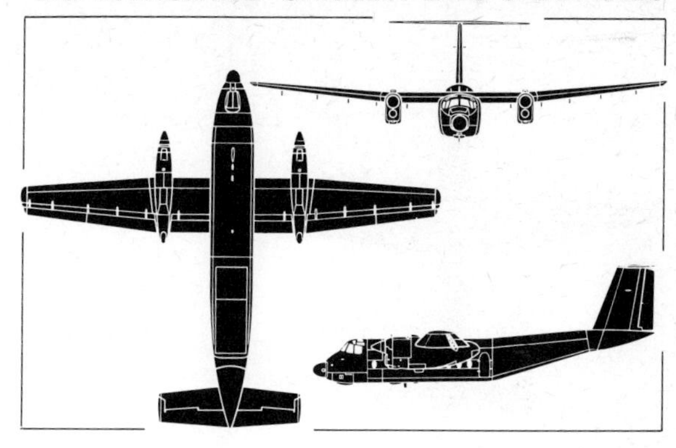

Both photograph (above) and general arrangement silhouette (right) depict the DHC-5 Buffalo, the former illustrating an example serving with the Brazilian Air Force

Country of Origin: Canada.
Type: STOL tactical transport.
Power Plant: Two 2,850 ehp General Electric (C-8A) T64-GE-10 turboprops or (CC-115) 3,055 ehp CT64-820-1 turboprops.
Performance (C-8A): Max speed, 271 mph (435 km/h) at 10,000 ft (3 050 m); high-speed cruise, 271 mph (435 km/h) at 10,000 ft (3 050 m); long-range cruise, 208 mph (335 km/h) at 10,000 ft (3 050 m); initial rate of climb, 1,890 ft/min (9,6 m/sec); service ceiling, 30,000 ft (9 150 m); range with max payload, 507 mls (815 km); range with max fuel, 2,170 mls (3 490 km) with 4,000 lb (1 815 kg) payload.
Weights: Operating weight empty, 23,157 lb (10 505 kg); max payload, 13,843 lb (6 279 kg); max fuel load, 13,598 lb (6 168 kg); max take-off, 41,000 lb (18 598 kg).
Dimensions: Span, 96 ft 0 in (29,26 m); length (C-8A) 77 ft 4 in (23,57 m); length (CC-118) 79 ft 0 in (24,08 m); height, 28 ft 8 in (8,73 m); wing area, 945 sq ft (87,8 m²).
Accommodation: Flight crew of three and provision for up to 41 troops, 35 paratroops or 24 stretchers plus six seats; or supplies and vehicles up to payload limit.

Status: Prototype first flown 9 April 1964. Deliveries began in April 1965. Production of 59 completed during 1972.
Notes: The DHC-5 was designed primarily to meet a US Army requirement, development costs being shared equally between the US Army, the Canadian government and the de Havilland company itself. The first four aircraft were delivered for US Army evaluation as CV-7As but a change of policy led to the Army programme being abandoned and these aircraft were transferred to the USAF as C-8As. The RCAF (now Canadian Armed Forces) purchased 15 (with the service designation CC-115) and these were delivered in 1967, with uprated engines and radar in the nose. Subsequently, the Brazilian Air Force ordered 24 in two equal batches, delivered in 1969 and 1970, and the Peruvian Air Force ordered 16, which were delivered in 1971/72 for service with the *Grupo Aereo* No 8. One of the original C-8As has been used as the basis for a NASA full-scale investigation into the augmentor-wing concept and a CC-115 has been fitted with a Bell ACLS (air-cushion landing system). The ACLS is based on the ground effect principle and, instead of wheels, the aircraft employs a cushion of air enclosed by an inflatable tube beneath the forward fuselage.

DE HAVILLAND CANADA DHC-4 CARIBOU

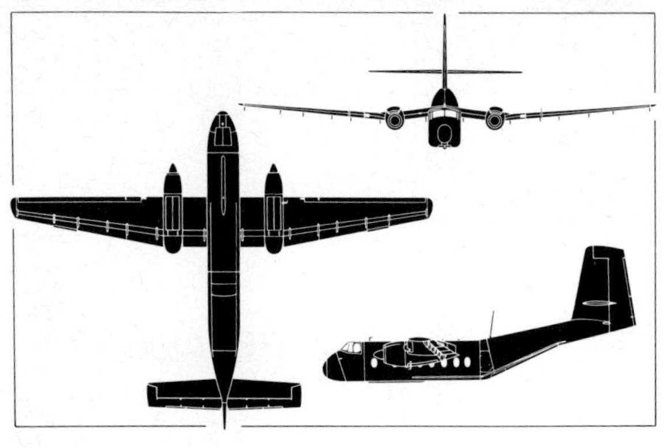

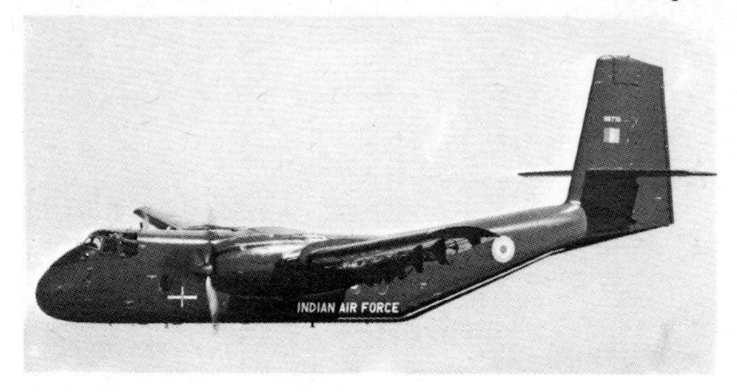

Both photograph (above) and general arrangement silhouette (left) illustrate the DHC-4 Caribou, the former depicting an example serving with the Indian Air Force

Country of Origin: Canada.
Type: Short range tactical transport.
Power Plant: Two 1,450 hp Pratt & Whitney R-2000-7M2 piston radials.
Performance: Max speed, 216 mph (347 km/h) at 6,500 ft (1 980 m); cruising speed, 182 mph (293 km/h) at 7,500 ft (2 285 m); initial rate of climb, 1,355 ft/min (6,9 m/sec); service ceiling, 24,800 ft (7 560 m); range with max payload, 242 mls (390 km); range with max fuel, 1,307 mls (2 103 km).
Weights: Basic operating, 18,260 lb (8 283 kg); max payload, 8,740 lb (3 965 kg); normal max take-off, 28,500 lb (12 928 kg); max overload, 31,300 lb (14 197 kg).
Dimensions: Span, 95 ft 7½ in (29,15 m); length, 72 ft 7 in (22,13 m); height, 72 ft 7 in (22,13 m); wing area, 912 sq ft (84,72 m²).
Accommodation: Crew of two and provision for 32 troops, 26 paratroops, 22 stretchers plus eight seats, two fully loaded jeeps or up to 3 tons of cargo.
Status: Prototype first flown 30 July 1958; service deliveries began 1959. US Type Approval (DHC-4) 23 December 1960 (DHC-4A) 11 July 1961. Total of 307 built by end of 1973.
Notes: The Caribou was designed primarily to meet a US Army requirement for a tactical transport able to operate from small, unprepared strips close to the front-line, carrying troops and supplies to the front and evacuating casualties on return flights. Following one prototype for company flight testing, de Havilland built five for US Army evaluation with the designation YAC-1 and one for the RCAF as CC-108. The US Army bought 159 AC-1s, later designated CV-2As (gross weight 26,000 lb–11 793 kg) and CV-2Bs (28,500 lb–12 928 kg); when transferred to the USAF in 1966 these aircraft were redesignated C-7A. The RCAF (now Canadian Armed Forces) acquired eight more CC-108s and other military users include the Royal Australian Air Force, Ghana Air Force, Kuwait Air Force, Zambian Air Force, Indian Air Force, Kenya Air Force, Royal Malaysian Air Force, Tanzanian Air Force, Spanish Air Force, Sultan of Oman's Air Force and Abu Dhabi Air Force. While manufactured primarily for military tasks, a small number of Caribou have been delivered to commercial operators, including AMOCO Ecuador, Ansett-MAL, Guyana Airways and Intermountain Aviation.

DOUGLAS C-47 SKYTRAIN (DAKOTA)

Despite its age the Douglas C-47 remains in the inventories of many of the world's air forces, the example illustrated above belonging to the Canadian Armed Forces

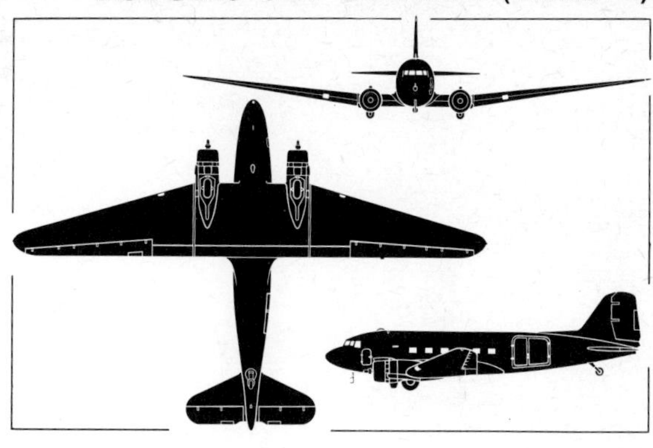

Country of Origin: USA.
Type: General-purpose transport.
Power Plant: Two Pratt & Whitney R-1830-90D or -92 radials of 1,200 hp or (C-117D) Wright R-1820-80 radials of 1,535 hp.
Performance (C-47A): Max speed, 229 mph (368,5 km/h) at 7,500 ft (2 286 m); cruising speed, 185 mph (298 km/h) at 7,500 ft (2 286 m); initial rate of climb, 1,500 ft/min (7,5 m/sec); max range, 2,125 mls (3 420 km).
Weights: Empty, 16,970 lb (7 700 kg); normal take-off, 26,000 lb (11 793 kg); max overload, 33,000 lb (14 970 kg).
Dimensions: Span, 95 ft 0 in (28,96 m); length, 64 ft 5½ in (19,64 m); height, 16 ft 11 in (5,16 m); wing area, 987 sq ft (91,69 m²).
Accommodation: Flight crew of three, and up to 28 troops.
Status: First flown (as commercial DST) 17 December 1935. First delivery to US Army Air Corps (as C-41) October 1938. First C-47 delivered 1 February 1942. Production totals (built on military contracts): two C-41; 965 C-47; 5,254 C-47A; 3,364 C-47B; 66 R4D-1; 2 R4D-2; 219 C-53; 159 C-53D; 17 C-117A, grand total of 10,048 in USA, plus additional production in Soviet Union (about 2,000 as Li-2) and Japan.
Notes: The C-47 Skytrain, C-53 Skytrooper and Dakota (RAF name for C-47) were all variants of the Douglas DC-3 (see *Civil* volume) and became the most widely known and used of any World War II aircraft. The USAF and USN continued to use special-purpose variants in 1973, new rôles having been found for the C-47 during the Vietnam war, particularly as a heavily-armed gunship (the AC-47D) and for electronic reconnaissance (the EC-47P and EC-47Q). The AC-47 gunship, or weapons platform, which is currently serving with the air forces of South Vietnam, Cambodia, Laos and Thailand, is intended to provide suppressive fire from three 7·62-mm Miniguns which are fired from side positions. Among the variants serving with the US Navy is the ski-equipped LC-47H for use in the Antarctic and the C-117D, similar to the commercial Super DC-3 with uprated engines and revised, more angular wing and tail outlines. Many other air forces continue to operate C-47 variants for transport, training and special duties, and a Soviet licence-built version, the Li-2 remains in service in some numbers with the Soviet Air Force and several Communist Bloc nations.

DOUGLAS C-54 SKYMASTER (AND C-118)

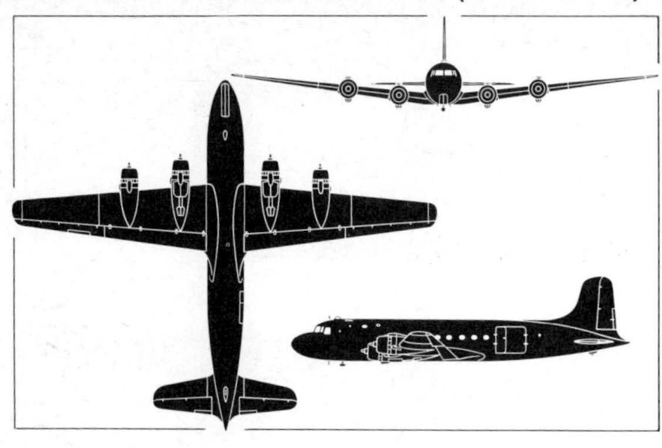

The photograph above depicts a C-54D of the Royal Danish Air Force and the general arrangement silhouette (left) is applicable to all remaining transport versions of the C-54

Country of Origin: USA.
Type: Medium-range transport.
Power Plant: Four Pratt & Whitney R-2000-7 or -11 radials of 1,350 hp or (C-54G, S and T) R-2000-9 radials of 1,450 hp.
Performance (C-54D): Max speed, 274 mph (441 km/h) at 14,000 ft (4 267 m); cruising speed, 239 mph (385 km/h) at 15,000 ft (4 572 m); max range, 3,900 mls (6 276 km).
Weights: Empty, 38,200 lb (17 327 kg); normal take-off, 62,000 lb (28 123 kg); max overload, 73,000 lb (33 112 kg).
Dimensions: Span, 117 ft 6 in (35,81 m); length, 93 ft 11 in (28,63 m); height, 27 ft $6\frac{1}{4}$ in (8,38 m); wing area, 1,463 sq ft (135,91 m^2).
Accommodation: Flight crew of five and up to 50 equipped troops or freight load of 32,000 lb (14 515 kg).
Status: First flown 26 March 1942, deliveries to US Army Air Corps began December 1942. Production totals: 24 C-54; 207 C-54A; 220 C-54B; 350 C-54D; 75 C-54E; 76 C-54G and 211 R5D variants for US Navy.
Notes: Douglas built a single DC-4E commercial transport prototype in 1938 but test flying which began on 21 June 1938 did not lead to large-scale commercial acceptance. Instead, a scaled-down DC-4A was projected, and a production batch of 24 was laid down in 1940 against airline orders. This entire batch was commandeered by the US Army Air Corps early in 1942, before the first had flown, and with the designation C-54, these aircraft became the first equipment for Air Transport Command, set up in December 1942 to ferry men and supplies around the world. Subsequent production versions were designated up to C-54G (see Status paragraph) except C-54C, which was a modified C-54A used by President Roosevelt and known as the "Sacred Cow". Post-war variants and redesignation of US Navy Skymasters led to designations up to C-54M. A number of examples remain in use with foreign air forces, including those of Argentina, Bolivia, Colombia, Denmark, Ethiopia, Guatemala, Honduras, Mexico, Nigeria, Paraguay, Peru, Portugal, South Africa, Spain, Thailand, and Zaïre. Developed from the C-54, the larger C-118 was the military equivalent of the DC-6 (see *Civil* volume), 166 being delivered to the USAF and USN, the version for the former service being the C-118A with accommodation for 76 fully-equipped troops, 60 casualty stretchers or 27,000 lb (12 247 kg) of freight, and that for the latter service, the C-118B.

DOUGLAS C-124 GLOBEMASTER II

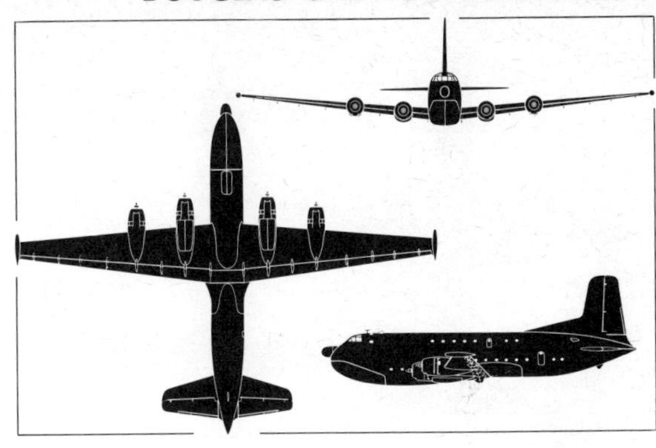

Both photograph (above) and general arrangement silhouette (right) illustrate the C-124C, the former depicting a C-124C of the Mississippi Air Guard

Country of Origin: USA.
Type: Long-range strategic transport.
Power Plant: Four Pratt & Whitney R-4360 (C-124A) -20WA radials of 3,500 hp or (C-124C)-63A radials of 3,800 hp.
Performance: Max speed, 304 mph (489 km/h) at 20,800 ft (6 340 m); high-speed cruise, 272 mph (437,7 km/h); long-range cruise, 230 mph (370 km/h); initial rate of climb, 760 ft/min (3,9 m/sec); service ceiling, 21,800 ft (6 645 m); range with max fuel, 6,820 mls (10 975 km); range with 56,000 lb (25 400 kg) payload, 1,232 mls (1 980 km).
Weights: Empty, 101,165 lb (45 887 kg); normal take-off, 185,000 lb (83 914 kg); max overload, 194,500 lb (88 223 kg).
Dimensions: Span, 174 ft 1½ in (53,08 m); length, 130 ft 5 in (39,75 m); height, 48 ft 3½ in (14,70 m); wing area, 2,506 sq ft (232,8 m²).
Accommodation: Flight crew of five and provision for up to 200 equipped troops or 127 stretchers plus 25 seats, or assorted cargo loads.
Status: Prototype (YC-124) first flown 27 November 1949. Production deliveries began May 1950. Production totals 204 C-124As, one YC-124B and 243 C-124Cs, final delivery May 1955.
Notes: The C-124 Globemaster II was evolved from the C-74 Globemaster I, from which it differed in having a new deep-section fuselage with nose-loading clam-shell doors and more powerful engines. The prototype YC-124 was converted from one of 15 C-74s ordered in 1945, and was later converted to the YC-124A with different mark of engines. The main production versions, C-124A and C-124C, differed primarily in engine power and operating weights; a nose radome and wing-tip fairings containing combustion heaters were introduced on the C-124C and later applied to the C-124A. At the time of its service introduction, the Globemaster II could accommodate 95 per cent of all types of US Army Field Force equipment without disassembly, the maximum cargo load being 74,000 lb (33 566 kg), and it was the largest production military transport in the world. One example, designated YC-124B, was experimentally fitted with four 5,500 ehp Pratt & Whitney YT34-P-6 turboprops, flying with these on 2 February 1954. The Globemaster IIs were progressively re-assigned to units of the Air Force Reserve and the Air National Guard, which had five C-124C Groups attached to MAC in 1973.

DOUGLAS C-133 CARGOMASTER

Both photograph (above) and general arrangement silhouette (left) depict the C-133A Cargomaster which currently equips two Military Airlift Command squadrons

Country of Origin: USA.
Type: Long-range strategic freighter.
Power Plant: Four Pratt & Whitney T34-P-9W turboprops rated at 7,500 shp each with water injection.
Performance: Max speed, 359 mph (578 km/h) at 8,700 ft (2 650 m); max cruise, 323 mph (520 km/h) at 26,850 ft (8 180 m); initial rate of climb, 1,280 ft/min (6,5 m/sec); service ceiling, 29,950 ft (9 140 m); range, 4,300 mls (6 920 km) with 44,000 lb (19 960 kg) payload.
Weights: Empty, 120,263 lb (54 550 kg); max take-off, 286,000 lb (129 727 kg).
Dimensions: Span, 179 ft 7¾ in (54,78 m); length, 157 ft 6½ in (47,01 m); height 48 ft 3 in (14,70 m); wing area, 2,673 sq ft (24,80 m²).
Accommodation: Flight crew of five (plus full reserve crew) and provision for up to 200 equipped troops, or up to 110,000 lb (49 895 kg) of varied stores.
Status: First C-133A flown 23 April 1956; first C-133B flown 31 October 1959. First delivery August 1957. Production totals, 34 C-133A and 15 C-133B. Production completed 1961.

Notes: The Cargomaster was developed to meet a USAF requirement for a specialised transport aircraft able to accommodate outsize loads, and was the first USAF transport specified to be powered by turboprop engines. The Douglas DTS-1333 design was accepted by the USAF to meet this requirement with a contract for 35 placed in 1954, no prototype being built. After the first seven C-133As had been built, the shape of the underside of the rear fuselage was changed to improve the airflow and a further change was made on the final three C-133As to introduce clam-shell rear loading doors in place of a combination of ramp and split doors. The same doors were featured in the C-133B, which had −9W engines replacing the dry-rated T34-P-7As of the earlier versions. The Cargomaster can accommodate 96 per cent of all US Army Field Force vehicles fully assembled, typical loads including 16 loaded jeep-type vehicles or two 40,000 lb (18 144 kg) earth-movers. The C-133B version carries a 52,000 lb (23 587 kg) load—10,000 lb (4 536 kg) more than the C-133A—over a distance of 4,000 mls (6 437 km) at 323 mph (520 km/h). Cargomasters equipped three squadrons of USAF Military Airlift Command until 1971, when the force was reduced to two squadrons.

EMBRAER EMB-110 BANDEIRANTE

The EMB-110 Bandeirante, illustrated by the photograph (above) and the general arrangement silhouette (right), was entering service with the Brazilian Air Force during 1973

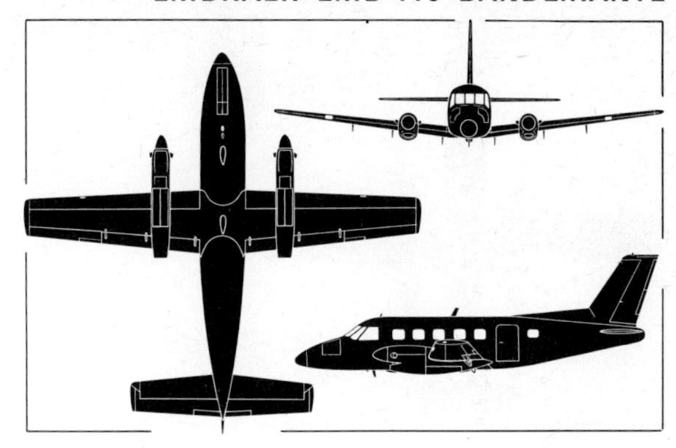

Country of Origin: Brazil.
Type: General purpose light transport.
Power Plant: Two 680 ehp Pratt & Whitney (UACL) PT6A-27 turboprops.
Performance: Max cruising speed, 260 mph (418 km/h) at 9,840 ft (3 000 m); initial rate of climb, 1,968 ft/min (10 m/sec); service ceiling, 27,950 ft (8 520 m); max range, 1,150 mls (1 850 km) with 30 min reserves.
Weights: Empty equipped, 6,437 lb (2 920 kg); max payload, 4,850 lb (2 200 kg); max take-off, 11,243 lb (5 100 kg); max landing, 10,692 lb (4 850 kg).
Dimensions: Span, 50 ft 2¼ in (15,30 m); length, 46 ft 8¼ in (14,22 m); height, 15 ft 6 in (4,73 m); wing area, 312·1 sq ft (29,0 m²).
Accommodation: Two pilots on flight deck and 6–12 seats in cabin, provision for four stretchers and two seats.
Status. Three YC-95 prototypes flown on 26 October 1968, 19 October 1969 and 25 June 1970 respectively. First of 80 C-95 production models for Brazilian Air Force flown on 15 August 1972. Deliveries began in late 1973. Total of 104 ordered by mid-1973 when planned production rate was to be two per month increasing to four per month in 1975, schedules calling for delivery of 17 by end of 1973, 36 during 1974 and 48 during 1975.
Notes: Development of the Bandeirante began under the design leadership of the well-known French designer Max Holste at the PAR—*Departamento de Aeronaves* of the *Centro Técnico Aeroespacial*, the original designation being IPD/PAR-6504. Following creation of the state-sponsored EMBRAER organisation in August 1969, the Bandeirante became an EMBRAER responsibility and a production contract was placed by the Brazilian Air Force, which designates the aircraft C-95 as a transport, able to accommodate up to 12 passengers or four stretchers with two medical attendants. The third prototype YC-95 was specially fitted with monitoring and recording equipment for use in connection with Brazilian space research activities. A pressurised version of the Bandeirante, designated EMB-120, was to be developed and some of the 80 ordered by the Brazilian Air Force may be delivered to this standard. An unspecified number of Bandeirantes are to be completed in navigational training configuration. Other versions are the EMB-110B for aerial surveillance and EMB-111 for maritime reconnaissance.

FAIRCHILD C-119 FLYING BOXCAR AND PACKET

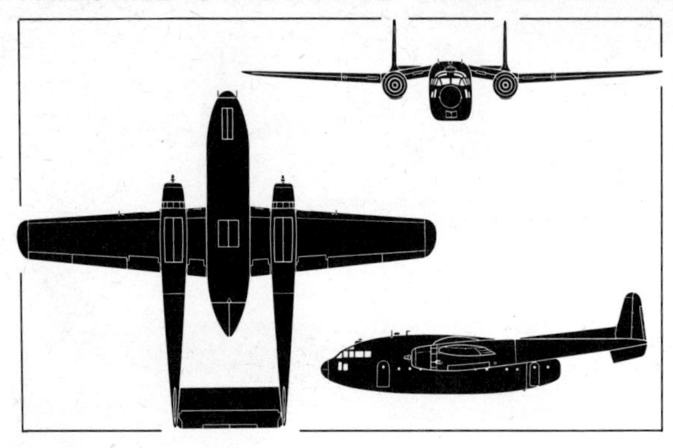

The photograph above illustrates a C-119G of the Nationalist Chinese Air Force, the general arrangement silhouette (left) also depicting this version of the elderly Fairchild transport

Country of Origin: USA.
Type: Medium-range tactical cargo and troop transport.
Power Plant: Two Wright R-3350-85 or -89A radials of 2,500 hp each or (C-119K) R-3350-999 TC18EA2 radials of 3,700 hp each with water injections and (Indian AF C-119 only) one HAL-built 4,700 lb st (2 132 kgp) Orpheus 701 booster turbojet or 3,400 lb st (1 542 kgp) Westinghouse J34 booster turbojet in dorsal pod.
Performance (C-119K): Max speed, 243 mph (391 km/h) at 10,000 ft (3 050 m); max cruising speed, 187 mph (300 km/h) at 10,000 ft (3 050 m); initial rate of climb (one engine out) 1,050 ft/min (5,3 m/sec); range with max payload 900 mls (1 595 km); ferry range 3,460 mls (5 570 km).
Weights (C-119K): Empty, 44,747 lb (20 300 kg); max payload, 20,000 lb (9 070 kg); max take-off and landing, 77,000 lb (34 925 kg).
Dimensions: Span, 109 ft 3 in (34,29 m); length, ·86 ft 6 in (26,36 m); height, 26 ft 6 in (8,07 m); wing area, 1,447 sq ft (134,43 m²).
Accommodation: Flight crew of four (two pilots, navigator and radio operator) plus loadmaster or jumpmaster and up to 62 fully-equipped troops or assorted items of cargo.
Status: First prototype (C-119A) flown in November 1947. Deliveries began December 1949. Production totalled 1,051, including 88 for Italy, India and Belgium and remainder for USAF/USN.
Notes: The C-119 Flying Boxcar was developed from the C-82 Packet and is sometimes referred to as the Packet also. Production variants were C-119B with R-4360-20 engines, C-119C with –20WA water injection engines, C-119F with –85 engines and 85,000 lb (38 555 kg) overload weight and C-119G, which was similar with Aeroproducts propellers. All Bs were later converted to Cs, all Fs to Gs, and 68 of the latter to C-119J with "beaver tail" door and ventral ramp openable in flight. Later modification programmes produced the AC-119G gunship with four 7·62-mm mini-gun pods in the fuselage, and the C-119K (unarmed) and AC-119K (armed) with underwing pods containing two 2,850 lb st (1 293 kgp) J85-GE-17 turbojets. The AC-119G and AC-119K operated in Vietnam until 1973 and three ANG units flew C-119Cs; other nations using the C-119 include Ethiopia, Brazil, Nationalist China, India, Italy and Morocco.

FAIRCHILD C-123 PROVIDER

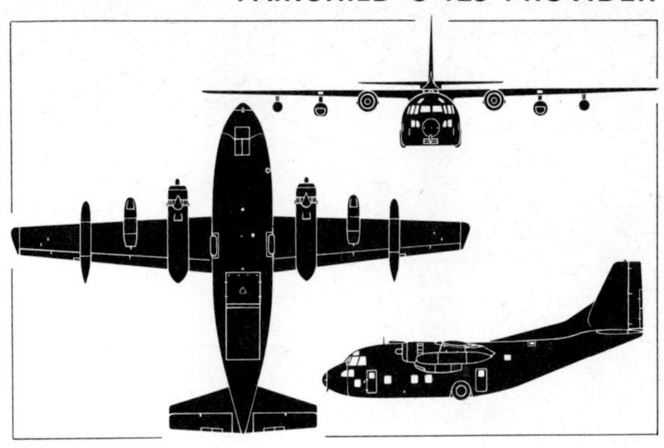

The photograph above depicts a C-123B of the Royal Thai Air Force and the general arrangement silhouette illustrates the turbojet-augmented C-123K with wider track undercarriage

Country of Origin: USA.
Type: Medium tactical assault transport.
Power Plant: Two 2,500 hp Pratt & Whitney R-2800-99W radials plus (C-123J) two 1,000 lb st (454 kgp) Fairchild J44-R-3 booster turbojets or (C-123K) two 2,850 lb st (1 293 kgp) General Electric J85-GE-17 booster turbojets.
Performance (C-123K): Max level speed, 228 mph (367 km/h); max cruising speed, 173 mph (278 km/h); initial rate of climb (one engine out) 1,220 ft/min (6,2 m/sec); service ceiling (one engine out) 21,100 ft (6 430 m).
Weights (C-123K): Empty, 35,366 lb (16 042 kg); basic operating 36,576 lb (16 590 kg); max payload, 15,000 lb (6 800 kg); max take-off and landing, 60,000 lb (27 215 kg).
Dimensions: Span, 110 ft 0 in (33,53 m); length, 76 ft 3 in (23,93 m); height, 34 ft 1 in (10,63 m); wing area, 1,223 sq ft (113,62 m²).
Accommodation: Flight crew of two and up to 60 equipped troops or 50 stretchers plus four seats or miscellaneous cargo loads.
Status: First flown (as XC-123 Avitruc) on 14 October 1949; first Fairchild-built C-123B flown on 1 September 1954; first C-123K flew on 27 May 1966. Production total, 5 by Chase and 302 by Fairchild.
Notes: The Provider originated in 1949 as an all-metal troop and cargo glider developed by Chase Aircraft and designated XG-20. One of two prototypes was fitted with two R-2800-83 engines in wing nacelles and tested as an assault transport, leading to a USAF production order for 300, placed with the Kaiser-Frazer Corporation, plus five pre-production models to be built by Chase. Fairchild took over the K-F commitment and built 302 C-123Bs between 1954 and 1958. Of this total, 24 were for MAP delivery to Venezuela and Saudi Arabia, and subsequently the USAF transferred some of its C-123s to Thailand and Vietnam. Later conversion programmes produced 10 C-123Hs with jet pods at the wing tips, serving with an ANG unit in Alaska in 1973, and the C-123K with underwing jet pods (specification above). Between 1966 and September 1968, 183 C-123Bs were converted to C-123Ks, primarily for service in Vietnam. A further variant is the UC-123B spray aircraft which, fitted with spray nozzles beneath the wings and tail, has been employed in support of US Department of Agriculture campaigns to control mosquitoes by the USAF Tactical Air Command.

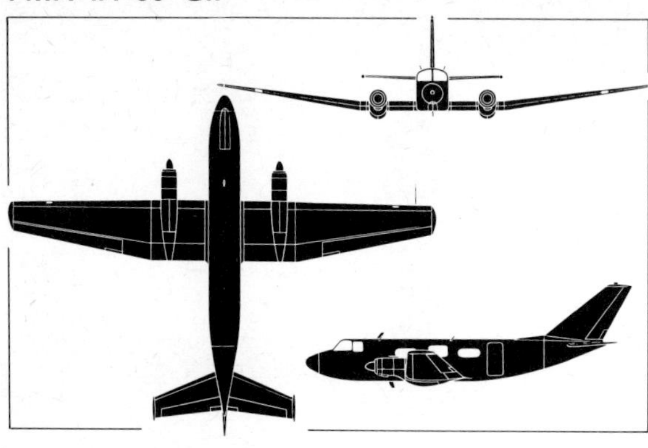

Both photograph (above) and general arrangement silhouette (left) illustrate the FMA GII, the former depicting an example serving with the I Brigada Aerea

Country of Origin: Argentina.
Type: General purpose transport.
Power Plant: Two 1,000 shp Turboméca Bastan VID turboprops.
Performance: Max speed, 310 mph (500 km/h); high speed cruise 305 mph (491 km/h); long-range cruise, 280 mph (450 km/h); initial rate of climb, 2,640 ft/min (13,4 m/sec); service ceiling, 41,000 ft (12 500 m); range with max payload, 1,240 mls (1,995 km); range with max fuel, 1,600 mls (2 575 km).
Weights: Empty equipped, 8,650 lb (3 924 kg); max payload, 3,307 lb (1 500 kg); max take-off, 16,204 lb (7 350 kg); max landing weight, 14,595 lb (6 620 kg).
Dimensions: Span, 64 ft $3\frac{1}{4}$ in (19,59 m); length, 50 ft $2\frac{1}{2}$ in (15,30 m); height, 18 ft 5 in (5,61 m); wing area, 450 sq ft (41,81 m^2).
Accommodation: Flight crew of two, alternative layouts for 10, 12 or 15 passengers as a transport or six stretchers plus two seats, or six seats as a navigation trainer.
Status: Prototype first flown on 23 April 1963; entered service early 1967. Total orders for 38 placed by 1973.
Notes: Originally built as the Guarani II, the IA 50 GII was developed from a single Guarani I prototype, which was in turn a transport derivative of the IA 35 Huanquero. New features of the G II were a swept-back single fin and rudder, shorter fuselage and more powerful engines. Two prototypes (with Bastan VIA engines) were built, plus a pre-production model, before the Argentine Air Force placed an initial order for 18, a second batch of 15 being ordered later. Included in the first batch of GIIs were 15 troop transports—one being fitted with skis for operation in the Antarctic—two aerial survey aircraft and one VIP transport for the Argentine Air Force and one staff transport for the Argentine Navy. One GII has been built for the use of the President of Argentina, four have been ordered by the Military Geographic Institute, and two have been completed for ground radio aids calibration. Several others delivered for civil tasks include one for an oil prospecting firm, one for the Federal Police and one for each of the provincial governments of Buenos Aires and Córdoba. A pressurised version of the GII has also been projected, together with variants for navigational training and photographic survey, but production of the basic IA 50 GII as described and illustrated here was expected to be phased out during the course of 1973.

GAF NOMAD 22

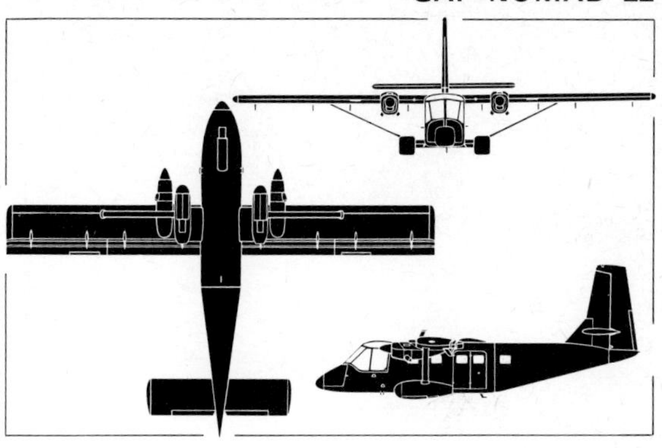

The Nomad 22, illustrated in prototype form by the photograph above and by the general arrangement silhouette (right) will enter service with the Australian Army during 1974

Country of Origin: Australia.
Type: STOL utility transport.
Power Plant: Two 400 ehp Allison 250-B17 turboprops.
Performance: Max cruise, 200 mph (322 km/h) at 10,000 ft (3 050 m); long-range cruise, 161 mph (259 km/h); initial rate of climb, 1,500 ft/min (7,6 m/sec); service ceiling 24,000 ft (7 315 m); range with max payload, 58 mls (93 km); range with 1,600 lb (726 kg) payload, 106 mls (172 km/h).
Weights: Typical operating empty, 4,630 lb (2 100 kg); max payload, 3,110 lb (1 410 kg); max take-off and anding, 8,000 lb (3 629 kg).
Dimensions: Span, 54 ft 0 in (16,46 m); length, 41 ft 3 in (12,57 m); height, 18 ft 1 in (5,45 m); wing area, 320 sq ft (29,7 m²).
Accommodation: Flight crew of one or two and up to 12 troops, or miscellaneous freight.
Status: First and second prototypes flown on 23 July and 5 December 1971 respectively. Initial production batch of 20 launched in May 1972, for first flight late 1973.
Notes: Design of the Nomad began in 1965 (as the Project N) to provide the indigenous Australian aircraft industry with a project having export potential in military and civil markets. The prototypes were to basic N22 configuration and in 1973 the first was equipped to the standard specified for production aircraft for the Australian Army, while the second was originally to have been converted to Nomad 24 configuration with lengthened fuselage to seat up to 15 passengers. Production of a batch of 20 Nomad 22s was authorised by the Australian government in May 1972, of which 11 were earmarked for use by the Australian Army Aviation Corps. A special feature of this version is the installation of Stanley ejection seats for the two pilots, and armour round the cockpit to give protection against ground fire. The first customer for a commercial Nomad was the Jack Brabham charter company, which ordered one early in 1973. To customer requirement, a version is available with hinged rear fuselage to allow straight-in loading of large freight items. Approval for the production of a further batch of 50 Nomads was given mid-1973, and it was anticipated that a half-dozen of these would be delivered to Australia's new Federal Coast Guard Service. Four Nomads modified for the coastal patrol task were to be supplied to the Indonesian Navy under an Australian defence aid programme announced during 1973.

HAWKER SIDDELEY HS 748 ANDOVER

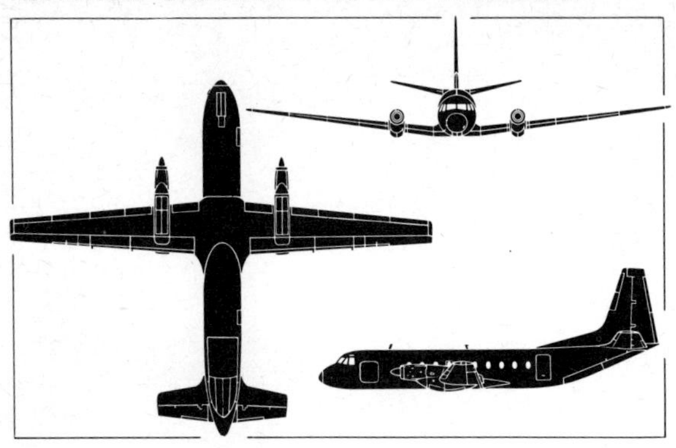

Both photograph (above) and general arrangement silhouette (left) depict the Andover C Mk 1 which equips one RAF squadron for tactical use

Country of Origin: United Kingdom.
Type: Troop and general purpose transport.
Power Plant: Two 3,245 ehp Rolls-Royce Dart 301 turboprops.
Performance: Max cruising speed, 265 mph (426 km/h) at 15,000 ft (4 570 m); initial rate of climb, 1,180 ft/min (6 m/sec); service ceiling, 24,000 ft (7 300 m); range with max payload, 282 mls (454 km); range with max fuel, 1,158 mls (1 865 km) with 8,530 lb (3 870 kg) payload.
Weights: Basic operating, 28,650 lb (12 996 kg); max payload, 15,350 lb (6 963 kg); max take-off, 50,000 lb (22 680 kg); max landing, 47,600 lb (21 590 kg).
Dimensions: Span, 98 ft 3 in (29,95 m); length, 78 ft 0 in (23,77 m); height, 30 ft 1 in (9,15 m); wing area, 831·4 sq ft (77,2 m²).
Accommodation: Flight crew of two or three and up to 58 equipped troops, 40 paratroops, 24 stretchers plus attendents or full payload as cargo.
Status: Prototype Andover (modified HS.748) first flown on 21 December 1963. First production Andover C Mk 1 flown on 9 July 1965. Deliveries to RAF began in 1966 and production of 31 completed in 1968.

Notes: The Andover was derived from the Hawker Siddeley HS.748 (see *Civil* volume) with the same wings and forward fuselage but a new rear fuselage and tail unit, uprated engines and a "kneeling" main undercarriage to facilitate loading through the rear doors. Only user of this version is the Royal Air Force, the principal unit being No 46 Squadron. The Andover name is also used for six HS.748s in RAF service, these being Andover CC Mk 2s, two serving with the Queen's Flight and four on VIP communications duties. Several other air forces have purchased standard HS.748s for training, transport or VIP duties, including the Royal Australian Air Force and Navy, the Brazilian Air Force, the Royal Thai Air Force, the Venezuelan Air Force, the Argentine Air Force, the Ecuadorian Air Force and the Sultan of Brunei. The Indian Air Force ordered a total of 93 from the Hindustan Aeronautics production line and 48 of these are specified to the standard of the HS.748 Series 2C with a large side-loading freight door with paradrop clearance. Hindustan Aeronautics has received orders for five HS.748 Series 1 and 40 Series 2 aircraft for the Indian Air Force for communications and training rôles, and these are to be followed by the 48 HS.748M (Series 2C) freighter and paratroop transports.

IAI-201 ARAVA

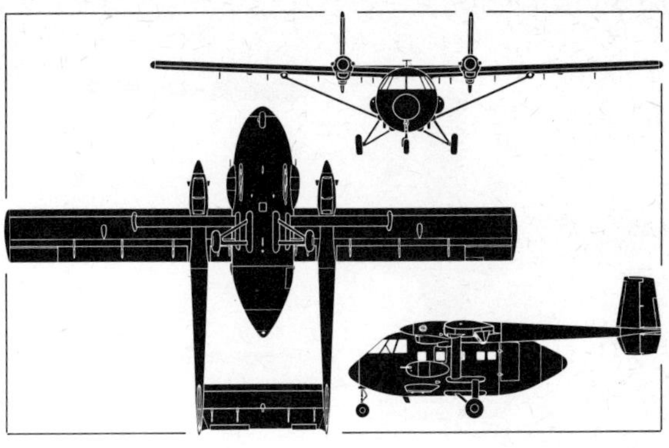

The Arava 201, illustrated by the photograph (above) and the general arrangement silhouette (right) is currently manufactured by Israel Aircraft Industries primarily for export

Country of Origin: Israel.
Type: General purpose STOL transport and gunship.
Power Plant: Two 783 shp Pratt & Whitney PT6A-34 turboprops.
Performance: Max speed, 203 mph (326 km/h) at 10,000 ft (3 050 m); high-speed cruise, 198 mph (319 km/h); long-range cruise, 193 mph (311 km/h); initial rate of climb, 1,564 ft (7·9 m/sec); service ceiling, 26,575 ft (8 100 m); range with max payload, 201 mls (323 km); range with max fuel, 806 mls (1 297 km).
Weights: Empty equipped (paratrooper) 7,787 lb (3 532 kg); max payload, 5,570 lb (2 526 kg); max take-off and landing, 15,000 lb (6 803 kg).
Dimensions: Span, 69 ft 6 in (20,88 m); length, 42 ft $7\frac{1}{2}$ in (12,99 m); height, 17 ft 1 in (5,21 m); wing area, 470·2 sq ft (43,68 m^2).
Accommodation: Flight crew of one or two plus 23 fully-equipped troops or 16 paratroops and two jumpmasters, or eight stretchers plus three seats. Alternative interior arrangements include an all-freight transport, a typical load comprising a jeep-mounted recoilless rifle and its four-man crew.
Armament: One 0·5-in (12·7-mm) machine gun on each side of fuselage with 250 rpg and provision for single aft-firing 0·5-in (12·7-mm) gun in flexible mount in the fuselage tail. Two hard-points on fuselage with 600-lb (272-kg) capacity each.
Status: Prototype IAI-101 first flown on 27 November 1970; second prototype flown in April 1971. Prototype IAI-201 first flown late 1971.
Notes: Development of the Arava began in 1966 as the first wholly indigenous product of Israel Aircraft Industries, and in the IAI-101 version was intended primarily as a civil light transport with STOL and rough field capability. A stretched version to seat 30 passengers, with PT6A-40 engines was also projected, but work on the civil variant was suspended in 1972 after prototypes had been tested, to allow IAI to concentrate on development and production of the IAI-201 for the Israeli Air Force and possible foreign military users. One of the initial Arava prototypes was completed to IAI-201 standard as described above and some eight production models having flown by mid-1973 when production rate was one per month. The first export deliveries were made in May of that year to Mexico, five being supplied for use by the Mexican Air Force. Eight more were then on order for other customers.

ILYUSHIN IL-14

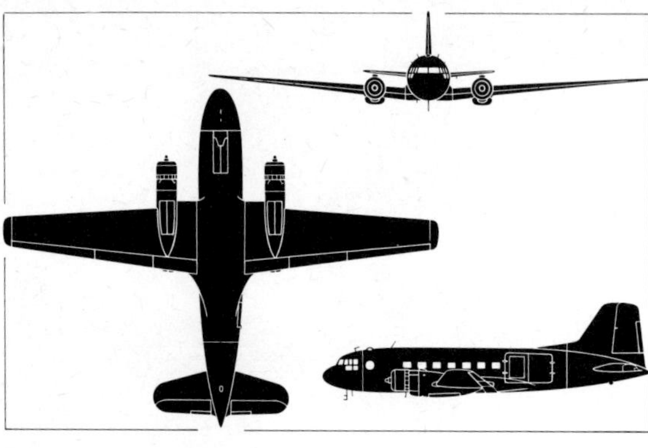

The photograph above depicts an Il-14 of the Egyptian Air Force and the general arrangement silhouette illustrates the basic military version of the transport

Country of Origin: USSR.
NATO Code Name: *Crate.*
Type: General purpose transport.
Power Plant: Two 1,900 hp Shvetsov ASh-82T piston radial engines.
Performance (Il-14M) : Max speed, 259 mph (417 km/h) ; high-speed cruise, 239 mph (385 km/h) ; long-range cruise, 193 mph (311 km/h) ; initial rate of climb, 1,220 ft/min (6,2 m/sec) ; service ceiling, 22,000 ft (6 705 m) ; range with max payload, 810 mls (1 304 km) ; range with max fuel, 1,988 mls (3 202 km).
Weights: Operational equipped, 27,776 lb (12 600 kg) ; max take-off, 39,683 lb (18 000 kg).
Dimensions: Span, 104 ft 0 in (31,69 m) ; length, 73 ft 2 in (22,30 m) ; height, 25 ft 11 in (7,90 m) ; wing area, 1,075 sq ft (99,7 m²).
Accommodation: Flight crew of three or four and ·standard accommodation for 24 or 28 passengers.
Status: Prototype Il-14 first flown in 1952 ; deliveries began (Il-14P) in 1954 and (Il-14M and Il-14T) in 1956.
Notes: The Il-14 was an outgrowth of the Il-12, a general purpose military transport developed during World War II as a successor to the Lisunov Li-2 (Douglas DC-3 built in the USSR). Large numbers of Il-12s were built for the Soviet Air Force and for commercial use by Aeroflot and other airlines. The Il-14 differed in having a refined structure, improved aerodynamics and uprated engines. Variants were the initial production Il-14P, also built in East Germany and Czechoslovakia, and the Il-14M, with fuselage lengthened by 3·3 ft (1 m) and carrying a bigger payload over a shorter range. The Czech-built version of the Il-14M, with 32 passenger seats, was known as the Avia 14-32 and a further improvement, the Avia 14 Salon, had a new fuel system using wing-tip tanks and other refinements, and the Il-14T was a commercial freighter variant of the military freight and paratroop transport version featuring a strengthened freight floor and large freight-loading doors in the fuselage portside, military examples also having observation blisters aft of the flight deck for the paradrop controller. In addition to commercial use, Il-14s entered service with the air forces of virtually all the Communist Bloc countries, and also survive in small numbers with the air forces of Afghanistan, Algeria, Cuba, Egypt, India, Indonesia, Iraq, Syria and Yugoslavia.

ILYUSHIN IL-76

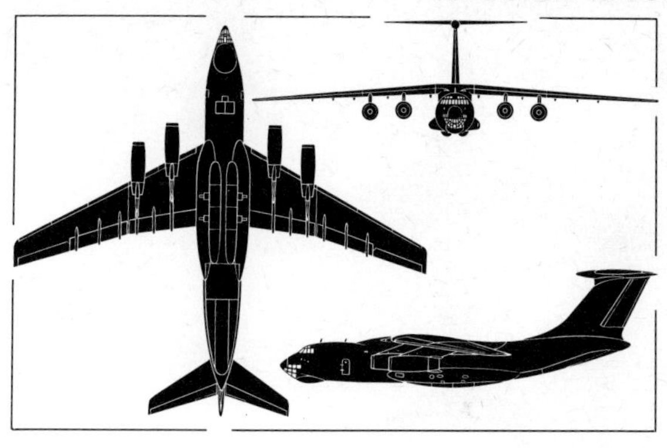

The Il-76, illustrated by both the photograph above and the general arrangement silhouette (right), is currently in production as both a military and civil freighter

Country of Origin: USSR.
NATO Code Name: *Candid.*
Type: Heavy freighter.
Power Plant: Four 26,455 lb st (12 000 kgp) Soloviev D-30-KP turbofans.
Performance: Max cruise, 528 mph (850 km/h) at 42,650 ft (13 000 m); range with max payload, 3,100 mls (5 000 km).
Weights: Max payload, 88,185 lb (40 000 kg); max take-off, 346,122 lb (157 000 kg).
Dimensions: Span, 165 ft 8½ in (50,50 m); length, 152 ft 10¼ in (46,59 m); height, 48 ft 5¼ in (14,76 m).
Accommodation: Normal flight crew of three (two pilots on flight deck and navigator in station in glazed nose) with optional provision for a fourth crew member on flight deck.
Status: Prototype first flown on 25 March 1971, with production underway in 1972 for deliveries to begin in 1973-74.
Notes: Although making its first public appearance (at the Paris Air Show, 1971) in civil markings and Aeroflot colours, the Il-76 appears to have been evolved primarily to meet a military requirement, and is generally similar in concept to the Lockheed C-141A StarLifter, being slightly larger, more powerful and heavier than its American counterpart. It employs a mechanised cargo-handling system, a high-flotation undercarriage—the main members of which comprise four individual units each of four parallel-mounted wheels—and extensive high-lift devices to achieve short-field performance. Clam-shell thrust reversers are fitted to all four power plants and straight-in freight loading is achieved through the rear fuselage which incorporates a loading ramp and clam-shell doors. According to an official Soviet statement, the Il-76 is intended to operate from short unprepared strips in Siberia and other underdeveloped areas of the Soviet Union during the period of the current five-year programme (1971–75), this indicating that the type will be operated by Aeroflot before entering the inventory of the Soviet Air Forces. Four prototype and pre-production examples of the Il-76 were undergoing trials in the Soviet Union by mid-1973, the second prototype being employed for commercial certification, and 100 Il-76s are to be delivered to Aeroflot from 1974, deliveries presumably being initiated prior to those to the Soviet Air Forces. It may be assumed that the Il-76 will supplant the An-12 in Soviet Air Force transport units.

KAWASAKI C-1A

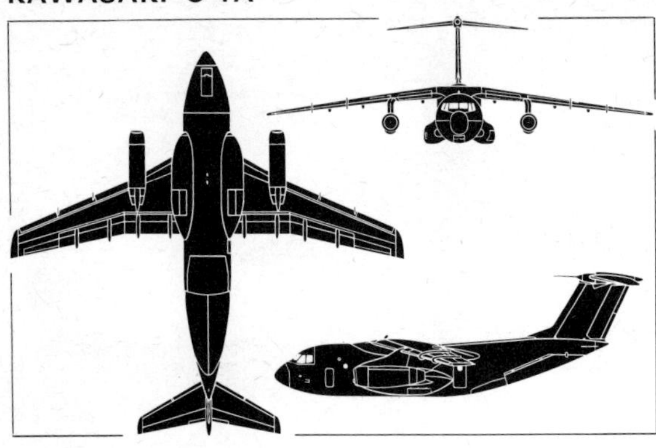

The Kawasaki C-1, the first prototype of which is illustrated by the photograph above, has been adopted as the standard medium-range transport of Japan's Air Self-Defence Force

Country of Origin: Japan.
Type: Medium-range transport.
Power Plant: Two 14,500 lb st (6 575 kg) Pratt & Whitney JT8D-9 turbofans.
Performance (estimated): Max level speed, 507 mph (815 km/h) at 25,000 ft (7 620 m); long-range cruise, 438 mph (704 km/h); initial rate of climb, 3,880 ft/min (19,5 m/sec); service ceiling, 39,375 ft (12 000 m); range with max payload, 807 mls (1 300 km); range with max fuel, 2,050 mls (3 300 km) with payload of 5,730 lb (2 600 kg).
Weights: Empty equipped, 53,130 lb (24 100 kg); payload, 17,640 lb (8 000 kg); max take-off, 85,320 lb (38 700 kg).
Dimensions: Span, 100 ft 4¾ in (30,60 m); length, 95 ft 1¾ in (29,00 m); height, 32 ft 9¼ in (9,99 m); wing area, 1,297 sq ft (120,5 m²).
Accommodation: Flight crew of four (two pilots, navigator and flight engineer) and loadmaster. Up to 60 equipped troops, 45 paratroops or 36 stretchers plus attendants, or miscellaneous freight on three pallets or loaded individually.
Status: First and second prototypes flown, respectively, on 12 November 1970 and 16 January 1971. Two pre-production C-1As ordered with first flown 19 September 1973, and production of 26 programmed in the Five-Year Defence Budget 1972–76.
Notes: The C-1A was developed to provide the JASDF with a replacement for its aged Curtiss C-46s which still provide the bulk of Japan's airlift capability in the early 'seventies. Initial design work on the new transport, which is Japan's first large jet-powered design, began in 1966 and was handled by NAMC, an industry consortium that had been responsible for designing and producing the YS-11 (see *Civil* volume). NAMC completed a mock-up but the two prototypes, under the designation XC-1A, were assembled by Kawasaki, which later assumed overall responsibility for the design. Two pre-production models were ordered early in 1972, for construction while the JASDF was completing an 800-hr test programme on the two prototypes. The two pre-production C-1As flew late in 1973, and the procurement of 26 was programmed at the beginning of 1974, with completion of deliveries during the 1977 Fiscal Year. A longer-range, larger-capacity derivative was under consideration in 1973 for possible development for the JASDF, together with an airborne early warning version.

LOCKHEED C-130 HERCULES

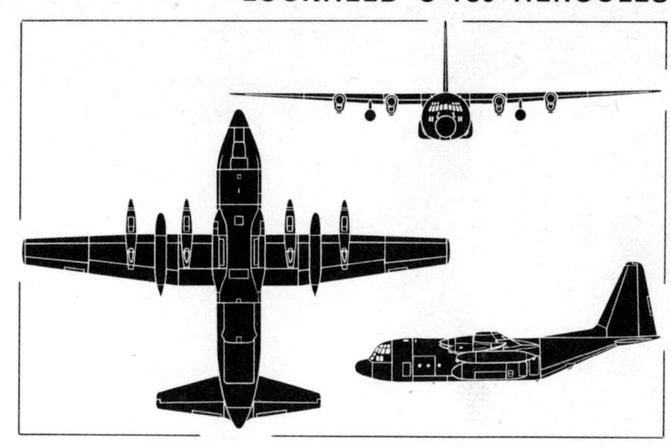

Both photograph (above) and general arrangement silhouette (right) depict the C-130H version of the Hercules, the former illustrating an example in service with the Brazilian Air Force

Country of Origin: USA.
Type: Medium/long-range military transport.
Power Plant: Four 4,050 ehp Allison T56-A-7A turboprops.
Performance (C-130E): Max speed, 384 mph (618 km/h); max cruising speed, 368 mph (592 km/h); economical cruising speed, 340 mph (547 km/h); initial rate of climb 1,830 ft/min (9,3 m/sec); service ceiling, 23,000 ft (7 010 m); range with max payload, 2,420 mls (3 895 km); range with max fuel, 4,700 mls (7 560 km).
Weights (C-130E): Empty equipped, 72,892 lb (33 063 kg); max payload, 45,000 lb (20 412 kg); normal take-off, 155,000 lb (70 310 kg); max overload, 175,000 lb (79 380 kg); max landing, 130,000 lb (58 970 kg).
Dimensions: Span, 132 ft 7 in (40,41 m); length, 97 ft 9 in (29,78 m); height, 38 ft 3 in (11,66 m); wing area, 1,745 sq ft (162,12 m²).
Accommodation: Flight crew of four (two pilots, navigator and systems manager) and provision for loadmaster; maximum provision for 92 equipped troops, 64 paratroops or 74 stretchers plus two seats, or assorted cargo loads up to payload limit, on six pallets or individually loaded.

Status: Prototype YC-130 first flown on 23 August 1954; first production C-130A flown on 7 April 1955; first C-130E flown on 25 August 1961; first HC-130H flown on 8 December 1964. Entered service with USAF December 1956. Total orders, over 1,300 for 28 different nations; production rate three a month at the beginning of 1974.
Notes: The Hercules entered production as the USAF's first tactical transport with turboprop engines, and has remained in front-line service in this rôle since 1956. Principal US versions are the C-130A with T56-A-1A or -9 engines and 155,000 lb (70 305 kg) gross weight; ski-equipped C-130D for Arctic operation; C-130E with increased fuel, higher weights and uprated engines; WC-130B and WC-130E for weather reconnaissance; AC-130E gunship; C-130F for USN and KC-130F tankers for USMC; HC-130H and HC-130N for rescue and recovery, HC-130P to air-refuel helicopters and EC-130Q for USN command communications. Major foreign users include the RAF (Hercules C Mk 1 and W Mk 2, the latter for weather reconnaissance), Belgium, Italy, Canada, Australia, New Zealand, Iran and Brazil, and many smaller air forces, orders having been placed in 1973 by Denmark, Zaïre, Abu Dhabi and Morocco.

LOCKHEED C-141 STARLIFTER

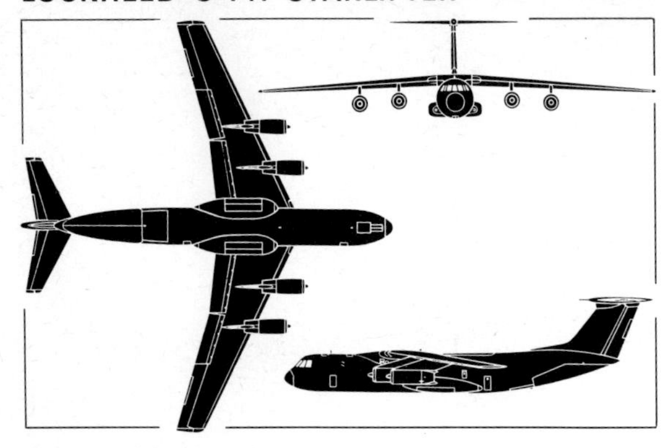

Currently the mainstay of the USAF Military Airlift Command, the C-141A StarLifter is illustrated by both the photograph (above) and the general arrangement silhouette (left)

Country of Origin: USA.
Type: Long-range Strategic Transport.
Power Plant: Four 21,000 lb (9 525 kgp) Pratt & Whitney TF33-P-7 turbojets.
Performance: Max speed, 571 mph (919 km/h) at 25,000 ft (7 600 m) high-speed cruise, 564 mph (908 km/h); long-range cruise, 495 mph (797 km/h); initial rate of climb, 3,100 ft/min (15,6 m/sec); service ceiling, 41,600 ft (12 680 m); range with max payload, 4 080 mls (6 565 km); range with max fuel, 6,140 mls (9 880 km) with payload of 31,870 lb (14 460 kg).
Weights: Empty equipped 133,773 lb (60 678 kg); max payload, 70 847 lb (32 136 kg); max take-off, 316,600 lb (143 600 kg); max landing, 257,500 lb (116 800 kg).
Dimensions: Span, 159 ft 11 in (48,74 m); length, 145 ft 0 in (44,20 m); height, 39 ft 3 in (11,96 m); wing area, 3,228 sq ft (299,9 m²).
Accommodation: Flight crew of four and 154 equipped troops or 123 paratroops or 80 stretchers plus 16 seats, or miscellaneous freight loaded on 10 pallets or stowed individually.
Status: No prototype. First C-141A flown on 17 December 1963. Entered service October 1964. Production total 285, completed in 1968.
Notes: Designed to meet a Specific Operational Requirement issued in May 1960, the C-141A StarLifter was the first jet-powered strategic freighter for the USAF, and became the mainstay of Military Airlift Command in place of interim C-135 jet transports and piston-powered C-97s and C-124s. The fuselage, including a rear-loading ramp and clam shell doors, was configured to allow the StarLifter to carry 90 per cent of all the items of air portable equipment in use with the US Army or Air Force. Dimensionally, the C-141A was able to accommodate a complete Minuteman ICBM, in container, but the 86,207 lb (39 103 kg) weight of this item was above the normal payload and a few C-141As were therefore specially modified, with beefed-up structure. Having a higher empty weight, those C-141As had a restricted gross weight of 300,600 lb (136 350 kg) and the range with a Minuteman was therefore reduced to 2,500 mls (4 025 km). The C-141A StarLifter remains the principal heavy transport of the USAF's Military Airlift Command, currently equipping a total of 14 squadrons, providing global-range airlift capability.

The world's largest military freighter, the C-5A Galaxy, which equips four squadrons of Military Airlift Command, is depicted by the photograph (above) and the silhouette (right)

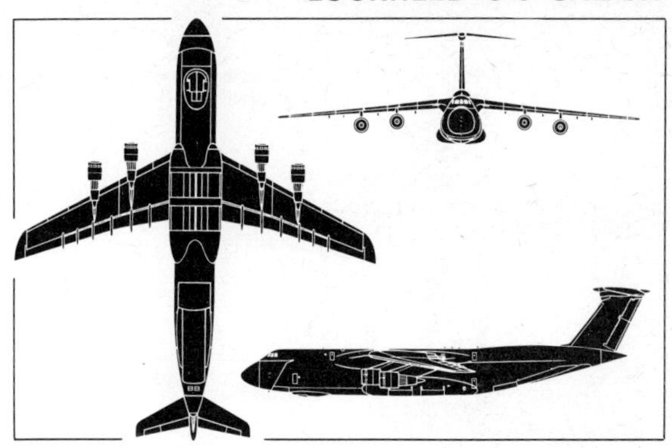

Country of Origin: USA.
Type: Long-range strategic freighter.
Power Plant: Four 41,000 lb st (18 600 kgp) General Electric TF39-GE-1 turbojets
Performance: Max speed, 571 mph (919 km/h) at 25,000 ft (7 620 m); high speed cruise, 541 mph (871 km/h) at 30,000 ft (9 150 m); long-range cruise, 537 mph (864 km/h) at 30,000 ft (9 150 m); initial rate of climb, 2,300 ft/min (11,7 m/sec); service ceiling, 34,000 ft (10 360 m); range with max payload, 2,950 mls (4 745 km); range with max fuel, 6,500 mls (10 460 km) with payload of 80,000 lb (36 287 kg).
Weights: Basic operational, 325,244 lb (147 528 kg); max payload, 265,000 lb (120 200 kg); max take-off, 764,500 lb (346 770 kg); max landing, 635,850 lb (288 416 kg).
Dimensions: Span, 222 ft $8\frac{1}{2}$ in (67,88 m); length, 247 ft 10 in (75,54 m); height, 65 ft $1\frac{1}{2}$ in (19,85 m); wing area, 6,200 sq ft (576 m²).
Accommodation: Flight crew of four (two pilots, navigator and system engineer) plus one loadmaster; 15 seats for relief crew, etc, and provision for up to 345 troops, of which total 75 are on upper deck and 270 on lower deck normally used to carry freight on 36 pallets or loaded individually. Typical loads include two M-60 tanks, an M-60 tank and two Iroquois helicopters, five M-113 personnel carriers, a 2·5-ton M-59 truck and a 0·5-ton M-151 truck, or 10 Pershing missiles with tow and launch vehicles.
Status: First of eight flight test and evaluation aircraft flown on 30 June 1968. Deliveries began 17 December 1969. Production of 81 completed May 1973.
Notes: Project definition of a very large strategic freighter to serve alongside the C-141 StarLifter (see page 128) began in 1963 and Lockheed was selected in October 1965 from three competing companies to build an aircraft in this category. The requirement was for a freighter that could carry 125,000 lb (56 700 kg) for 8,000 mls (12 875 km), or a maximum of twice that load for a shorter distance. It had to be able to take-off from an 8,000 ft (2 440 m) runway when fully loaded and to land on a 4,000 ft (1 220 m) semi-prepared runway. Because of cost escalation, USAF plans to buy enough C-5As to equip six squadrons were modified to provide a four-squadron force in MAC each squadron possessing 16 aircraft, and this was achieved during 1973.

NORD N.2501 NORATLAS

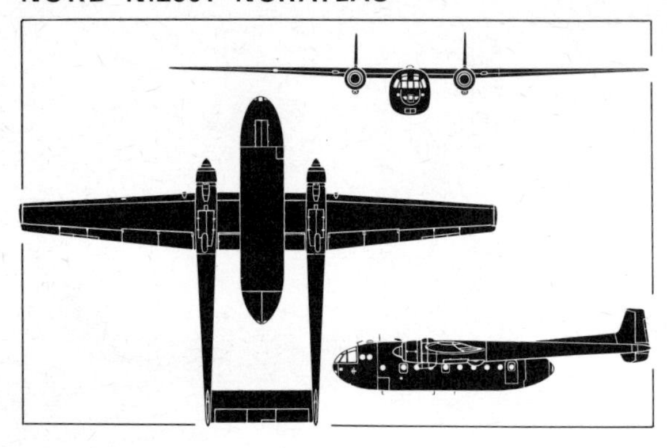

The photograph above illustrates an N.2501D (former Luftwaffe) of the Israeli air arm and the general arrangement silhouette (left) is applicable to all non-jet boosted versions

Country of Origin: France.
Type: Medium-range tactical transport.
Power Plant: Two 2,090 hp SNECMA-Bristol Hercules 738 or 758 piston radials plus (N.2502 and N.2504 only) two 882 lb st (400 kgp) Turboméca Marboré II turbojets.
Performance: Max speed, 251 mph (404 km/h) at 10,000 ft (3 050 m); long-range cruise, 200 mph (322 km/h) at 10,000 ft (3 050 m); initial rate of climb, 1,180 ft/min (5,9 m/sec); service ceiling 23,300 ft (7 100 m); range, 1,710 mls (2 750 km) with payload of 14,994 lb (6 800 kg).
Weights: Empty, 29,327 lb (13 300 kg); payload, 13 227 lb (6 000 kg); max take-off, 48,500 lb (22 000 kg).
Dimensions: Span, 106 ft 7½ in (32,48 m); length, 72 ft 0½ in (21,95 m); height, 19 ft 8¼ in (6,0 m); wing area, 1,089 sq ft (101,0 m²).
Accommodation: Flight crew of four or five and 45 equipped troops, 36 paratroops or 18 stretchers plus medical attendants.
Status: Prototype N.2500 first flown on 10 September 1949; N.2501 prototype first flown on 30 November 1950; N.2502 flown on 1 June 1955; N.2504 flown on 17 November 1958; N.2508 flown on 29 May 1957. Production deliveries began 1951; quantities comprised N.2500, one; N.2501, 267 in France and 161 in Germany; N.2502, 21; N.2503, one; N.2504, 5; N.2506, one; N.2508, two.
Notes: The Noratlas was evolved in the period immediately following World War II to provide the *Armée de l'Air* with a troop and paratroop transport and as such it entered service in 1951, continuing as front-line equipment until the advent of the Transall C.160 some 16 years later. The *Armée de l'Air* acquired 200 N.2501s, and the French *Aéronavale* used five N.2504s to train anti-submarine radar operators; the N.2501 was also exported to Israel (30), Portugal (12) and Germany (25), with another 161 built for the *Luftwaffe* in Germany. Surplus Noratlas transports from Germany have entered service with the air forces of Niger, Nigeria, and the Chad Republic, but current planning calls for the Noratlas *Armée de l'Air* Noratlas transports have been supplied to various former French territories, including the Central African Republic and the Chad Republic, but current planning calls for the Noratlas to remain in service with the *Armée de l'Air* until at least the early 'eighties.

PIAGGIO P.166M

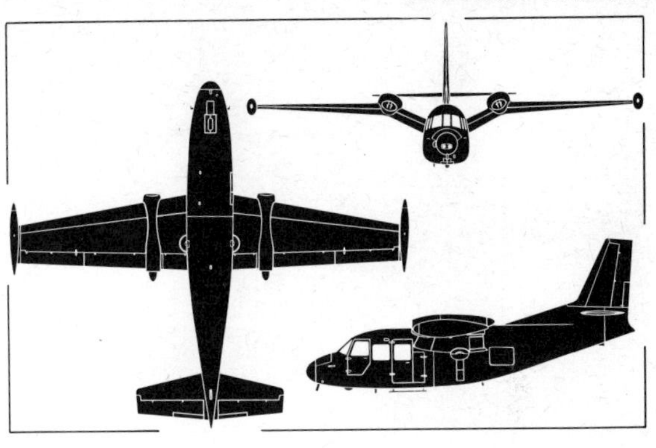

The photograph above depicts a P.166M of the Italian Air Force and the general arrangement silhouette is applicable to this version. The South African P.166S has larger wingtip tanks

Country of Origin: Italy.
Type: General purpose transport.
Power Plant: Two 340 hp Lycoming GSO-480-B1C6 six-cylinder horizontally-opposed engines.
Performance: Max speed, 222 mph (357 km/h) at 9,500 ft (2 900 m); high speed cruise, 207 mph (333 km/h); long-range cruise, 174 mph (280 km/h); initial rate of climb, 1,240 ft/min (6,3 m/sec); service ceiling, 25,500 ft (7 770 m); max range, 1,200 mls (1 930 km).
Weights: Empty, 5,180 lb (2 350 kg); max take-off, 8,115 lb (3 680 kg).
Dimensions: Span, 46 ft 9 in (14,25 m); length, 38 ft 1 in (11,60 m); height, 16 ft 5 in (5,0 m); wing area, 286 sq ft (26,56 m²).
Accommodation: Flight crew of two pilots and accommodation for 6–8 passengers.
Status: Prototype (P.166) first flown on 26 November 1957; certification and deliveries in 1958. First P.166S flown October 1968; certification in 1969.
Notes: The P.166 evolved from the amphibious P.136, being of similar configuration with two engines on a high wing, but somewhat larger and designed only for land operations. Of more than 100 P.166s built, about half have been for military use, including 51 P.166Ms for the Italian Air Force, which operates the type for communications and light transport duties. The South African Air Force acquired nine of the P.166S variant, which is especially equipped for search and surveillance duties, with nose radar, individual pilot's doors, cargo door and emergency escape hatch in the cabin roof. The fuel capacity is increased and extensive nav/com equipment is fitted. With the name Albatross, the P.166S serves with No 27 Squadron, SAAF, at Ysterplaat, Cape Town, on short-range patrol and fishery surveillance duties, and nine more were on order in 1973. Several military variants of the basic design were proposed in 1973, these including the P.166-BL2 PAR para-dropping version accommodating seven paratroops; the P.166-BL2 ASP submarine search and detection version with nose-mounted anti-snorkel radar; the P.166-BL2 MAR maritime reconnaissance and long-range patrol version, and the P.166-BL2 AML armed military version. The Piaggio company was also working on a proposed derivative of the P.166 with turboprop engines, two aircraft being under conversion in 1973.

PIAGGIO PD-808

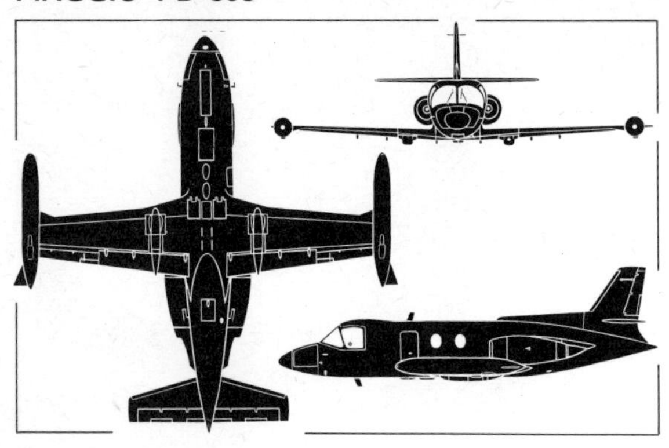

Both photograph (above) and general arrangement silhouette (left) illustrate the PD-808 as currently serving with the Italian Air Force in a variety of rôles

Country of Origin: Italy.
Type: Staff and personnel transport and trainer.
Power Plant: Two 3,360 lb st (1 524 kgp) Rolls-Royce (Piaggio) Viper 526 turbojets.
Performance: Max level speed, 529 mph (852 km/h) at 19,500 ft (5 945 km); high speed cruise, 497 mph (800 km/h) at 36,000 ft (11 000 m); long-range cruise, 449 mph (722 km/h) at 41,000 ft (12 500 m); initial rate of climb, 5,400 ft/min (27,5 m/sec); service ceiling, 45,000 ft (13 715 m); range with max fuel, 1,322 mls (2 128 km) with payload of 840 lb (381 kg).
Weights: Empty equipped, 10,650 lb (4 830 kg); max payload, 1,600 lb (726 kg); max take-off, 18,000 lb (8 165 kg); max landing, 16,000 lb (7 257 kg).
Dimensions: Span, 43 ft 3½ in (13,20 m); length, 42 ft 2 in (12,80 m); height, 15 ft 9 in (4,80 m); wing area, 225 sq ft (20,9 m²).
Accommodation: Flight crew of two: six (VIP) or nine (communications) passenger seats in cabin.
Status: First and second prototypes flown on 29 August 1964 and 14 June 1966 respectively; civil certification on 29 November 1966. Production of 25 completed in 1973.
Notes: Originally known as the Vespa-Jet, the PD-808 was designed basically by the Douglas Aircraft Company's El Segundo Division, with detail design and manufacture undertaken by Piaggio. Intended for the civil market, the PD-808 was ordered only by the Italian Air Force, which financed construction of the two prototypes (powered by 3,000 lb st/1 360 kgp Viper 525 engines) and in 1965 ordered a batch of 25. Included in the first 13 delivered to the Air Force were four with VIP interiors, three for electronic countermeasures with special equipment and five-man crew, and six nine-seat communications aircraft, the remaining 12 being completed for checking Italy's airways and navigational systems. The nine-seat version may also be employed in the navigational training rôle, with one main student station in the co-pilot's seat and either two or three more students' stations in the main cabin. Two executive models have been built for civil use, with Viper 526 engines and a version with AiResearch TFE-731-2 turbofans has been projected as the PD-808TF. An aeromedical version has also been proposed, this having accommodation for two casualty stretchers, two seated casualties and a medical attendant.

SHORT BELFAST C.MK 1

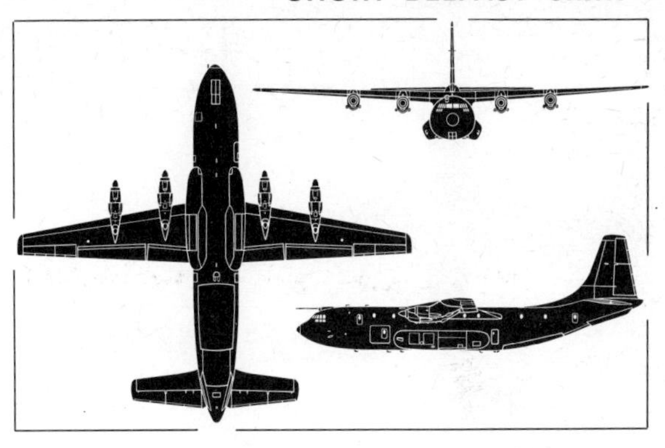

The photograph (above) and general arrangement silhouette (right) depict the standard Belfast C Mk 1 which is currently operated by the RAF's No 53 Squadron

Country of Origin: United Kingdom.
Type: Long-range strategic freighter.
Power Plant: Four 5,730 ehp Rolls-Royce Tyne 101 turboprops.
Performance: Max cruise 352 mph (566 km/h) at 28,000 ft (8 535 m); economical cruise, 336 mph (540 km/h) at 24,000 ft (7 315 m); initial rate of climb, 1,060 ft/min (5,4 m/sec); service ceiling, 30,000 ft (9 145 m); range with max payload, 1,000 mls (1 610 km); range with max fuel, 5,300 mls (8 530 km).
Weights: Basic operating, 127,000 lb (57 600 kg); max payload, 78,000 lb (35 400 kg); max take-off, 230,000 lb (104 300 kg); max landing, 215,000 lb (97 520 kg).
Dimensions: Span, 158 ft 9½ in (48,42 m); length, 136 ft 5 in (41,69 m); height, 47 ft 0 in (14,30 m); wing area, 2,466 sq ft (229 m²).
Accommodation: Flight crew of four, plus air quartermaster. Miscellaneous freight loads carried individually, with alternative provision to carry up to 250 equipped troops.
Status: First and second aircraft first flown on 5 January and 1 May 1964. Production total 10, first delivery in January 1966, final delivery end 1967.

Notes: Built for the RAF to Specification C.203, the Belfast was the outcome of a long series of design studies based on the Bristol Britannia, the RAF version of which was produced at Shorts' Belfast works. Early designs in the series were identified as the S.C.5 Britannic, and an RAF intention to order was indicated early in 1958, with a prospective need for 30. A completely new fuselage was planned, capable of carrying a Blue Streak missile, married to Britannia wings and tail unit and Tyne engines. Following the defence cuts in 1958, the RAF requirement dropped to 10 aircraft and the design was revised during 1959 to the S.C.5/10 Belfast. Flight trials with the first five Belfasts revealed a major drag problem and modifications were introduced during production of the final five, which became the first to enter service, with No 53 Squadron RAF. The other Belfasts were cycled through an extensive modification and overhaul programme before delivery to the same squadron. The modifications, which resulted in a 20 per cent improvement in long-range cruise performance, and enabled the Belfast to carry a 30,000-lb (13 608-kg) payload over 3,985 mls (6 413 km) with full RAF fuel reserves, included wing/fuselage intersection fairings and ventral anti-drag strakes.

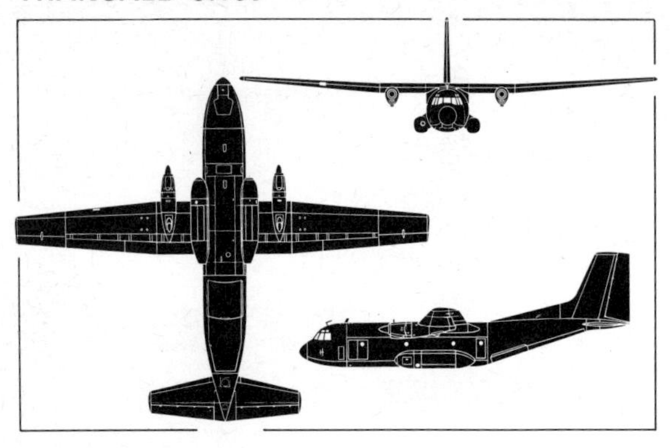

Both the photograph (above) and the general arrangement silhouette (left) illustrate the C-160, the former depicting a C-160F of the 61e Escadre of the Armée de l'Air

Country of Origin: France and Federal Germany.
Type: Medium-range tactical transport.
Power Plant: Two 6,100 ehp Rolls-Royce Tyne RTy 20 Mk 22 turboprops.
Performance: Max speed, 333 mph (536 km/h) at 14,760 ft (4 500 m); max cruise, 306 mph (492 km/h) at 26,250 ft (8 000 m); initial rate of climb, 1,440 ft/min (7,3 m/sec); service ceiling, 27,900 ft (8 500 m); range with max payload, 730 mls (1 175 km); range with 50 per cent payload, 2,832 mls (4 558 km).
Weights: Operating weight empty, 63,815 lb (28 946 kg); max payload, 35,270 lb (16 000 kg); normal take-off, 97,450 lb (44 200 kg); max overload, 108,250 lb (49 100 kg).
Dimensions: Span, 131 ft 3 in (40,0 m); length, 106 ft 3½ in (32,40 m); height, 38 ft 5 in (11,65 m); wing area, 1,723 sq ft (160,1 m²).
Accommodation: Flight crew of four (two pilots, navigator and flight engineer); up to 93 troops, 81 paratroops or 62 stretchers plus four seats or miscellaneous supplies individually loaded.
Status: First of three prototypes flown on 25 February 1963; first of six C-160A pre-production aircraft flown on 21 May 1965; first production C-160D flown on 2 November 1967 and delivered in April 1968; first C-160F flown on 13 April 1967 and delivered in October 1967; first C-160Z flown on 28 February 1969 and delivered in July 1969. Production comprised 110 C-160D for Germany, 50 C-160F for France and nine C-160Z for South Africa. Production ended October 1972.
Notes: The Transall project began in 1959 as a joint Franco-German programme to produce a tactical transport to replace the Nord Noratlas in service with the *Luftwaffe* and a proportion of the aircraft of this type serving with the *Armée de l'Air*. VFW-Fokker became the overall project manager, with additional final assembly lines at MBB (originally HFB) and Aérospatiale (originally Nord). The Transall is operated by three *Luftwaffe* transport groups, *Lufttransportgeschwader* 61, 62 and 63, three *Armée de l'Air* squadrons of the 61e *Escadre de Transport* and No 28 Squadron of the SAAF at Waterkloof. During 1971, Germany transferred to the Turkish Air Force 20 Transalls which were surplus to *Luftwaffe* requirements. During the course of 1973, four Transalls were modified by SOGERMA under the designation C-160P for use by the French Postal Service.

BOEING KC-135 STRATOTANKER and C-137

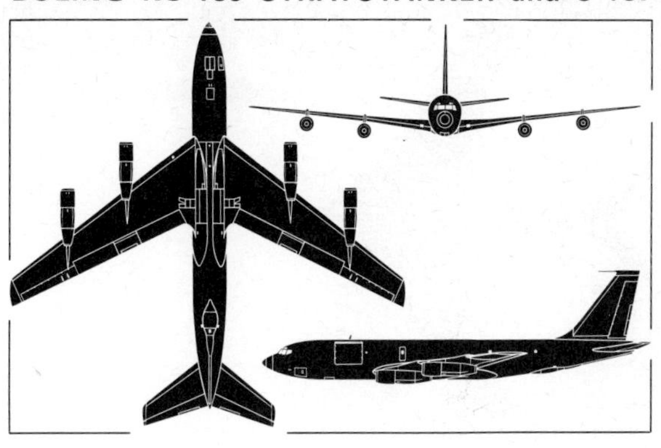

The photograph above depicts a C-135F tanker of the Armée de l'Air and the general arrangement silhouette (right) illustrates the KC-135A tanker and transport of the USAF

Country of Origin: USA.
Type: In-flight refuelling tanker and transport.
Power Plant: Four 13,750 lb st (6 237 kgp) (with water injection) Pratt & Whitney J57-P-59W turbojets.
Performance: Max speed, 585 mph (941 km/h) at 30,000 ft (9 144 m); cruising speed, 530 mph (853 km/h); initial rate of climb, 2,000 ft/min (6 096 m/sec); service ceiling, 50,000 ft (15 240 m); range, 1,150 mls (1 850 km) carrying 120,000 lb (54 430 kg) of transfer fuel.
Weights: Empty, 98,446 lb (44 654 kg); fuel load, 120,000 lb (54 430 kg); max take-off, 297,000 lb (134 717 kg).
Dimensions: Span, 130 ft 10 in (39,87 m); length, 136 ft 3 in (41,53 m); height, 38 ft 4 in (11,68 m); wing area, 2,433 sq ft (225,7 m²).
Accommodation: Flight crew of six; optional provision for up to 160 troops or 83,000 lb (37 648 kg) of cargo.
Status: First flown 31 August 1956. Entered service 18 June 1957. Production of 732 KC-135A completed in January 1965, plus 17 EC-135C, 4 RC-135A, 10 RC-135B, 15 C-135A, 30 C-135B and 12 C-135F (for France).

Notes: The Stratotanker was selected by the USAF as its first jet-powered in-flight refuelling tanker in October 1954, after Boeing had demonstrated the concept with the basic Dash-80 prototype jet transport. The initial KC-135A variant remained in service as the principal SAC and TAC tanker in 1973, in addition to 12 similar C-135Fs (with probe-and-drogue refuelling) in service in France. Some specially-modified versions were redesignated KC-135Q, R and T. For service as airborne command posts the USAF bought 17 EC-135Cs (with TF33 engines) and later modification programmes produced the EC-135G, H, J, K, L, N and P variants. The RC-135A and B were bought for photo-reconnaissance duties, the RC-135C, D, E, M, S and U being later modifications. The C-135A and C-135B (TF33 engines) were built as troop and cargo transports for MAC, having no refuelling boom, and some later became VC-135B and WC-135B for weather reconnaissance. Versions of the larger commercial Boeing 707 (see *Civil* volume) have been bought for military service, by the USAF (as VC-137A, B and C), by the Luftwaffe and Portuguese Air Force (as personnel transports) and by the Canadian Defence Force (CC-137) and Iranian Imperial Air Force (as tanker/transports).

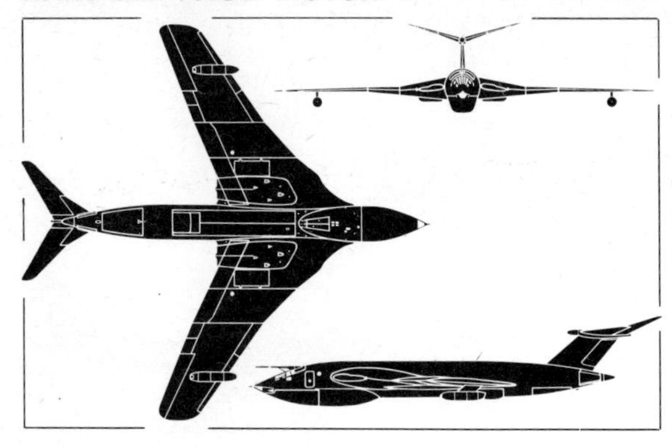

The photograph above depicts a Victor SR Mk 2 of No 543 Sqdn, RAF, and the general arrangement silhouette (left) illustrates the Victor K Mk 1A tanker

Country of Origin: United Kingdom.
Type: Flight refuelling tanker.
Power Plant: Four 20,600 lb st (9 345 kg) Rolls-Royce Conway 201 (R Co 17) turbojets.
Performance: Max speed, approximately 650 mph (1 045 km/h) above 40,000 ft (12 192 m); service ceiling, over 55,000 ft (16 764 m); unrefuelled range, over 2,000 mls (3 220 km).
Weights: Gross weight, 223,000 lb (101 150 kg).
Dimensions: Span, 120 ft 0 in (36,58 m); length, 114 ft 11 in (35,02 m); height, 30 ft 1½ in (9,18 m); wing area, 2,597 sq ft (240·27 m²).
Accommodation: Flight crew of four.
Status: Prototype Victor bomber first flown 24 December 1952; first production B Mk 1 flown on 1 February 1956; first B Mk 2 flown on 20 February 1959; first B(K) Mk 1A tanker conversion flown on 28 April 1965; first K Mk 2 tanker conversion flown on 1 March 1972. Production totals, 50 B Mk 1, 30 B Mk 2, completed in 1962.
Notes: In its original guise as a strategic bomber, the Victor was part of Britain's V-Force which provided the nation's main deterrent in the 'fifties and 'sixties. Variants were the initial production B Mk 1 with 11,050 lb st (5 012 kgp) Armstrong Siddeley Sapphire 202/207 turbojets: B Mk 1A conversions with tail warning radar and other modifications; the K Mk 1 and K Mk 1A refuelling tanker conversions; the B Mk 2 with extra wing span, Conway 103 (RCo 11) engines and increased weights; the B Mk 2R conversions with uprated Conway 201s; the SR Mk 2 conversions for photographic reconnaissance and the K Mk 2 tankers. All conversion programmes but that of the K Mk 2 were handled by Handley Page before its failure in 1970; the K Mk 2 programme covering 20 aircraft, was taken over by Hawker Siddeley at Woodford, and these aircraft, delivered to the RAF in 1973/74, are progressively supplementing and will eventually replace the Victor K Mk 1A tanker to provide the RAF's sole aerial refuelling force. Pods under each wing contain retractable hose and drogue equipment, allowing two aircraft to refuel simultaneously when required. In addition to the tankers, the RAF operated one squadron of Victor SR Mk 2s in 1973, but these are to be superseded by specially-equipped Vulcans. The SR Mk 2 carries containers which, hung in the weapons bay, can accommodate a wide variety of camera combinations.

AERITALIA-AERMACCHI AM.3C

Both photograph (above) and general arrangement silhouette (right) depict the production AM.3C, the former illustrating one of the first aircraft for South Africa

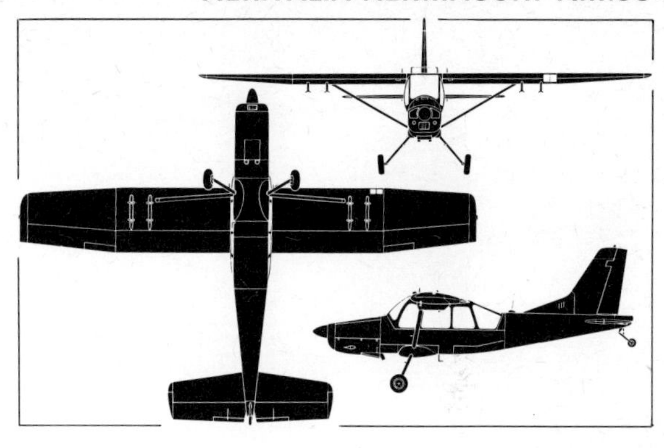

Country of Origin: Italy.
Type: Battlefield surveillance and forward air control aircraft.
Power Plant: One Piaggio-Lycoming GSO-480-B1B6 six-cylinder horizontally-opposed engine.
Performance (at normal loaded weight): Max speed, 173 mph (278 km/h) at 8,000 ft (2 440 m); max cruise, 153 mph (246 km/h) at 8,000 ft (2 440 m); initial rate of climb, 1,378 ft/min (7,0 m/sec); service ceiling, 27,560 ft (8 400 m); max range, 615 mls (990 km) with 30-min fuel reserve; max endurance 4·5 hrs.
Weights: Empty equipped, 2,548 lb (1 156 kg); normal loaded (2 crew) 3,307 lb (1 500 kg); max take-off (with underwing stores), 3,858 lb (1 750 kg).
Dimensions: Span, 41 ft $5\frac{1}{2}$ in (12,64 m); length, 29 ft $5\frac{1}{2}$ in (8,98 m); height, 8 ft 11 in (2,72 m); wing area, 219·15 sq ft (20,36 m^2).
Accommodation: Three seats in tandem, with dual controls and rear seat removable to allow stores to be carried; alternative layouts with pilot only and one stretcher or cargo.
Armament: Four underwing stores stations, stressed for (inner) 375 lb (170 kg) each and (outer), 200 lb (91 kg) each. Armament options include four LAU-32A or MATRA 181 rocket launchers, 12 80-mm SURA rockets, two Nord AS.11 missiles, two 7·62 mm Minigun pods, two Matra 7·62 mm twin machine gun pods or bombs, in various combinations.
Status: First and second prototypes (with Continental GTSIO-520-C engines) first flown on 12 May 1967 and 22 August 1968. Both re-engined with Lycoming GSO-480 and flown as AM.3C during 1969. First production AM.3C flown late 1972, from batch of 100.
Notes: Original design was Aermacchi MB.335, developed jointly by Aermacchi and Aerfer as AM.3 and based on the wing of the original Aermacchi–Lockheed AL.60. After extended evaluation, the AM.3C lost out to the SIAI–Marchetti SM.1019 (see page 153) as the Italian Army's new AOP, but 20 were ordered by the Army for special duties, from an initial batch of 100 laid down by Aeritalia. Other orders were placed by South Africa (for 40) and Rwanda (for 3). Initial production deliveries to the South African Air Force began in the spring of 1973, and it is anticipated that assembly and part-manufacture will be undertaken in South Africa by the Atlas Aircraft Corporation.

BEECH U-21 UTE/QUEEN AIR

Both the photograph (above) and general arrangement silhouette (left) illustrate the U-21A which is extensively used by the US Army as a staff transport and utility aircraft

Country of Origin: USA.
Type: Special reconnaissance, transport and utility aircraft.
Power Plant: Two (U-21A, D, E and G) 550 shp Pratt & Whitney T74-CP-700 or (U-21F) 680 shp PT6A-28 or (RU-21B and C) 750 shp T74-CP-702 turboprops.
Performance: (U-21A and U-21G) Max speed, 249 mph (401 km/h) at 11,000 ft (3 350 m); max cruising speed, 245 mph (395 km/h) at 10,000 ft (3 050 m); economical cruising speed, 205 mph (328 km/h) at 10,000 ft (3 050 m); initial rate of climb, 2,000 ft/min (10,1 m/sec); service ceiling, 25,500 ft (7 775 m); range with max fuel, 1,676 mls (2 697 km); range with max payload, 1,167 mls (1 878 km).
Weights: Empty equipped (U-21A and U-21G) 5,464 lb (2 478 kg); (U-21F) 6,728 lb (3 051 kg); max payload (U-21A and U-21G) 4,186 lb (1 898 kg); max take-off weight (U-21A, D, E, G), 9,650 lb (4 377 kg); (RU-21B and C) 10,900 lb (4 944 kg) (U-21F) 11,500 lb (5 216 kg).
Dimensions: Span, 45 ft 10½ in (13,98 m); length (except U-21F) 35 ft 6 in (10,82 m); length (U-21F), 39 ft 11¼ in (12,17 m); height (except U-21F) 14 ft 2½ in (4,33 m); height (U-21F), 15 ft 4¼ in (4,68 m); wing area, 279·7 sq ft (25,98 m²).
Accommodation: Crew of two with dual control; up to ten combat-equipped troops, or three stretchers plus three seated; or six passengers, or cargo, or special equipment operators.
Status: Prototype (NU-8F) first flown in May 1963. First production U-21A flown in March 1967. Deliveries to US Army began on 16 May 1967. Approximately 184 delivered, in various models, by 1971.
Notes: The U-21 was derived from the commercial Queen Air (used by the US Army as the U-8) through installation of turboprop engines, producing the NU-8F prototype in 1963. Deliveries have included 124 U-21As and 17 similar U-21Gs (with updated equipment) for staff and utility transport duties, plus a series of special electronic reconnaissance versions. Designated RU-21A, RU-21B, RU-21C, RU-21D and RU-21E according to equipment and power plant, these carried large aerials above and below the wings and tailplane and were used extensively over Vietnam. The five U-21Fs are commercial model King Air A100s acquired by the Army in 1971 and a single VC-6A is a commercial King Air C90 acquired by the USAF as a VIP transport.

CESSNA O-1 BIRD DOG

The photograph above depicts an O-1A Bird Dog of Japan's Ground Self-Defence Force and the general arrangement silhouette (right) is applicable to all current versions

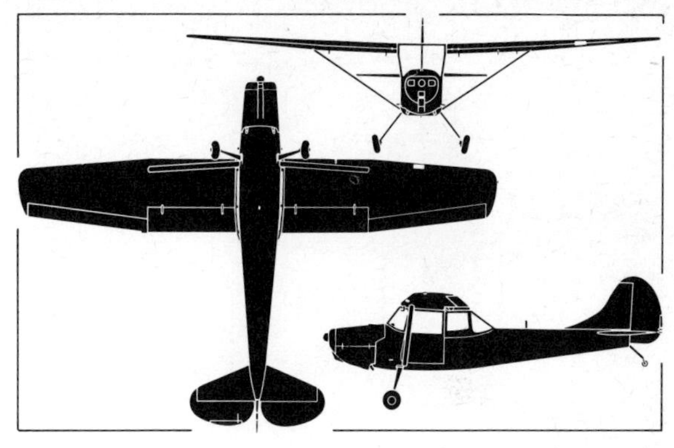

Country of Origin: USA.
Type: Observation and forward air control aircraft.
Power Plant: One 213 hp Continental O-470-11 six-cylinder horizontally-opposed engine.
Performance (O-1E): Max speed, 115 mph (184 km/h); economical cruise (29 per cent power) 104 mph (166,5 km/h) at 5,000 ft (1 525 m); initial rate of climb, 1,150 ft/min (5,8 m/sec); service ceiling, 18,500 ft (5 640 m); range, 530 mls (848 km).
Weights: Empty, 1,614 lb (680 kg); max take-off (O-1E), 2,400 lb (1 090 kg), (O-1F) 2,800 lb (1 270 kg).
Dimensions: Span, 36 ft 0 in (10,9 m); length, 25 ft 10 in (7,89 m); height, 7 ft 4 in (2,23 m); wing area, 174 sq ft (16,16 m²).
Accommodation: Two seats in tandem, optional dual control.
Status: Prototype (Model 305) first flown December 1949; first production L-19A (O-1A) rolled out November 1950. First deliveries of TL-19D (TO-1D), second half of 1956. First deliveries of L-19E (O-1E), November 1956. Production totals, O-1A, 2,499; L-19A-IT, 66; O-1B, 62; XL-19B, one; XL-19C, two; TO-1D, 307; O-1E, 494, grand total, 3,431. Approximately 100 also built by Fuji in Japan.

Notes: The Cessna Model 305, built as a private venture, won a competition in 1950 to become the new standard liaison and observation monoplane for the US Army. Originally designated L-19, the series was redesignated O-1 in 1962. Production versions were the L-19/O-1A; the OE-1/O-1B which was similar for the US Marine Corps; the L-19A-IT instrument trainer and similar TL-19D/TO-1D, and the L-19E/O-1E with new equipment and higher gross weight. For special duties as forward air control aircraft in Vietnam, the USAF took over many of the Army's Bird Dogs, the TO-1D and O-1A becoming O-1F and O-1G respectively with USAF equipment fitted. Other major users of the Bird Dog included Japan (where it was built under licence), France, Canada, Italy and Pakistan. During 1972, new Bird Dogs were being produced by a Pakistan Army workshop using a combination of US spares and indigenous new components, and as a result of tests with two re-manufactured Bird Dogs re-engined with a turboprop, this type is now being manufactured in Italy as the SM 1019A (see page 153) for the Italian Army. The Bird Dog has been manufactured in larger numbers than any other post-WWII air observation post and forward air control aircraft.

CESSNA O-2

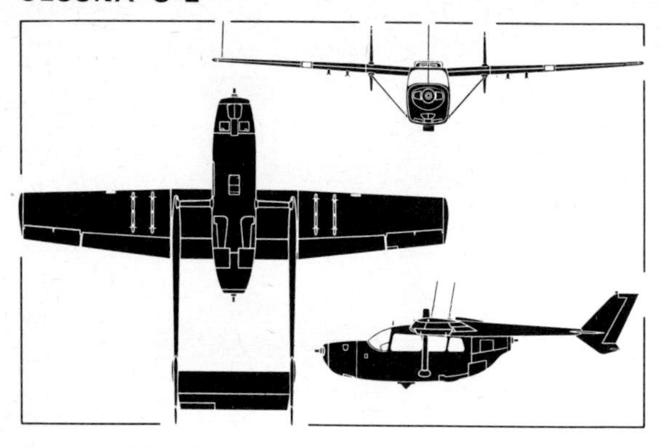

The photograph above illustrates the O-2B for psychological warfare and the general arrangement silhouette (left) depicts the O-2A forward air control version

Country of Origin: USA.
Type: Forward air control and observation aircraft.
Power Plant: Two 210 hp Continental IO-360-C/D six-cylinder horizontally-opposed engines.
Performance: Max speed, 199 mph (320 km/h) at sea level; cruising speed, 144 mph (232 km/h) at 10,000 ft (3 050 m); initial rate of climb, 1,100 ft/min (5,6 m/sec); service ceiling, 18,000 ft (5 490 m); range, 1,060 mls (1 705 km).
Weights: Empty, 2,848 lb (1 291 kg); normal take-off, 4,630 lb (2 100 kg); max overload, 5,400 lb (2 450 kg).
Dimensions: Span, 38 ft 2 in (11,63 m); length, 29 ft 9 in (9,07 m); height, 9 ft 4 in (2,84 m); wing area, 202·5 sq ft (18,81 m^2).
Accommodation: Pilot and observer, side-by-side; two passengers optional in cabin.
Armament (O-2A only): Four wing strong points permit carriage of 7·62 mm Mini-gun pods, rockets, flares and similar ordnance loads.
Status: Deliveries of O-2B began 31 March 1967; deliveries of O-2A began April 1967. Total of 510 delivered by end 1970.
Notes: This is the military version of the distinctive "push-pull" Cessna Skymaster (see *Civil* volume), first ordered by the USAF in December 1966 to provide a specialised Forward Air Control aircraft for use in Vietnam in place of the O-1 Bird Dog (see page 139). The O-2A variant (346 delivered) introduced wing strong points to carry marker flares, rockets or gunpods and a special electronic installation, and had a higher gross weight than the commercial version. The O-2B was closer to its civil counterpart, the special features being a dispenser for leaflet dropping and loud speakers for air-to-ground broadcasting; these aircraft were used for psychological warfare in Vietnam. Twelve O-2As were acquired by the Iranian Imperial Air Force in 1970 for training, liaison and observation duties. A military version of the Skymaster has also been developed in France by Reims Aviation, which builds Cessna aircraft under licence. First flown on 26 May 1970, this variant is known as the FTMA Milirole and like the O-2A has four wing strongpoints. It has new high-lift trailing edge flaps which give a 40 per cent improvement in field performance, and is designed to carry, in addition to two crew, four passengers or two stretchers or cargo. Small numbers of the commercial Skymaster have been delivered to the Ecuadorian Air Force and the Venezuelan Navy.

CESSNA U-17 AND SKYWAGON

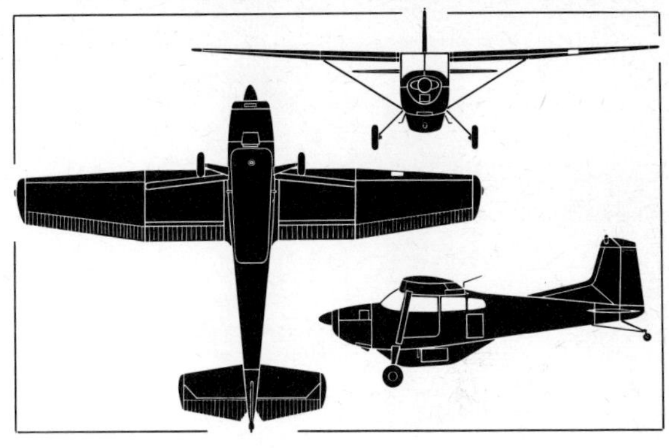

Both photograph (above) and general arrangement silhouette (right) illustrate the U-17 utility monoplane, the latter depicting the glass-fibre Cargo-Pack fitted

Country of Origin: USA.
Type: General utility monoplane.
Power Plant: One 300 hp Continental IO-520-D six-cylinder horizontally-opposed engine.
Performance: Max speed, 178 mph (286 km/h) at sea level; max cruise, 169 mph (272 km/h) at 7,500 ft (2 285 m); economical cruise, 129 mph (208 km/h) at 10,000 ft (3 050 m); initial rate of climb, 1,010 ft/min (5,1·m/sec); service ceiling, 17,150 ft (5 230 m); range, 660 mls (1 062 km) at max cruise (75 per cent power), standard tankage; range, 1,075 mls (1 730 km) at economical cruise, long-range tanks.
Weights: Empty equipped, 1,585 lb (719 km); max take-off, 3,350 lb (1 519 kg).
Dimensions: Span 35 ft 10 in (10,92 m); length, 25 ft 9 in (7,85 m); height, 7 ft 9 in (2,36 m); wing area, 174 sq ft (16,16 m²).
Accommodation: Pilot and optional arrangements for up to five passengers or cargo in lieu of seats.
Status: Prototype (Model 185) flown in July 1960. First U-17A delivered in 1963. Over 300 built.
Notes: The Cessna Model 185 Skywagon is one of a family of Cessna utility aircraft which serve in both military and civil guise. Essentially a multi-purpose derivative of the earlier Model 180 with a strengthened structure to render it more suitable for cargo hauling and other heavy duty tasks. Among its special features are removable passenger seats to allow the whole cabin to be used for cargo-carrying; ability to operate on skis, floats or as an amphibian, and provision for a glass-fibre Cargo-Pack to be carried under the fuselage, with a capacity of 300 lb (136 kg). In 1963, the USAF selected the Model 185 for supply to foreign nations receiving US aid through MAP, and the designation U-17 was allocated, although no examples were acquired for use by the USAF itself. Procurement totals included 169 U-17A and 136 U-17B, plus some U-17Cs with less powerful O-470-L engines, and nations supplied with U-17s included Bolivia, Costa Rica, Laos and South Vietnam. The South African Air Force acquired similar Cessna 185As and 185Ds on direct contract, for use by No 41 Squadron at Grand Central, Johannesburg. Some Model 185s have been purchased on direct contract by other air forces. Examples of other high wing Cessnas, as described in the *Civil* volume, have also been acquired for military use by various nations.

DE HAVILLAND CANADA DHC-2 BEAVER

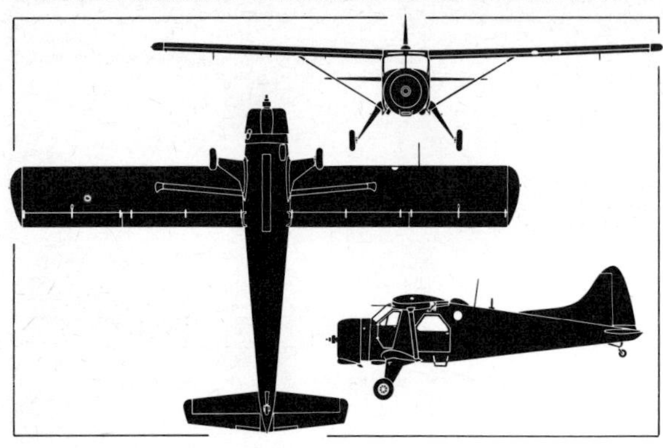

The standard production DHC-2 Beaver is illustrated by both the photograph (above) and the general arrangement silhouette (left), the former depicting a Finnish Air Force aircraft

Country of Origin: Canada.
Type: Light utility transport.
Power Plant: One 450 hp Pratt & Whitney R-985-AN Wasp Junior radial.
Performance: Max speed, 140 mph (225 km/h); max cruising speed, 135 mph (217 km/h) at sea level; economical cruising speed, 125 mph (201 km/h); initial rate of climb, 1,020 ft/min (5,2 m/sec); service ceiling, 18,000 ft (5 490 m); range with max payload, 483 mls (777 km); range with max fuel, 778 mls (1 252 km).
Weights: Basic operating, 3,000 lb (1 361 kg); max take-off, 5,100 lb (2 313 kg).
Dimensions: Span, 48 ft 0 in (14,64 m); length, 30 ft 4 in (9,24 m); height, 9 ft 0 in (2,75 m); wing area, 250 sq ft (23·2 m²).
Accommodation: Pilot and up to seven passengers (provision for dual control if second pilot carried). Floor stressed for freight with lightweight collapsible bush seats interchangeable with cargo attachments.
Status: Prototype first flown August 1947. Certification (civil version) on 12 March 1948. Production deliveries began 1954; total production over 1,657 including 968 for US Army as U-6A (originally L-20A). Single Mk II (Leonides engine) first flown in 1953. DHC-2 Mk III Turbo-Beaver first flown on 30 December, 1963. Volpar/DHC Model 4000 first flown April 1972.
Notes: The Beaver was the second design of the Canadian de Havilland company, and the first of the company's series of "bush" aircraft. It was produced in both civil and military guise, the largest single user being the US Army, which selected the Beaver in 1951. Initially with the description L-20A, later changed to U-6A, 968 Beavers were supplied to the US services, about one quarter going to the USAF and a few to the USN. The Beaver was selected for service with the British Army in 1960, 46 being acquired as Beaver AL Mk 1s. One experimental Beaver II had a Leonides engine; the Beaver III, of which a few examples were produced, had a 579 ehp Pratt & Whitney PT6A-6 turboprop, and in 1972 Volpar produced an alternative turboprop conversion scheme using the 715 ehp Garrett AiResearch TPE-331 engine. Beavers entered service in about 70 countries around the world and many air forces acquired examples for use on general transport, communications and liaison duties, current operators including Argentina, Austria, Chile, Finland, UK, Kenya, S. Korea, Oman, Netherlands, Peru, USA and S. Vietnam.

DORNIER DO 27

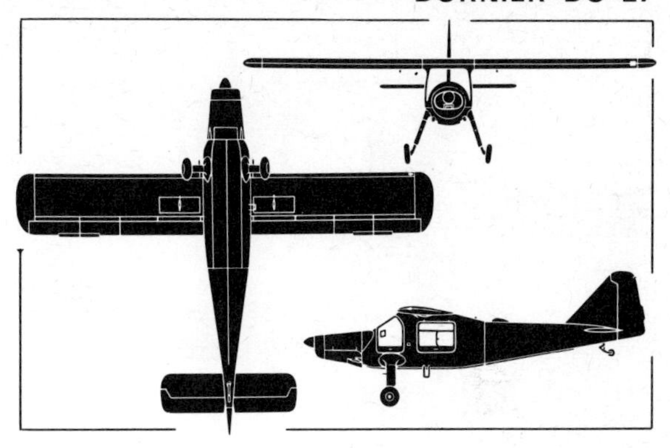

Both the photograph (above) and the general arrangement silhouette (right) illustrate the Do 27A, the former depicting a CASA-built example (C-127) of the Spanish Air Force

Country of Origin: Federal Germany.
Type: General utility aircraft and AOP.
Power Plant: (A-4) One 270 hp Lycoming GO-480-B1A6 or (H-2) 340 hp Lycoming GSO-480-B1B6 six-cylinder horizontally-opposed engine.
Performance (A-4): Max speed, 141 mph (227 km/h) at 3,280 ft (1 000 m); cruising speed (75 per cent power) 130 mph (210 km/h); cruising speed (65 per cent power) 109 mph (175 km/h); initial rate of climb, 650 ft/min (3,3 m/sec); service ceiling, 10,825 ft (3 300 m); range with max fuel (60 per cent power), at 3,280 ft (1 000 m), 685 mls (1 100 km).
Performance (H-2): Max speed, 152 mph (245 km/h) at 3,280 ft (1 000 m); cruising speed (75 per cent power) 132 mph (212 km/h); cruising speed (65 per cent power) 112 mph (180 km/h); initial rate of climb, 965 ft/min (4,9 m/sec); service ceiling, 22,000 ft (6 700 m); range with max fuel, 845 mls (1 360 km).
Weights: Empty equipped, 2,365 lb (1 072 kg); max take-off, 4,070 lb (1 850 kg).
Dimensions: Span, 39 ft 4½ in (12,0 m); length, 31 ft 6 in (9,6 m); height, 9 ft 2 in (2,8 m); wing area, 208·8 sq ft (19,4 m²).

Accommodation: Two seats side-by-side in cockpit and provision for four to six seats in cabin aft.
Status: First of three prototypes flown on 27 June 1955. First production Do 27A flown on 17 October 1956. Production totals: Do 27A-1, 195; Do 27A-3, 75; Do 27A-4, 52; Do 27B-1, 88; Do 27B-3, 18; Do 27H-1, 1; Do 27H-2, 12; Do 27Q-3, 1; Do 27Q-4, 34; Do 27Q-5, 1; total 571, completed in 1966. In addition, 50 C-127 built by CASA, first example flown on 3 December 1959.
Notes: First post-war design by Professor Claude Dornier was the Do 25, designed and built in Spain to meet a Spanish Air Force requirement for a general purpose light aircraft. Two prototypes were built, the first flight being on 25 June 1954, with a 150 hp ENMA Tigre G-IVB engine, and CASA subsequently built 50 of the production Do 27A version as C-127s for the Spanish Air Force (nomenclature L.9). Principal German production, following three Do 27 prototypes, was for the Luftwaffe, which received all the A versions listed above (with only detail differences) and the B versions, which had dual control—a total of 428. Other variants listed were for the civil market, but a few were also acquired by other air forces.

DORNIER DO 28D SKYSERVANT

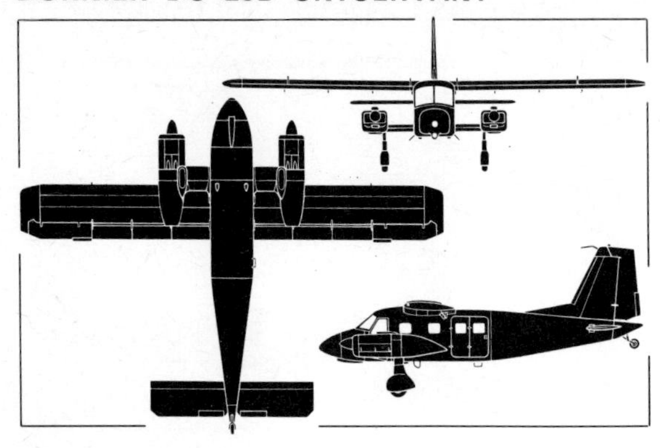

The photograph above depicts a Do 28D-1 Skyservant of Federal Germany's Marineflieger but this type is most widely used by the Luftwaffe in the utility/communications rôle

Country of Origin: Federal Germany.
Type: Light STOL utility aircraft.
Power Plant: Two 380 hp Lycoming IGSO-540-A1E six-cylinder horizontally-opposed engines.
Performance: Max speed, 199 mph (320 km/h) at 10,000 ft (3 050 m); 75 per cent power cruise, 178 mph (286 km/h) at 10,000 ft (3 050 m); 50 per cent power cruise, 143 mph (230 km/h), initial rate of climb, 1,180 ft/min (6 m/sec); service ceiling, 24,280 ft (7 400 m); range with max fuel, no reserves, 1,143 mls (1 837 km).
Weights: Empty, 4,775 lb (2 166 kg); max take-off and landing, 8,050 lb (3 650 kg).
Dimensions: Span, 50 ft 10¼ in (15,50 m); length, 38 ft 0¾ in (11,60 m); height, 13 ft 2 in (4,01 m); wing area, 308·9 sq ft (28,70 m²).
Accommodation: Flight crew of one or two and up to 13 passengers, or five stretchers and five seats, or miscellaneous freight.
Status: Prototype first flown on 23 February 1966; civil certification (Do 28D) on 24 February 1967 and (Do 28D-1) 6 November 1967. Deliveries began in mid-1967; production rate 6–8 per month in 1973.

Notes: Although resembling the Do 28 (see *Civil* volume) in general configuration and size, the Do 28D Skyservant was a completely new design with better performance and payload. The first seven aircraft were Do 28Ds, with a slightly lesser wing span and lower gross weight, the main production variant being the Do 28D-1 as described above. Although offered on the civil market, the Skyservant found its largest market among the German armed forces, with orders for 101 for the *Luftwaffe* and 20 for the *Marineflieger* in the general utility/communications role, and four for the *Flugbereitschaft*, a unit of the *Luftwaffe* providing VIP transport services. Each of 17 *Luftwaffe Geschwader* has four Skyservants assigned for its own use. The Somalia Police Air Wing has two Skyservants, some serve with the Zambian Air Force, four were ordered during 1973 for the Nigerian Air Force for IFR training, VIP transportation, aeromedical tasks and freight transportation, and the Turkish Geographical Institute has five in civil guise. The Skyservant may be fitted with wheel–ski gear or floats, and the current Do 28D-2 is being offered with special equipment for environmental control, oceanological reconnaissance and scientific environmental surveillance, as well as for other specialised rôles.

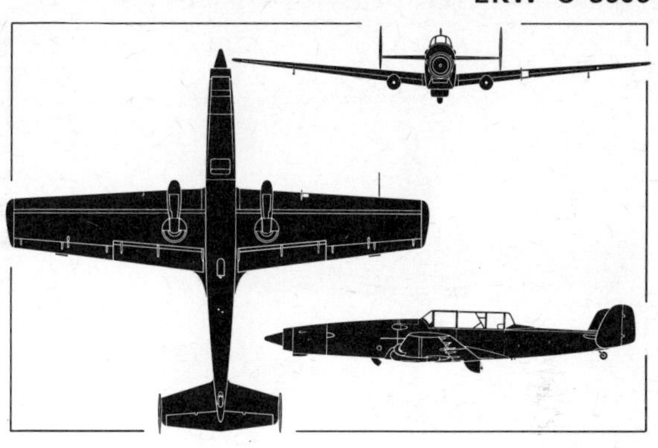

The C-3605, illustrated by the photograph above and the general arrangement silhouette (right) is a conversion of the piston-engined C-3603

Country of Origin: Switzerland.
Type: Target towing aircraft.
Power Plant: One 1,150 ehp Avco-Lycoming T5307A (LTC 1) turboprop.
Performance: Max speed, 268 mph (432 km/h) at 10,000 ft (3 050 m); max cruise, 261 mph (420 km/h) at 10,000 ft (3 050 m); economical cruise, 217 mph (350 km/h) at 20,000 ft (6 100 m); initial rate of climb, 2,470 ft/min (12,54 m/sec); service ceiling 32,810 ft (10 000 m); range (internal fuel only) 609 mls (980 km).
Weights: Empty equipped, 5,806 lb (2 634 kg); normal take-off, 7,275 lb (3 300 kg); max overload, 8,192 lb (3 716 kg).
Dimensions: Span, 45 ft 1 in (13,74 m); length, 39 ft 5¾ in (12,03 m); height, 13 ft 3½ in (4,05 m); wing area, 308·9 sq ft (28,7 m²).
Accommodation: Pilot and observer/winch operator in tandem.
Status: Prototype (C-3603) first flown on 23 November 1941. Total of 150 delivered to Swiss Air Force by 1948. Prototype conversion to C-3605 flown on 19 August 1968. Conversion of 23 C-3605s completed by end-1972.
Notes: In service with the Swiss Air Force since 1971 as target tugs, the C-3605s are converted from C-3603 reconnaissance bombers, following successful testing of a prototype conversion in 1968/69. The modification, primarily, concerns installation of a Lycoming turboprop in place of the original Hispano-Suiza piston engine, and fitting a winch and target stowage in the fuselage beneath the rear seat. A third, central fin was added after initial testing of the C-3605 prototype with the standard twin fins and rudders of the original design. The C-3605 is not a re-manufactured aircraft, the aim having been to change the engine with minimum alteration to existing structure, an additional section being introduced ahead of the forward fuselage mainframe in order to maintain the existing CG position after installation of the lower-weight turboprop. Both the C-3603 and C-3605 derive from the C-3601 and C-3602 prototypes which were ordered from the EKW concern at Emmen in 1936 and first flew in 1939. A production order was placed in 1940, leading to the appearance of the first C-3603 in November 1941. In addition to 144 C-3603s, the EKW built 10 C-3604s, with more powerful engine and other improvements. Six more C-3603s were assembled from spares in 1947/48 but none now remain in service in the original rôle.

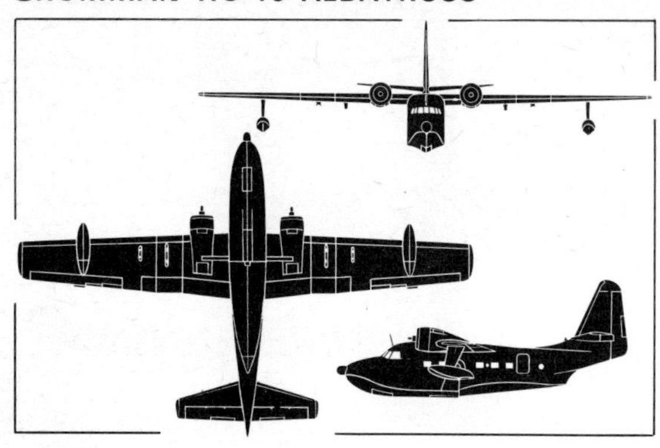

Both photograph (above) and general arrangement silhouette (left) illustrate the HU-16B version of the Albatross, the former depicting an example operated by the Spanish Air Force

Country of Origin: USA.
Type: Air-sea rescue and reconnaissance amphibian.
Power Plant: Two 1,425 hp Wright R-1820-76A or -76B or (CSR-110) 1,525 hp R-1820-82 radial engines.
Performance (HU-16B): Max speed, 236 mph (379 km/h); max cruising speed, 224 mph (362 km/h); long range cruise, 124 mph (200 km/h); initial rate of climb, 1,450 ft/min (7,3 m/sec); service ceiling, 21,500 ft (6 550 m); range with max fuel, 2,850 mls (4 587 km) with 30 min plus 5 per cent reserve; max endurance (with external tanks) 22·9 hrs.
Weights (HU-16B): Empty equipped, 22,883 lb (10 380 kg); normal take-off, 30,353 lb (13 768 kg); max overload, 37,500 lb (17 010 kg).
Dimensions: Span (HU-16A) 80 ft 0 in (24,38 m) (HU-16B) 96 ft 8 in (29,46 m); length, (HU-16A) 60 ft 8 in (18,49 m) (HU-16B) 62 ft 10 in (19,18 m); height (HU-16A) 24 ft 10 in (7,57 m), (HU-16B) 25 ft 10 in (7,87 m); wing area (HU-16A) 833 sq ft (77,39 m²), (HU-16B) 1,035 sq ft (96,2 m²).
Accommodation: Crew of five or six and up to 12 stretchers (for rescue duty) or up to 22 seats (for transport rôle). Crew of four in ASW rôle.
Armament (ASW versions only): Four Mk 43 torpedoes or two Mk 43 torpedoes and two Mk 54-2 or Mk 101 depth charges.
Status: Prototype (XJR2F-1) first flown 24 October 1947; first SA-16B conversions flown 16 January 1956. Total production about 460, completed by 1961.
Notes: Origin of the Albatross was a US Navy requirement for a general purpose utility amphibian, the prototype being designated XJR2F-1 in this category. Production models for the US Navy were UF-1, UF-1L (for Antarctic operation) and UF-1T (trainers), later being redesignated HU-16C, LU-16C and TU-16C. The USAF bought 305 SA-16As (later HU-16As) for air-sea rescue duties, most being later converted to SA-16B (later HU-16B) with bigger wing span and taller tail, as indicated above. Navy aircraft similarly modified became UF-2 (later HU-16D) and the US Coast Guard bought 34 UF-1G (UF-2G when modified, then re-designated HU-16E). Through MAP and other programmes, the Albatross was supplied to some 12 other nations, and a special ASW version was supplied to Norway and Spain. The 16 Norwegian aircraft were later passed on to Greece and Spain.

HAL HAOP-27 KRISHAK

The Krishak, illustrated by the photograph (above) and the general arrangement silhouette (right), is currently the standard Indian Army AOP and light utility aircraft

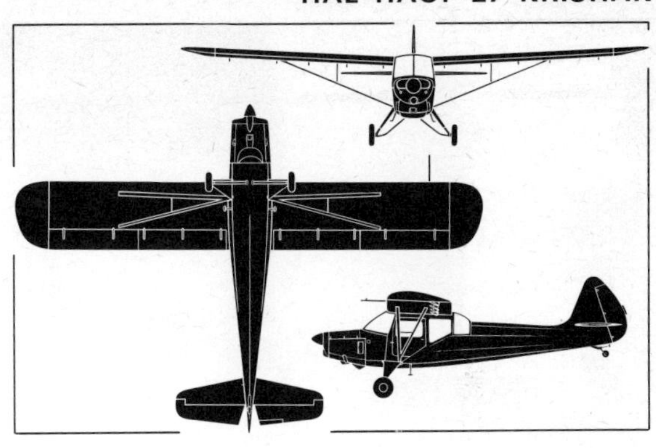

Country of Origin: India.
Type: Air observation post and light utility aircraft.
Power Plant: One 225 hp Continental O-470-J six-cylinder horizontally-opposed air-cooled engine.
Performance: Max speed, 130 mph (209 km/h) at sea level, 116 mph (187 km/h) at 5,000 ft (1 500 m); initial rate of climb, 900 ft/min (4,6 m/sec); service ceiling, 19,500 ft (5 940 m); range, 500 mls (805 km) with auxiliary tankage fitted.
Weights: Empty, 1,970 lb (894 kg); max take-off, 2,800 lb (1 270 kg).
Dimensions: Span, 37 ft 6 in (11,43 m); length, 27 ft 7 in (8,41 m); height, 7 ft 9 in (2,36 m); wing area, 200 sq ft (18,58 m²).
Accommodation: Two seats side-by-side and optional third seat behind.
Status: First and second prototypes (Krishak Mk I) flown in November 1959 and November 1960 respectively. Prototype Krishak Mk II flown in 1965. Production of 63 primarily for Indian Army completed in 1969.
Notes: The Bangalore Division of Hindustan Aeronautics Ltd. began development of a two-seat ultra-light high wing monoplane, the HUL-26, in 1958, completing a prototype in time for a first flight to be made on 28 September 1958. This type entered production as the Pushpak, for flying club use, and it was taken as the basis for development of a slightly larger multi-purpose variant. Two prototypes were built as four-seaters, but this version did not go into production. Instead, a three-seat Mk II was evolved to meet Indian Army requirements for an air observation post and liaison monoplane. The prototype of this Mk II version flew in 1965 and deliveries to Army Air OP Squadrons and Flights as replacements for Auster AOP Mk 9s, were completed in 1969 and about 50 were in service in 1973. Stressed to a limiting load factor of 3·8*g*, the HAOP-27 employs an essentially similar structure to that of the two-seat Pushpak, with fabric-covered metal wings and a welded steel-tube fuselage also covered by fabric. Fuel is carried in two wing tanks, with provision for an auxiliary tank under the fuselage. A stretcher can be accommodated in the rear of the cabin, in place of the third seat. As a result of a policy decision taken in 1973, all Indian Army air observation post and forward air control duties are to be undertaken by helicopters, and from 1974 the Krishak was being progressively replaced by the HAL-built SA 315B Cheetah (Lama).

HELIO AU-24A STALLION

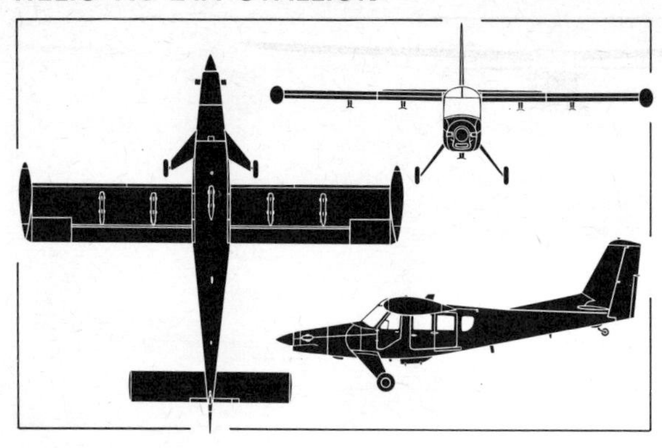

The AU-24A Stallion, illustrated by the photograph (above) and the general arrangement silhouette (left) is currently serving with the Khmer Air Force as a "mini-gunship"

Country of Origin: USA.
Type: Armed light utility aircraft.
Power Plant: One 715 ehp Pratt & Whitney (UACL) PT6A-27 turboprops.
Performance (at gross weight of 5,100 lb–2 313 kg) : Max speed, 216 mph (348 km/h) at 10,000 ft (3 050 m) ; max cruising speed, 206 mph (332 km/h) at 10,000 ft (3 050 m) ; economical cruising speed, 160 mph (257 km/h) at 10,000 ft (3 050 m) ; initial rate of climb, 2,200 ft/min (11,2 m/sec) ; service ceiling, 28,000 ft (8 530 m) ; range with max payload, 445 mls (716 km) ; range with max fuel, 1,090 mls (1 755 km).
Weights: Empty, 2,860 lb (1 297 kg) ; normal take-off, 5,100 lb (2 313 kg) ; max overload, 6,300 lb (2 857 kg).
Dimensions: Span, 41 ft 0 in (12,50 m) ; length, 39 ft 7 in (12,07 m) ; height, 9 ft 3 in (2,81 m) ; wing area, 242 sq ft (22,48 m²).
Accommodation: Two crew side-by-side, and up to nine passengers.
Armament: One 20-mm Mini gun in fuselage, firing through cabin door. Strong point under fuselage with 500-lb (227-kg) capacity ; four underwing strong points with total capacity of 1,800 lb (816 kg).
Status: Prototype (commercial HST-550) first flown on 5 June 1964 ; FAA certification in August 1965. Production model H-550A certificated August 1969. AU-24A deliveries made in 1972.
Notes: The Stallion originated as a civil utility STOL aircraft with the ability to carry up to nine passengers in addition to crew of two. The prototype was designated HST-550 and like the earlier Helio designs (see *Civil* volume) it incorporated full-span automatic leading-edge slats, an augmented lateral control system and slotted flaps. The prototype Stallion was the first Helio design to make use of a turboprop engine and the production version, designated H-550A, appeared in 1966. An armed version of the Stallion was developed to meet requirements for a light armed utility aircraft suitable for operation by the South Vietnam Air Force and elsewhere in SE Asia. Engineering changes included restressing to increase the gross weight by 1,200 lb (544 kg) and to provide five external strong points for bombs, rockets, etc. After demonstration of the prototype, 15 similar aircraft were procured by the USAF with the designation AU-24A, 14 of these being transferred to the Khmer (Cambodian) Air Force in 1973.

MAX HOLSTE M.H.1521M BROUSSARD

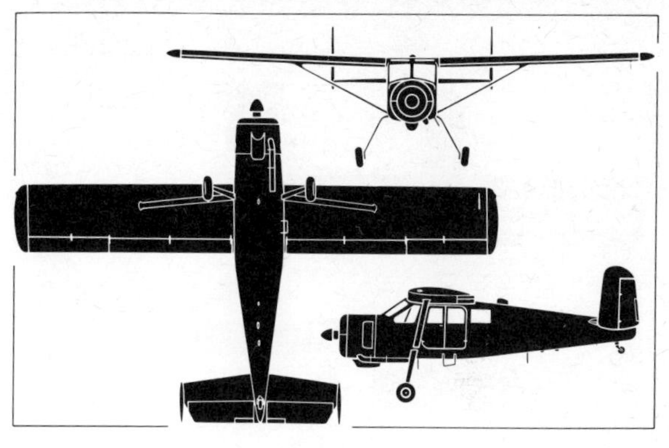

Both the photograph (above) and the general arrangement silhouette (right) illustrate the M.H.1521M Broussard which has now been supplied to a number of small African air arms

Country of Origin: France.
Type: Light utility transport and AOP.
Power Plant: One 450 hp Pratt & Whitney R-985-AN-1 Wasp radial engine.
Performance: Max speed, 168 mph (270 km/h) at 3,280 ft (1 000 m); max cruise, 152 mph (245 km/h); economical cruise, 143 mph (230 km/h); initial rate of climb, 1,082 ft/min (5,5 m/sec); service ceiling, 18,045 ft (5 500 m); time to 4,920 ft (1 500 m) 5 min; range, 745 mls (1 200 km) at 132 mph (212 km/h) at 3,280 ft (1 000 m).
Weights: Empty, 3,373 lb (1 530 kg); normal take-off, 5,511 lb (2 500 kg); max overload, 5,953 lb (2 700 kg).
Dimensions: Span, 45 ft 1 in (13,75 m); length, 28 ft 2½ in (8,60 m); height, 9 ft 2 in (2,79 m); wing area, 273·4 sq ft (25,40 m²).
Accommodation: One pilot and five passengers. Rear four seats removable for freight carrying or for the ambulance rôle, in which two casualty stretchers and two sitting casualties may be accommodated.
Status: First of two prototypes flown on 17 November 1952; first production (civil) flown on 16 June 1954; first production (military) flown on 24 June 1954. Production total 363 completed in 1959.
Notes: The Broussard (Bushman) was developed from the M.H.152 prototypes, which were smaller and were powered by the 240 hp Salmson 8AS in-line engine. The M.H.1521 prototypes were also built as private ventures but of the initial production batch of 24, 18 were ordered by the French Army and subsequent orders from the Army and the Air Force kept the Broussard in production for five years. The basic military model was identified as the M.H.1521M; other variants were the M.H.1521A for civil agricultural use, the M.H.1521C commercial light utility freighter, the M.H.1522 with experimental double slotted flaps and full-span wing leading-edge slots (first flown on 11 February 1958) and the M.H.153, one of the prototypes re-engined with a Turboméca Astazou turboprop. Through the French Army and Air Force, examples of the Broussard were supplied to the air arms of several of the former French territories in North Africa, including Cameroon, Ivory Coast, Mauritania, Niger, Senegal, Togo and Upper Volta, and other military users included Portugal and Madagascar. Examples of the Broussard remained in service with most of these in 1973.

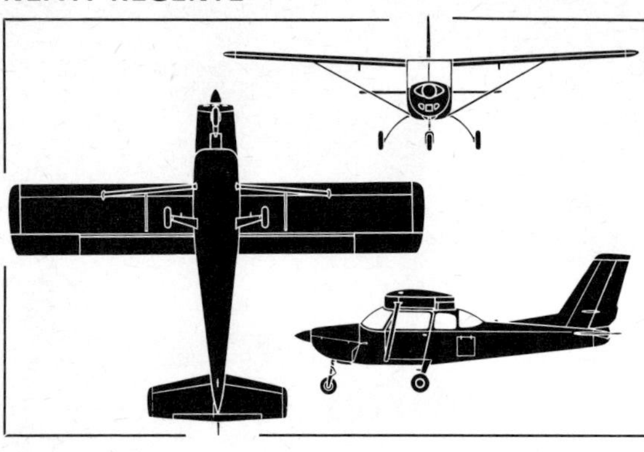

The photograph above depicts the C-42 Regente liaison aircraft and the general arrangement silhouette (left) illustrates the L-42 Regente air observation post

Country of Origin: Brazil.
Type: Air observations post and liaison aircraft.
Power Plant: One (C-42) 180 hp Lycoming O-360-A1D four-cylinder or (L-42) 210 hp Continental IO-360-D six-cylinder horizontally-opposed engine.
Performance (C-42): Max speed, 137 mph (220 km/h) at sea level; cruising speed, 132 mph (212 km/h) at 5,100 ft (1 550 m); initial rate of climb, 690 ft/min (3,5 m/sec); service ceiling, 11,800 ft (3 600 m); range with max payload, 651 mls (904 km); range with max fuel, 576 mls (928 km).
Performance (L-42): Max speed, 153 mph (246 km/h) at sea level; max cruising speed, 142 mph (229 km/h) at sea level; cruising speed (75 per cent power) 134 mph (216 km/h) at 5,000 ft (1 525 m); initial rate of climb, 918 ft/min (4,7 m/sec); service ceiling, 15,810 ft (4 820 m); range with max payload, 547 mls (925 km); range with max fuel, 590 mls (950 km).
Weights (C-42): Empty equipped, 1,410 lb (640 kg); max take-off, 2,293 lb (1 040 kg).
Weights (L-42): Empty equipped, 1,622 lb (736 kg); max take-off, 2,293 lb (1 040 kg).

Dimensions: Span, 29 ft 11½ in (9,13 m); length, 23 ft 7¾ in (7,21 m); height, 9 ft 7¼ in (2,93 m); wing area, 144·8 sq ft (13,45 m²).
Accommodation (L-42): Two seats side-by-side for pilot and co-pilot or observer and third seat in rear of cabin for navigator or observer.
Status: Prototype (Model 360C) first flown on 7 September 1961; first production C-42 flown in February 1965 and production of 80 completed in 1968. Prototype YL-24 flown in October 1967; first production L-42 flown in June 1969 and production completed in March 1971.
Notes: Brazil's Neiva company has produced a series of light aircraft of conventional high-wing design, the first with an all-metal airframe being the Model 360C Regente in 1961. This type was ordered into production for the Brazilian Air Force, which assigned it the designation U-42 (later changed to C-42) as a light utility aircraft. Eighty were built, these aircraft being four-seaters. A second variant, the Regente 420L, was evolved more specifically for AOP duties, having a cut-down rear fuselage and "all-round" cabin transparencies and is designated L-42 in service.

PILATUS/FAIRCHILD TURBO-PORTER

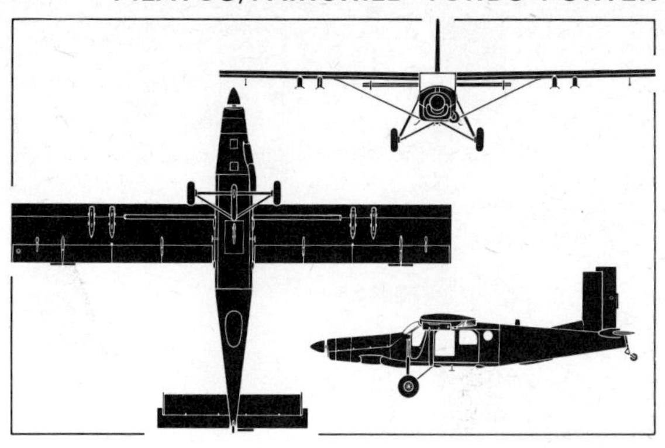

The photograph above illustrates the PC-6/B-1 Porter which is operated by the Australian Army and the general arrangement silhouette depicts the AU-23A Peacemaker armed version

Country of Origin: Switzerland/USA.
Type: Light utility and armed observation aircraft.
Power Plant: One (PC-6/A) 523 shp Turboméca Astazou IIG; (PC-6/A1) 573 shp Astazou X11; (PC-6/A2) 573 shp Astazou X1VE; (PC-6/B) 550 shp Pratt & Whitney PT6A-6A; (PC-6/B1) 550 shp PT6A-20; (PC-6/B2) 550 shp PT6A-27; (PC-6/C); 575 shp AiResearch TPE 331-25D or (PC-6/C1 and AU-23A) 576 shp TPE 331-1-100 turboprop engine.
Performance (PC-6/C1 and AU-23A): Max cruising speed, 164 mph (264 km/h) at 10,000 ft (3 050 m); economical cruising speed, 144 mph (231 km/h) at 10,000 ft (3 050 m); initial rate of climb, 1,607 ft/min (8,2 m/sec); service ceiling, 27,875 ft (8 500 m); range with max internal fuel, 683 mls (1 100 km); range with two external fuel tanks, 1,044 mls (1 680 km).
Weights: Empty equipped, 2,612 lb (1 185 kg); max take-off and landing, 4,850 lb (2 200 kg).
Dimensions: Span, 49 ft 8 in (15,13 m); length, (Astazou engine) 36 ft 4½ in (11,08 m); length (PT6A engine), 36 ft 1 in (11,00 m); length (TPE 331 engine) 35 ft 9 in (10,90 m); height, 10 ft 6 in (3,20 m); wing area, 310 sq ft (28,80 m²).

Accommodation: Pilot and up to nine other occupants.
Armament (AU-23A only): One hardpoint under fuselage with capacity of 590 lb (268 kg); four underwing strong points, with capacity of 510 lb (231 kg) inner and 350 lb (159 kg) outer; max external load 2,000 lb (906 kg). Provision in cabin for two 7·62-mm miniguns or one 20-mm cannon.
Status: Prototype turboprop version (PC-6/A) first flown 2 May 1961; first PC-6/B flown on 1 May 1964; first PC-6/C (by Fairchild) flown October 1965, (by Pilatus) flown 4 March 1966.
Notes: The Turbo-Porter was evolved from the piston-engined Porter (see *Civil* volume) and was first certificated in December 1963 in Switzerland. About 250 had been built by 1973, including a number for military use. The Australian Army bought a total of 14 PC-6/B-1s, the Sudan bought eight, Israel at least two and Peru one. In 1966, Fairchild in the USA began production of a batch of 100 Porters, both the PC-6/B and PC-6/C version being offered, and in 1970, an armed version was developed with the characteristics indicated above. The USAF bought 15 of this armed variant, the AU-23A, for evaluation, supplying 13 of these to the Royal Thai Air Force, the Thai Police acquiring a further five.

ROCKWELL (NORTH AMERICAN) OV-10 BRONCO

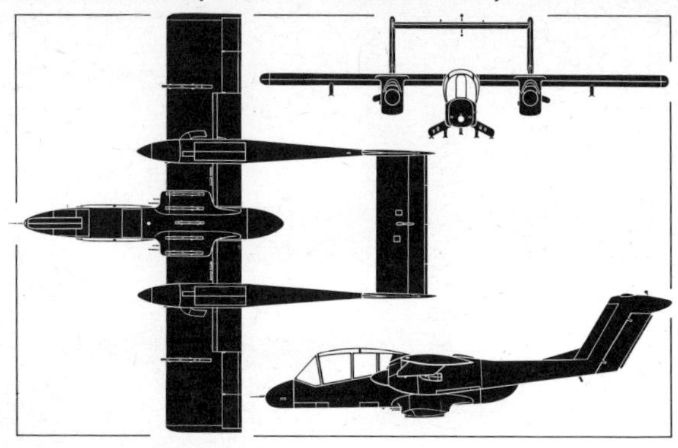

The OV-10C of the Royal Thai Air Force depicted by the photograph above is externally similar to the OV-10A illustrated by the general arrangement silhouette (left)

Country of Origin: USA.
Type: Forward air control, target tug and utility aircraft.
Power Plant: Two 715 shp Garrett AiResearch T76-G-10 (left) and -12 (right) turboprops.
Performance: Max speed, 281 mph (452 km/h) at 10,000 ft (3 048 m); cruising speed, 224 mph (361 km/h) at 18,000 ft (5 486 m); initial rate of climb, 2,800 ft/min (14,2 m/sec); radius of action, typical close support mission, 190 mls (306 km) with 2,800 lb (1 270 kg) ordnance load and 1 hr loiter at 5,000 ft (1 524 m); radius of action, typical FAC mission, 167 mls (269 km) with 3-hr loiter; ferry range, 1,430 mls (2 300 km).
Weights: Empty, 7,190 lb (3 260 kg); normal loaded, 12,500 lb (5 670 kg); max take-off, 14,444 lb (6 550 kg).
Dimensions: Span, 40 ft (12,19 m); length (excluding nose probe) 39 ft 9 in (12,12 m); overall height, 15 ft 1 in (4,62 m); wing area, 291 sq ft (27,03 m²).
Accommodation: Two in tandem.
Armament: Four 7·62 mm M60 C machine guns in sponsons. Seven external store stations (one on fuselage, four on sponsons, two on wings) with total combined capacity of 4,600 lb (2 086 kg).

Status: First of seven YOV-10A prototypes flown on 16 July 1965; first flight with Hoerner wing-tips (No 5 YOV-10A), 15 August 1966; first flight with Pratt & Whitney T74 engines (No 7 YOV-10A), 7 October 1966; first flight in production configuration, March 1967; first production OV-10A flown on 6 August 1967; first OV-10C flown 9 December 1970; first YOV-10D flown on 9 June 1970. Production totals: YOV-10A, 7; OV-10A (USAF), 157; OV-10A (USMC), 114; OV-10B, 17; OV-10B (Z), 1; OV-10C, 32; OV-10E, 16.
Notes: The OV-10 was built to joint USAF/USMC/USN requirements for an armed forward air control and light armed reconnaissance aircraft and was deployed to Vietnam in 1968 for use by all three services. Export models are the OV-10B for Germany, OV-10C for Thailand and OV-10E for Venezuela. The German OV-10B(Z) version had an additional J85 turbojet above the fuselage to boost speed to 393 mph (632 km/h) but only one was completed. Two YOV-10D were Night Observation Gun-Ships converted from OV-10As for Marine Corps evaluation, having a three-barrel 20-mm gun in a remotely-controlled ventral turret and IR sensor in the nose.

SIAI-MARCHETTI SM 1019A

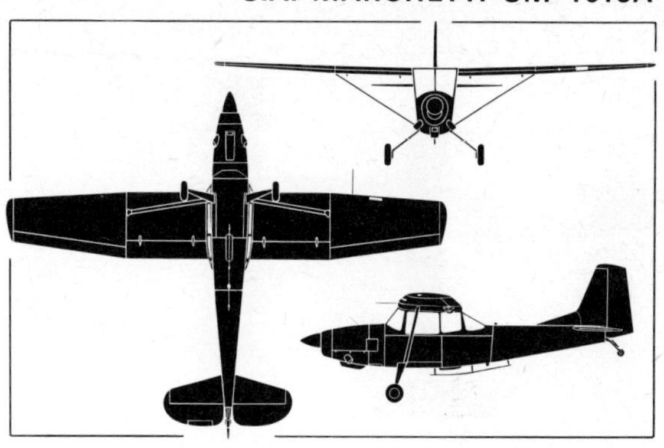

The SM 1019, illustrated in prototype form by the photograph above and the general arrangement silhouette (right), is currently in production for the Italian Army

Country of Origin: Italy.
Type: Light utility and observation aircraft.
Power Plant: One 317 shp Allison Model 250-B15G turboprop engine.
Performance: Max speed, 188 mph (302 km/h) at 6,000 ft (1 830 m); max cruise, 173 mph (278 km/h) at 6,000 ft (1 830 m); economical cruise, 135 mph (217 km/h) at 10,000 ft (3 050 m); initial rate of climb, 1,625 ft/min (8,25 m/sec); range with external stores, 320 mls (515 km); range with max fuel (10 min reserve) 765 mls (1 230 km).
Weights: Empty equipped, 1,480 lb (672 kg); max payload, 1,300 lb (590 kg); max take-off and loading, 2,800 lb (1 270 kg).
Dimensions: Span, 36 ft 0 in (10,97 m); length, 27 ft $11\frac{1}{2}$ in (8,52 m); height, 7 ft $9\frac{3}{4}$ in (2,38 m); wing area, 174 sq ft (16,16 m^2).
Accommodation: Two seats in tandem for pilot and observer or co-pilot, with optional dual controls.
Armament: Two stores stations under wings capable of carrying minigun pods, rockets, etc, up to a maximum external load of 500 lb (227 kg).
Status: First and second prototypes flown on 24 May 1969 and 18 February 1971 respectively. Production of 100 began in 1973, with first production deliveries scheduled to commence mid-1974.
Notes: To meet an Italian Army requirement for an aircraft to replace its Cessna L-19 (O-1) Bird Dogs, SIAI–Marchetti evolved this refined version, using much of the basic Cessna design but introducing a turboprop engine, enlarged tail surfaces and a modified structure. The extra power derived from the turboprop engine, without any increase in gross weight compared with that of the final O-1 variants (see page 139), bestows on the SM 1019A a STOL performance, the take-off run when fully loaded being less than 200 ft (65,6 m). Several differences distinguished the second prototype, designated SM 1019A, from the first, including a revised fuel system with provision for a long-range tank under the fuselage to supplement the four wing tanks which have a total capacity of 68 Imp gal (310 l), and a modified and lengthened cabin. After competitive evaluation between the AM.3C (see page 137) and the SM 1019, the latter was chosen in 1972 by the *Aviazione Leggera dell' Esercito* (Italian Army Aviation) as its new light battlefield surveillance and forward air control aircraft and production of an initial batch began during the course of 1973.

UTVA-60 AND UTVA-66

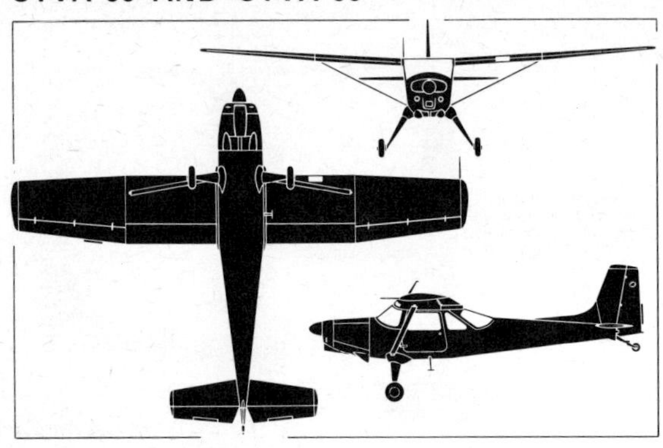

The UTVA-60, illustrated by the photograph above, and the UTVA-66, depicted by the general arrangement silhouette (left), are used in numbers by the Yugoslav armed forces

Country of Origin: Yugoslavia.
Type: Light utility and observation aircraft.
Power Plant: One (UTVA-60) 270 hp Lycoming GO-480-B1A6, (UTVA-60H) 296 hp Lycoming GO-480-G1H6 or (UTVA-66) 270 hp Lycoming GSO-480-B1J6 six-cylinder horizontally-opposed engine.
Performance (UTVA-60-AT1): Max speed, 157 mph (252 km/h); max cruise, 143 mph (230 km/h); economical cruise, 114 mph (184 km/h); initial rate of climb, 1,260 ft/min (6,4 m/sec); service ceiling, 17,060 ft (5 200 m); range, 485 mls (780 km).
Performance (UTVA-66): Max speed, 155 mph (250 km/h); max cruise, 143 mph (230 km/h); initial rate of climb, 885 ft/min (4,5 m/sec); service ceiling, 22,000 ft (6 700 m); range, 466 mls (750 km).
Weights (UTVA-60-AT1): Empty equipped, 2,100 lb (952 kg); normal take-off, 3,192 lb (1 448 kg); max take-off, 3,571 lb (1 620 kg).
Weights (UTVA-66): Empty equipped, 2,756 lb (1 250 kg); max take-off, 4,000 lb (1 184 kg).
Dimensions: Span, 37 ft 5 in (11,40 m); length (UTVA-60) 26 ft $11\frac{1}{2}$ in (8,22 m); length (UTVA-66) 27 ft 6 in (8,38 m), height, (UTVA-60) 8 ft 11 in (2,72 m); height (UTVA-66) 10 ft 6 in (3,20 m); wing area, 194·5 sq ft (18,08 m^2).
Accommodation: Pilot and three passengers or two stretchers.
Status: Prototype (UTVA-56) flown 22 April 1959; production of UTVA-60 began 1960 and UTVA-66 flown in 1966.
Notes: The series of UTVA high-wing light aircraft were designed by Branislav Nikolic and Dragoslav Petkovic and began in 1959 with the UTVA-56. This was "productionised" as the UTVA-60, and put into production for the Yugoslav Air Force. Several versions were designated, including the UTVA-60-AT1 basic four-seater for utility use; the 60-AT2 with dual controls for use as a trainer; the 60-AG equipped for agricultural duties; the 60-AM ambulance version with accommodation for two stretchers and the 60-H floatplane with twin Edo floats (first flown on 29 October 1961). A further development of the same basic design emerged in 1966 as the UTVA-66, the principal new features being fixed slots on the wing leading edge, larger tail unit and strengthened undercarriage, these modifications allowing the UTVA-66 to operate at lower speeds and from shorter runways or landing strips.

AERMACCHI M.B.326

The photograph above depicts the single-seat M.B.326K which was entering South African service in 1973, the general arrangement silhouette illustrating the two-seat M.B.326G

Country of Origin: Italy.
Type: Basic trainer and light strike aircraft.
Power Plant: One 2,500 lb st (1 134 kgp) Rolls-Royce (M.B.326B, D and F) Viper 11 or (M.B.326G, GB, H and M) 3,410 lb st (1 547 kgp) Viper 20 Mk 540 or (M.B.326K) 4,000 lb st (1 814 kgp) Viper 632-43 turbojet.
Performance (M.B.326G, clean): Max speed, 539 mph (867 km/h); max cruising speed, 495 mph (797 km/h); initial rate of climb, 6,050 ft/min (31,5 m/sec); service ceiling, 47,000 ft (14 325 m); range, internal and tip tanks, 1,150 mls (1 850 km); range, with underwing tanks, 1,520 mls (2 445 km).
Performance (M.B.326K, full weapon load): Max speed, 426 mph (686 km/h) at 30,000 ft (9 145 m); initial rate of climb, 3,750 ft/min (19,05 m/sec); time to reach 35,000 ft (10 670 m), 23 min; typical combat radius, 4,000 lb (1 814 kg) external load, 57 mls (91 km); max ferry range, 2,072 mls (3 334 km).
Weights (M.B.326G, unarmed trainer): Basic operating, 5,920 lb (2 685 kg); max take-off (clean), 10,090 lb (4 577 kg).
Weights (M.B.326K): Empty equipped, 6,240 lb (2 830 kg); operating weight empty, 6,500 lb (2 948 kg); take-off (clean), 9,680 lb (4 390 kg); max take-off, 12,000 lb (5 443 kg).
Dimensions: Span (over tip tanks), 35 ft 7 in (10,85 m); length, 34 ft 11 in (10,64 m); height, 12 ft 2½ in (3,72 m); wing area, 208·3 sq ft (19,4 m²).
Accommodation: Two in tandem or (M.B.326K) pilot only.
Armament (M.B.326F, G): Provision for up to 4,000 lb (1 814 kg) of bombs, gun pods, rockets, etc on six wing pylons. (M.B.326K): Two 30-mm DEFA or Aden cannon with 150 rpg. Six wing stations of which four stressed for 1,000 lb (454 kg) each and two for 750 lb (340 kg) each, with max external load of 4,500 lb (2 040 kg).
Status: Prototype M.B.326 flown on 10 December 1957; first production M.B.326 flown on 5 October 1960; first M.B.326K (Viper 540) flown on 22 August 1970 and second (Viper 632) in May 1971. About 300 built by Aermacchi by 1973; 85 built by CAC in Australia, over 160 built by Atlas in South Africa (named Impala) and 112 ordered from Embraer in Brazil (as AT-26 Xavante, first flown 3 September 1971). 40 M.B.326K in production in 1973.
Notes: Over 100 M.B.326 supplied to Italian Air Force for basic pilot trainers, eight M.B.326B to Tunisia; seven M.B.326F to Ghana; 17 M.B.326G to Zaïre; eight M.B.326G to Argentine Navy.

AERO L 29 DELFIN

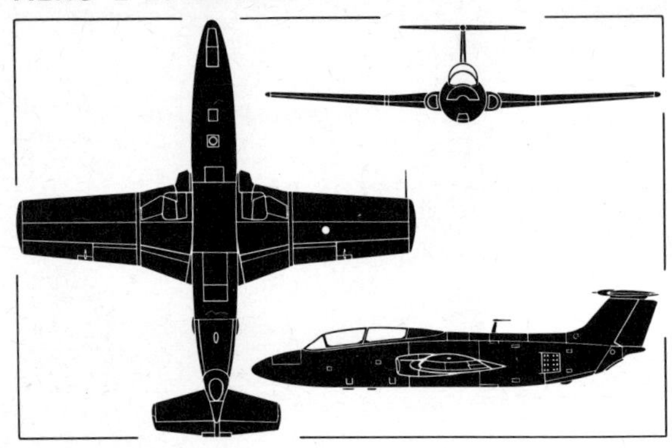

Both photograph above and general arrangement silhouette (left) illustrate the standard L 29 Delfin, the former depicting an example serving with the Hungarian Air Force

Country of Origin: Czechoslovakia.
NATO Code Name: *Maya*
Type: Basic trainer.
Power Plant: One 1,960 lb st (890 kgp) M-701 VC-150 or S-50 turbojet.
Performance: Max speed, 382 mph (615 km/h) at sea level and 407 mph (655 km/h) at 16,400 ft (5 000 m); normal cruising speed, 340 mph (547 km/h) at 16,400 ft (5 000 m); initial rate of climb, 2,755 ft/min (14 m/sec); service ceiling, 36,100 ft (11 000 m); max range with internal fuel, 397 mls (640 km); range with drop tanks, 555 mls (894 km).
Weights: Empty, 5,027 lb (2 280 kg); normal take-off, 7,231 lb (3 280 kg); max overload, 7,804 lb (3 540 kg).
Dimensions: Span, 33 ft 9 in (10,29 m); length, 35 ft $5\frac{1}{2}$ in (10,81 m); height, 10 ft 3 in (3,13 m); wing area, 213·1 sq ft (19,80 m^2).
Accommodation: Two in tandem, full dual control.
Armament: Wing strong points can carry (in place of drop tanks) two bombs of up to 220 lb (100 kg) weight, two 7·62 mm gun pods or eight air-to-ground rockets.

Status: First prototype flown on 5 April 1959 and second in July 1960. First production model flown in April 1963. Production continuing in 1973, with more than 2,500 built.
Notes: A specification for a basic jet trainer to serve in the air forces of the Soviet Bloc nations was drawn up in 1957, leading to development of suitable designs in Czechoslovakia, Poland (see page 200) and the Soviet Union. Pending development of a suitable engine in Czechoslovakia, the Czech prototype, designated XL 29, was fitted with a Bristol Siddeley Viper engine; the second prototype, with the Czech M701 engine, was designated L 229 and after being adjudged the winner of the design contest, the type entered production as the L 29. Deliveries to the Soviet Union began in May 1963, concurrently with the first deliveries to the Czech Air Force, and the air forces of East Germany, Bulgaria, Hungary and Rumania also received early deliveries. Syria, Indonesia and Egypt also placed orders for Delfins, and small numbers have been supplied to Uganda and Nigeria, some being used in an attack rôle during the war against Biafra. A version of the Delfin specially modified for aerobatic use is designated the L 29A Akrobat and has the rear seat removed and the canopy faired over.

AERO L 39 ALBATROSS

The L 39, illustrated by the photograph above and the general arrangement silhouette (right), is currently in production as a successor to the L 29 Delfin

Country of Origin: Czechoslovakia.
Type: Basic and advanced trainer.
Power Plant: One 3,307 lb st (1 500 kgp) Walter Titan (Ivchenko AI-25W) turbofan.
Performance: Max speed, 379 mph (610 km/h) at sea level and 454 mph (730 km/h) at 16,400 ft (5 000 m); initial rate of climb, 3,740 ft/min (19 m/sec); service ceiling, 37,225 ft (11 350 m); range on internal fuel only, 680 mls (1 100 km) with 5% reserves; range with full tip-tanks, no reserves, 930 mls (1 500 km).
Weights: Empty, 6,283 lb (2 850 kg); normal loaded, 8 377 lb (3 800 kg); max take-off, 9,480 lb (4 300 kg).
Dimensions: Span, 29 ft 10¾ in (9,11 m); length, 39 ft 9 in (12,11 m); height, 14 ft 4¼ in (4,38 m); wing area, 202·4 sq ft (18,8 m²).
Accommodation: Two in tandem, full dual control.
Armament: Provision for underwing bombs or rocket pods.
Status: First of five prototypes flown on 4 November 1968; first of pre-production batch of 10 completed in 1971; production deliveries began in 1973. Production of 700 planned, to provide 300 for Soviet Air Force, 300 for Czech Air Force and 100 for export.

Notes: Development of the L 39 was initiated at the Vodochody works to provide a successor for the L 29 Delfin (see page 156), although its performance represents only a modest improvement over that of the earlier type. A protracted flight test programme was undertaken, using at least five flying prototypes (in addition to two built for static testing) and modifications were made to the intakes before the L 39 entered production. One prototype was tested in the weapon training rôle with underwing stores, and in 1973 a light strike version of the L 39 was also under development with the designation L 39Z. This has wing and fuselage hardpoints for gun pods, ASMs and bombs, and in its definitive production version is intended to be powered by an afterburning version of the AI-25V engine, rated at 4,410 lb st (2 000 kgp). In addition to the Czech Air Force, the Iraqi Air Force placed orders for the L 39, for deliveries starting in 1974, and was expected to acquire examples of the L 39Z also. Experience derived from the L 39 programme is being utilised in the development of the Aero A 159, a light single-seat twin-engined attack fighter with a thrust-to-weight ratio of the order of 1:1, a built-in cannon armament and four wing hardpoints for ordnance.

AEROTEC T-23 UIRAPURU

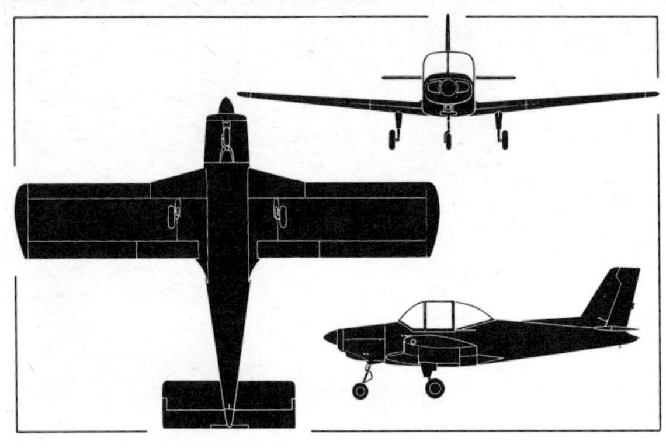

The Aerotec T-23 Uirapuru, illustrated by the photograph above and the general arrangement silhouette (left) is now standard Brazilian primary training equipment

Country of Origin: Brazil.
Type: Primary trainer.
Power Plant: One 160 hp Lycoming O-320-B2B four-cylinder horizontally-opposed engine.
Performance: Max speed, 140 mph (225 km/h); max cruising speed, 115 mph (185 km/h) at 5,000 ft (1 525 m); initial rate of climb, 787 ft/min (4 m/sec); service ceiling, 14,760 ft (4 500 m); max range, 495 mls (800 km); endurance 4 hrs.
Weights: Empty, 1,190 lb (540 kg); max take-off, 1,825 lb (840 kg).
Dimensions: Span, 27 ft 10¾ in (8,50 m); length, 21 ft 8 in (6,60 m); height, 8 ft 10 in (2,70 m); wing area, 145·3 sq ft (13,50 m²).
Accommodation: Two seats side by side with full dual control.
Status: Prototype first flown on 2 June 1965; first pre-production T-23 flown on 23 January 1968. Production deliveries began late 1968; 90 built by 1973.
Notes: The original Uirapuru was a private venture prototype designed by two Brazilian engineers and flown in 1965 with a 108 hp Lycoming O-235-C1 engine. A second prototype was completed subsequently, with a 150 hp Lycoming O-320-A. Early in 1968 the Brazilian Air Force adopted the Uirapuru as a replacement for its locally-built Fokker S.11 and S.12 Instructor primary trainers, and placed an order for 30. Two pre-production models flown in 1968 had the same 150 hp engine as the second prototype but production models, with the Brazilian Air Force designation T-23, have the 160 hp engine as indicated above. Further orders were placed in 1969 to bring the total Brazilian Air Force purchase of T-23s to 70, these serving primarily at the Military Pilot Training Centre (*Centro de Formaçâo de Pilotos Militares*) at Natal. The Paraguayan Air Force ordered 20 in 1971. A civil version was developed, differing from the T-23 in having a modified cockpit canopy and additional wing fuel tanks to provide a maximum endurance of 6½ hrs and a range of 805 mls (1 300 km). A four-seat version of the Uirapuru was also developed during 1972. The Uirapuru is of all-metal construction, with some components of glass-fibre. The fixed tricycle undercarriage has a steerable nosewheel and hydraulic brakes. Provision is made for wing-tip tanks to be fitted, supplementing an integral tank in the leading edge of each wing. Uirapuru production was continuing at two per month in 1973.

AEROSPACE CT/4 AIRTRAINER

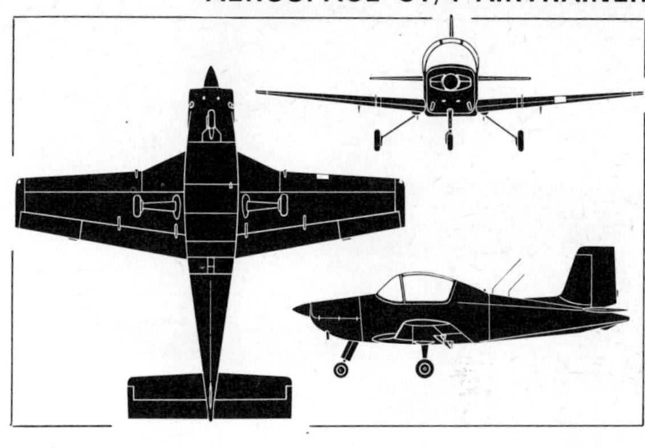

Both photograph (above) and general arrangement silhouette (right) illustrate the Airtrainer scheduled to enter service with the Australian and Thai air forces in 1974.

Country of Origin: New Zealand.
Type: Basic trainer.
Power Plant: One 210 hp Continental IO-360-D six-cylinder horizontally-opposed engine.
Performance: Max speed, 180 mph (290 km/h) at sea level; cruising speed at 75 per cent power, 155 mph (250 km/h) at sea level and 142 mph (228 km/h) at 10,000 ft (3 050 m); cruising speed at 65 per cent power, 145 mph (233 km/h) at sea level and 139 mph (224 km/h) at 5,000 ft (1 524 m); initial rate of climb, 1,345 ft/min (6,8 m/sec); range with internal fuel only, 10 per cent reserve, 808 mls (1 300 km) at 145 mph (233 km/h).
Weights: Empty equipped, 1,520 lb (690 kg); max take-off, 2,350 lb (1 070 kg).
Dimensions: Span, 26 ft 0 in (7,92 m); length, 23 ft 5½ in (7,15 m); height, 8 ft 6 in (2,59 m); wing area, 129 sq ft (12,00 m²).
Accommodation: Two seats side-by-side with full dual control.
Status: Prototype first flown on 21 February 1972. First delivery to R Thai AF on 23 October 1973.
Notes: The origins of the Airtrainer can be traced back to 1953, in which year a design for a two-seat light aircraft by Henry Millicer (then chief aerodynamicist of the Australian Government Aircraft Factories) won a competition organised by the British Royal Aero Club. This design eventually was put into production by Victa in Australia, and was subsequently taken over by Aero Engine Services Ltd (AESL) of New Zealand (see *Civil* volume). Originally of wooden construction, the Airtourer was later redesigned to have a metal structure, and then became the basis for a four-seat development, the Victa Aircruiser. Using the same revised structure as the Aircruiser, restressed for *g* limits of +6 to −3, the Airtrainer is dimensionally similar to the Airtourer but has a number of modifications to suit it to the military training rôle. These include detachable spoiler strips on the leading edges of the wing outboard panels, fitted only when the aircraft is engaged on spinning exercises; and provision for 13-Imp gal (59-l) wingtip auxiliary tanks. A 200 hp Lycoming IO-360-B engine is an optional alternative to the Continental specified above. Initial orders for the Airtrainer were placed by the Royal Thai Air Force (for 24) and the RAAF (for 37). During the course of 1973, AESL amalgamated with Air Parts (NZ) to form New Zealand Aerospace Industries, integrating production of the Airtrainer with the Fletcher agricultural aircraft.

AISA I-115

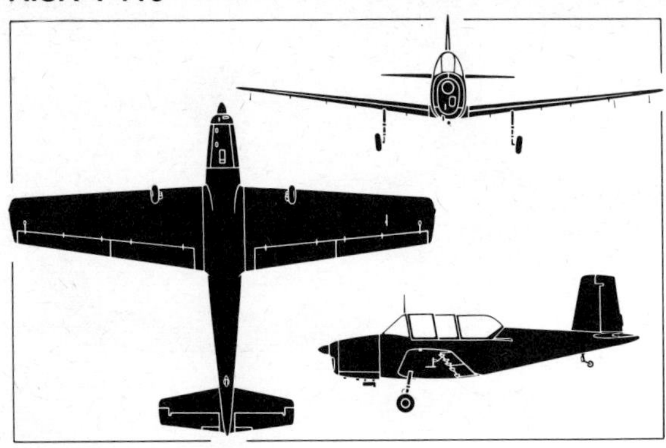

The I-115, illustrated by the photograph above and the general arrangement silhouette (left), was built in larger numbers than any aircraft of indigenous Spanish design

Country of Origin: Spain.
Type: Primary trainer.
Power Plant: One 150 hp ENMA Tiger G-IV-B four-cylinder-in-line engine.
Performance: Max speed, 143 mph (230 km/h); cruising speed, 127 mph (204 km/h); initial rate of climb, 686 ft/min (3,7 m/sec); service ceiling, 14,050 ft (4 300 m); range, 620 mls (1 000 km); endurance, 3½ hrs.
Weights: Empty, 1,500 lb (680 kg); max take-off (normal operation) 2,182 lb (990 kg); max take-off (aerobatic) 1,875 lb (850 kg).
Dimensions: Span, 31 ft 3 in (9,54 m); length, 24 ft 1 in (7,35 m); height, 6 ft 10 in (2,1 m); wing area, 150,6 sq ft (14,0 m²).
Accommodation: Two in tandem with full dual controls.
Status: Prototype first flown on 16 July 1952. Production of 450 for the Spanish Air Force between 1956 and 1961.
Notes: The I-115 was one of several light aircraft originated in the immediate post-war period by Iberavia SA, its production being undertaken by Aeronautica Industrial SA at Cuatro Vientos. Construction was of wood throughout, with plywood covering and the I-115 was chosen by the Spanish Air Force in 1954 to replace its Bücker Jungmeister biplanes, being the first post-war type of wholly indigenous design to go into large scale production. In air force service, the I-115 carried the nomenclature E.9, and went into use at the various pilot training schools and the *Academia General del Aire*. It remained a principal primary trainer of the Spanish Air Force in 1973. It currently serves with the *Escuela General del Aire* at San Javier where No 792 Squadron employs this type of elementary navigational training, and it is operated by the Training Centre of the University Air Militia. A substantial number of I-115s are operated for both liaison and training tasks by the various base flights, and others are included in the equipment of government-sponsored flying clubs. A similar design evolved by AISA as the I-11B had side-by-side instead of tandem seating, and a 90 hp Continental C90-12F engine. This was also used by the Spanish Air Force in small numbers, primarily as a liaison type rather than a trainer, as indicated by the service nomenclature L.8C. Production of the I-11B totalled 180, of which the first 70 had only basic flying instruments and the remainder had full blind-flying panels. Some aircraft of this type are powered by the 93 hp ENMA Flecha engine.

BAC 167 STRIKEMASTER AND JET PROVOST

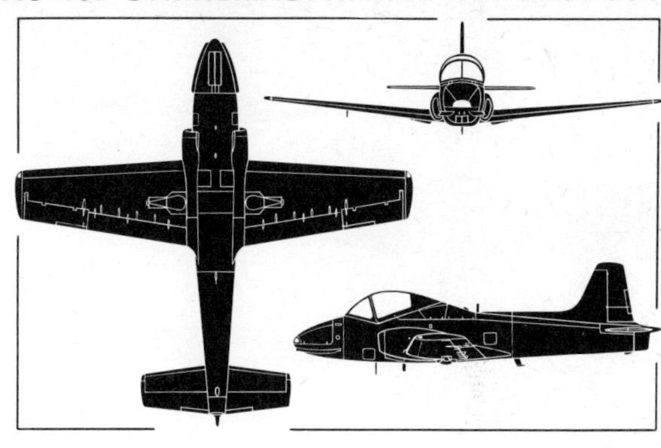

Both the photograph above and the general arrangement silhouette (right) illustrate the Jet Provost T Mk 5, the Strikemaster being externally similar

Country of Origin: United Kingdom.
Type: Basic trainer and light strike aircraft.
Power Plant: One 1,750 lb st (794 kgp) Rolls-Royce Viper (Jet Provost and 3 and 51) Mk 102 or (Jet Provost, 4, 5, 52 and 55) 2,500 lb st (1 134 kgp) Viper 201 or (Strikemaster) 3,410 lb st (1 547 kgp) Viper 535.
Performance (Jet Provost 5) : Max speed, 440 mph (708 km/h) at a weight of 6,400 lb (2 900 kg) at 25,000 ft (7 620 m) ; initial rate of climb, 3,550 ft/min (18 m/sec) ; service ceiling, 34,500 ft (10 500 m) ; max range with tip tanks, 900 mls (1 450 km) at 35,000 ft (10 670 m).
Performance (Strikemaster) : Max speed (clean), 472 mph (760 km/h) at 20,000 ft (6 100 m) ; initial rate of climb, 5,300 ft/min (26 m/sec) ; radius of action (max armament) 250 mls (400 km) ; radius of action (max fuel) 735 mls (1 200 km) ; ferry range, 1,450 naut mls (2 700 km).
Weights (Jet Provost) : Normal take-off (internal fuel only), 7,629 lb (3 460 kg) ; max take-off (Mk 55, with tip tanks), 8,524 lb (3 866 kg) ; max overload, 9,200 lb (4 173 kg).
Weights (Strikemaster) : Empty equipped, 6,270 lb (2 844 kg) ; normal take-off (trainer), 9,200 lb (4 170 kg) ; normal take-off (strike), 11,500 lb (5 215 kg).
Dimensions: Span, 35 ft 4 in (10,77 m) ; span (over tip tanks), 36 ft 11 in (11,25 m) ; length, 33 ft 8½ in (10,27 m) ; height, 10 ft 2 in (3,10 m) ; wing area, 213·7 sq ft (19,80 m²).
Armament: Provision for two 7·62-mm machine guns in fuselage. Four (Jet Provost 55) or eight (Strikemaster) wing strong points, for maximum ordnance load of 3,000 lb (1 360 kg).
Status: Prototype Jet Provost first flown 26 June 1954 ; first JP3 flown on 22 June 1958 ; first JP5 flown on 28 February 1967 ; first Strikemaster flown on 26 October 1967. Production totals : Jet Provost 1, 10 ; JP2, 2 ; JP3, 201 ; JP4, 198 ; JP5, 110 ; JP51, 22 ; JP52, 43 ; JP55, 5 ; Strikemaster, about 115 up to end of 1973.
Notes: Jet Provost was RAF's first jet trainer, Mks 3, 4 and 5 being for RAF service and Mks 51, 52 and 55 for export. Strikemaster (and Jet Provost 5) have pressurised cockpit and users comprise Saudi Arabia (35 Mk 80) ; South Arabia (4 Mk 81) ; Oman (20 Mk 82) ; Kuwait (12 Mk 83) ; Singapore (16 Mk 84) ; Kenya (6 Mk 87) ; New Zealand (10 Mk 88) and Ecuador (8 Mk 89). The BAC 145 Jet Provost T Mk 5 and BAC 167 Strikemaster are externally similar.

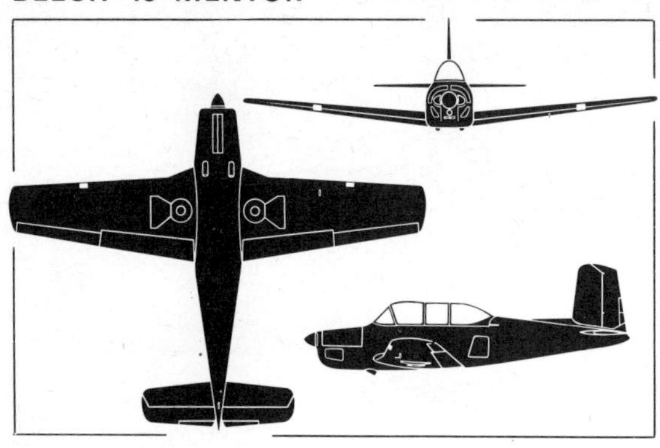

The Beech Mentor, illustrated by the photograph above and the general arrangement silhouette (left), remains in wide service, the photo depicting a Spanish Air Force example

Country of Origin: USA.
Type: Primary trainer.
Power Plant: One 225 hp Continental O-470-13A six-cylinder horizontally-opposed engine.
Performance: Max speed, 189 mph (304 km/h) at sea level, 185 mph (297 km/h) at 7,500 ft (2 286 m); max cruising speed 173 mph (277 km/h) at 10,000 ft (3 050 m); economical cruise, 170 mph (274 km/h) at 7,500 ft (2 286 m); initial rate of climb, 1,230 ft/min (6,2 m/sec); service ceiling, 20,000 ft (6 100 m); range, 737 mls (1 186 km) at 7,500 ft (2 286 m).
Weights: Empty (T-34A), 2,156 lb (978 kg); empty (T-34B) 2,254 lb (1 022 kg); max take-off (T-34A), 2,950 lb (1 338 kg); max take-off (T-34B), 2,985 lb (1 354 kg).
Dimensions: Span, 32 ft 10 m (10 m); length, 25 ft 11 in (7,9 m); height, 10 ft $0\frac{1}{4}$ in (3,04 m); wing area, 177·6 sq ft (16,49 m^2).
Accommodation: Two in tandem, dual controls.
Armament: Optional provision for one 0·30-in (7,62-mm) machine gun in each wing plus practice bombs or rockets.
Status: Prototype (Model 45) first flown 2 December 1948. First pre-production YT-34 flown in May 1950; production deliveries of T-34A began in 1954. Production totals: 350 T-34A and 423 T-34B and 170 export Model B45s by Beech, 100 T-34A and 25 Model 45A by Canadian Car and Foundry, 176 by Fuji and 75 in Argentina, final deliveries 1962.
Notes: Mentor prototype was a derivative of the civil Bonanza (see *Civil* volume). Design was adopted by USAF on 4 March 1953 and by US Navy on 17 June 1954, designation being T-34A and T-34B respectively, the latter with O-470-4 engine. US production was supplemented by a production line in Canada which contributed 100 T-34As to the USAF and 25 to the RCAF; subsequently, the latter transferred 23 to Turkey and one to Greece, while the US government supplied others through MAP to Spain (as E-17) and Saudi Arabia. Beech exported similar Model B45s direct to Argentina (15), Chile (66), Colombia (41), Mexico (4), El Salvador (3) and Venezuela (41). In Japan, Fuji built 140 Model B45s for the JASDF and 36 for the Philippine Air Force, 50 of this total being assembled from Beech components, and another 75 sets by the parent company were assembled at Cordoba for the Argentine Air Force. Two T-34Bs were being experimentally fitted with 400 shp PT6A-25 turboprops during 1973.

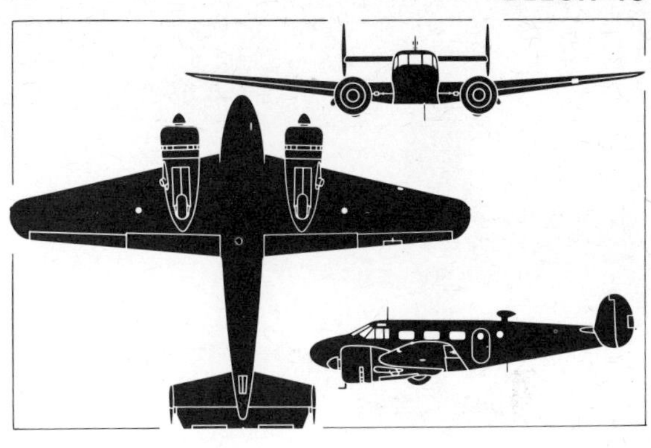

Illustrated by the photograph above and the general arrangement silhouette (right), the Beech 18 remains in widespread service in the communications rôle

Country of Origin: USA.
Type: Multi-engine crew trainer and light transport.
Power Plant: Two 450 hp Pratt & Whitney R-985-AN-1 Wasp Junior seven-cylinder radial engines.
Performance: Max speed, 230 mph (368 km/h); cruising speed, 211 mph (340 km/h); initial rate of climb, 1,280 ft/min (6,5 m/sec); service ceiling, 22,000 ft (6 710 m); range, 1,300 mls (2 585 km).
Weights: Typical empty weight, 5,610 lb (2 546 kg); normal take-off, 8,500 lb (3 860 kg); max loaded, 9,000 lb (4 082 kg).
Dimensions: Span, 47 ft 7 in (14,5 m); length, 31 ft $11\frac{1}{2}$ in (10,4 m); height, 9 ft $2\frac{1}{2}$ in (2,8 m); wing area, 349 sq ft (32,4 m^2).
Accommodation: Two crew with dual controls. Up to seven passenger seats in cabin.
Status: Prototype (commercial model) first flown on 15 January 1937. Production deliveries (military and civil) began 1938. Total production, 7,091, including 5,204 in military versions, final delivery 26 November 1969.
Notes: The Beech 18 enjoyed an uninterrupted production life of 32 years, from 1937 to 1969, which is believed to be a world record. In this period, it evolved through many versions, for both military and civil use, but all but the final Super 18 model retained essentially the same characteristics. Large numbers were used by the USAF for communications (C-45 variant) and for training (AT-7 and AT-11 versions) but these are no longer in service, the last having been retired at the beginning of 1964. Similar models are operated by the US Navy (RC-45J and TC-45J), and by numerous foreign air forces, particularly in South America. One of the largest post-war users of the Beech 18 was the RCAF, which at one time had nearly 400 on strength and in 1973 still used about 60 under the name Expeditor. Canadian versions carry special designation according to rôle, including Mk 3T, Mk 3N, Mk 3TM, Mk 3NM and Mk 3 NMT. The Japanese armed forces are also among the largest remaining users of Beech 18s. Several air arms currently operate the Super 18 version in the utility rôle. Introduced in 1954 as a refined version of the D18S, the Super 18 embodied a number of design refinements, the G18 of 1959 introducing an enlarged windshield, the 1961 model embodying further refinement in the form of enlarged so-called panoramic windows, this version being supplanted by the H18 in 1962, a nosewheel undercarriage being offered as an optional feature in 1964.

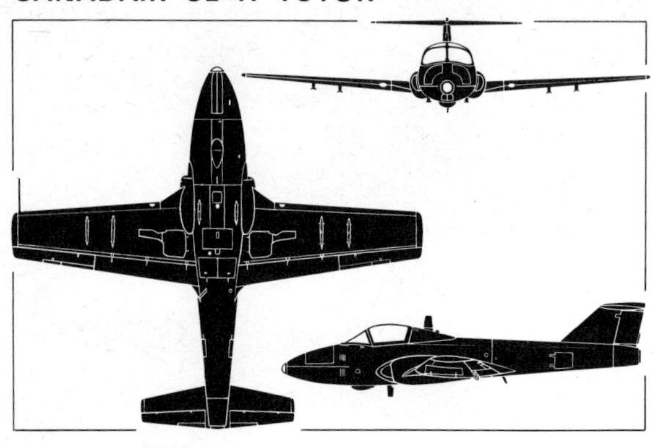

The photograph above and the general arrangement silhouette (left) depict the CL-41G, the armed version of the Tutor used by the Royal Malaysian Air Force

Country of Origin: Canada.
Type: Basic trainer and light attack aircraft
Power Plant: One 2,950 lb st (1 340 kgp) General Electric J85-J4 (CL-41G) or 2,633 lb st (1 195 kgp) J85-CAN-40 turbojet.
Performance (CL-41A): Max speed, 498 mph (801 km/h) at 28,500 ft (8 700 m); initial rate of climb, 4,220 ft/min (21,4 m/sec); service ceiling, 43,000 ft (13 100 m); range, 944 mls (1 519 km).
Performance (CL-41G): Max speed, 480 mph (774 km/h) at 28,500 ft (8 700 m); service ceiling, 42,200 ft (12 800 m); range (with six external tanks) 1,340 mls (2 157 km).
Weights (CL-41A): Empty, 4,895 lb (2 220 kg); max take-off, 7,397 lb (3 355 kg).
Weights (CL-41G): Empty, 5,296 lb (2 400 kg); max take-off, 11,288 lb (5 131 kg).
Dimensions: Span, 36 ft 6 in (11,13 m); length, 32 ft 0 in (9,75 m); height, 9 ft 3$\frac{3}{4}$ in (2,84 m); wing area, 220 sq ft (20,44 m²).
Accommodation: Two side by side, with dual controls.
Armament (CL-41G): Six wing hardpoints for up to 3,500 lb (1 590 kg) of bombs, rockets, mini-gun pods etc.

Status: Prototype first flown on 13 January 1960; first production CL-41A flown in October 1963. CL-41R flown on 13 July 1962. First CL-41G flown in June 1964. Production totals: two prototypes, 190 CL-41A and 20 CL-41G, completed in 1968.
Notes: Canadair built a prototype of its CL-41A basic jet trainer as a private venture, powered by a 2,400 lb st (1 088 kgp) Pratt & Whitney JT12A-5 turbojet. After evaluation, this design was selected by the RCAF as its new basic trainer and 190 were built, with the first delivery on 29 October 1963 and the last in 1966. These aircraft carry the RCAF designation CT-114. The second prototype was converted to the CL-41R configuration this being an experimental systems trainer with provision for NASARR system in a lengthened nose. The 15th production CL-41A was modified to prototype CL-41G light strike and weapon trainer configuration, with four underwing strong points. The production CL-41G has, in addition, two strong points under the centre section. Twenty were delivered to the Royal Malaysian Air Force in 1967–68, taking the local name of Tebuan, and these are used at Kuantan by No 9 Squadron for training and No 6 for light strike duties. The CL-41A remains the standard basic trainer of the Canadian Armed Forces.

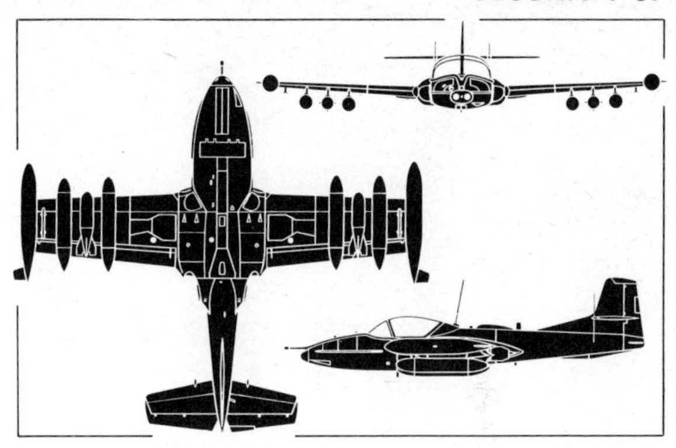

The photograph above depicts the T-37A and the general arrangement silhouette (right) illustrates the A-37B light attack aircraft derivative

Country of Origin: USA.
Type: Basic trainer and light attack aircraft.
Power Plant (T-37): Two 1,025 lb st (465 kgp) Continental T69-T-25 turbojets or (A-37A) two 2,400 lb st (1 090 kgp) General Electric J85-GE-5 turbojets or (A-37B) two 2,850 lb st (1 293 kgp) J85-GE-17A turbojets.
Performance (T-37B): Max speed, 425 mph (684 km/h) at 20,000 ft (6 100 m); normal cruising speed, 360 mph (579 km/h) at 35,000 ft (10 670 m); initial rate of climb, 3,370 ft/min (17,1 m/sec); service ceiling, 39,200 ft (11 948 m); range, 932 mls (1 500 km) at 333 mph (536 km/h) at 35,000 ft (10 670 m) on standard tankage with 5 per cent reserve.
Performance (A-37B): Max speed, 507 mph (816 km/h) at 16,000 ft (4 875 m); max cruising speed, 489 mph (787 km/h) at 25,000 ft (7 620 m); initial rate of climb, 6,990 ft/min (33,5 m/sec); service ceiling, 41,765 ft (12 730 m); range with 4,100 lb (1 860 kg) armament, 460 mls (740 km); max fuel range, 1,012 mls (1 628 km).
Weights (T-37B): Max take-off, 6,574 lb (2 982 kg); (T-37C), max take-off, 8,007 lb (3 632 kg); (A-37B), empty equipped, 6,211 lb (2 817 kg); max take-off, 14,000 lb (6 350 kg).

Dimensions: Span, 33 ft 9¼ in (10,3 m); span over tip tanks, 35 ft 10½ in (10,93 m); length, 29 ft 3½ in (8,93 m); height, 9 ft 2 in (2,8 m); wing area, 183·9 sq ft (17,09 m²).
Accommodation: Two side-by-side with dual controls.
Armament: T-37A and T-37B usually unarmed. T-37C has two underwing strong points for gun pods, rockets, bombs, etc. A-37 has one 7·62-mm Minigun in fuselage and four pylon stations on each wing with combined capacity of 5,680 lb (2 576 kg).
Status: Prototype XT-37 first flown on 12 October 1954; first T-37A flown on 27 September 1955; first YAT-37D flown on 22 October 1963. Production total over 1,700 by early 1973 comprising 3 XT-37, 12 T-37, 537 T-37A, 447 T-37B, 252 T-37C, 39 A-37A and 416 A-37B.
Notes: T-37B is standard USAF basic trainer (all T-37As, with J69-T-9 engines, converted to T-37Bs) and also used by Luftwaffe (in USA), Royal Thai Air Force, Chilean, Pakistan, Greek, Turkish and Peruvian Air Force. Armed T-37C used by air forces of Brazil, Portugal, Cambodia, Chile and Columbia. More heavily armed A-37 was used by USAF in Vietnam, later supplied to ANG and South Vietnamese Air Force, Peruvian Air Force (24) and Guatamala.

CESSNA T-41 MESCALERO

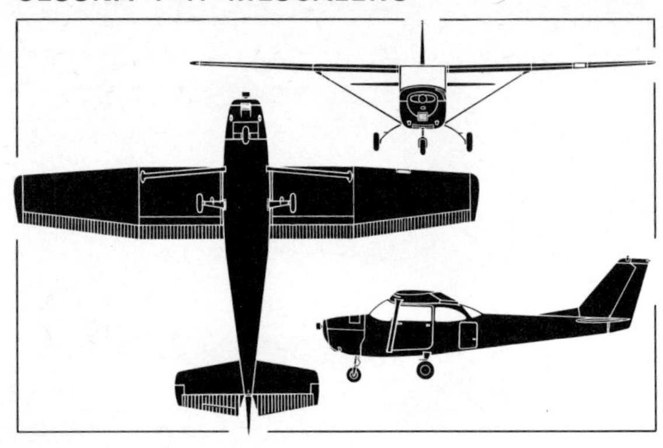

The photograph above depicts a T-41D which has been supplied to various foreign air forces, the general arrangement silhouette (left) illustrating the lower-powered T-41A

Country of Origin: USA.
Type: Primary trainer.
Power Plant: One (T-41A) 150 hp Lycoming O-320-E2D four-cylinder or (T-41B/C/D) 210 hp Continental IO-360-D six-cylinder horizontally-opposed engine.
Performance: (T-41B) Max speed, 153 mph (246 km/h); max cruising speed, 105 mph (169 km/h) at 10,000 ft (3 050 m); initial rate of climb, 880 ft/min (4,5 m/sec); service ceiling, 17,000 ft (5 180 m); range, 1,010 mls (1 625 km).
Weights: Empty equipped, 1,405 lb (637 kg); max take-off and landing, 2,550 lb (1 156 kg).
Dimensions: Span, 35 ft 10 in (10,92 m); length, 26 ft 11 in (8,20 m); height, 8 ft 9½ in (2,68 m); wing area, 174 sq ft (16,16 m^2).
Accommodation: Two crew side-by-side with dual control and two passenger seats.
Status: Commercial Model 172 introduced in November 1955. First USAF order for T-41A announced July 1964 with delivery of the first batch between September 1964 and July 1965. Production total 237 including 8 for Ecuador and 26 for Peru. Total of 255 T-41B delivered to US Army in 1966/67. Total of 52 T-41C delivered to USAF in 1967/68. Total of 238 T-41D built by 1973.
Notes: The USAF selected the commercial Cessna Model 172 in July 1964 when it was decided to abandon the "all-through" jet training programme and to provide 30 hrs on a piston-engined primary trainer for all *ab initio* students. These T-41As are operated under contract to the USAF by civilian flying schools, a unique feature of the aircraft being that they bear civil registrations chosen to duplicate the USAF serial numbers. Ecuador and Peru also acquired T-41As. For use at the USAF Academy in Colorado, the USAF purchased the T-41C, a version of the commercial R172E with more powerful engine, as indicated above. The US Army had previously ordered a similar model as the T-41B, for use on training and site support duties. Differing from the T-41C in having a constant-speed propeller and 28-volt electrical system, the T-41D has been supplied to various overseas air forces including those of Colombia, Ecuador and Turkey. The Royal Saudi Air Force purchased eight F172Gs from the Reims production line in France, these being similar to the T-41B, and eight Model 172Ks were delivered to the Singapore Air Defence Command.

COMMONWEALTH CA-25 WINJEEL

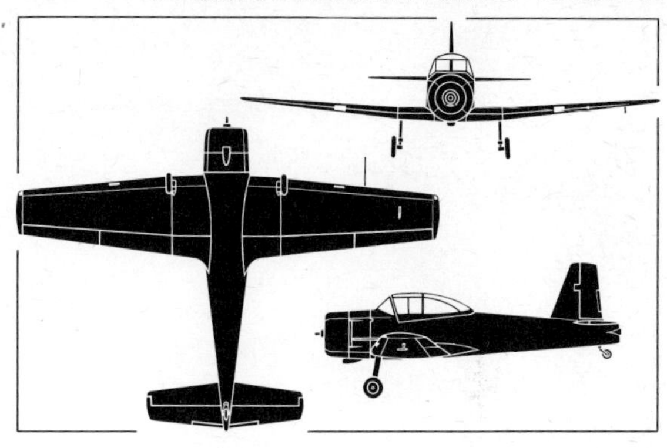

The CA-25 Winjeel, illustrated by the photograph above and the general arrangement silhouette (right), was expected to be phased out by the RAAF during the course of 1975

Country of Origin: Australia.
Type: Basic trainer.
Power Plant: One 445 hp Pratt & Whitney R-985-AN-2 Wasp Junior radial engine.
Performance: Max speed, 188 mph (303 km/h); max cruising speed, 165 mph (265 km/h) at 8,500 ft (2 590 m); economical cruising speed, 158 mph (253 km/h) at 5,000 ft (1 525 m); initial rate of climb, 1,500 ft/min (7,6 m/sec); service ceiling, 15,800 ft (4 815 m); cruising endurance 3½ hrs.
Weights: Empty, 3,289 lb (1 492 kg); max fuel, 542 lb (246 kg); max take-off, 4,265 lb (1 935 kg).
Dimensions: Span, 38 ft 7½ in (11,77 m); length, 28 ft 0½ in (8,55 m); height, 9 ft 1 in (2,77 m); wing area, 249 sq ft (23,13 m²).
Accommodation: Two side-by-side with full dual control; optional third seat in rear.
Status: First of two prototypes flown on 3 February 1951. First production model flown on 23 February 1955. Production of 62 for RAAF completed early 1958.
Notes: The RAAF issued a specification during 1948 for a basic trainer, to cover those phases of the training programme then being covered by the Tiger Moth (*ab initio*) and Wirraway. To meet this requirement, which specified a three-seat aircraft of simple and robust construction, with easy maintenance, economy of operation and safe handling characteristics, Commonwealth Aircraft Corporation designed the CA-22. The RAAF ordered two prototypes, and after prolonged testing, a production order was placed. At one stage, it was planned to develop a suitable engine in Australia, the CAC Cicada, but this did not reach the flight test stage. The production model, CA-25, had some differences, especially around the tail unit, to improve the spinning characteristics, and entered service with the RAAF's No 1 Basic Flying Training School in September 1955. Delivery was completed in 1958 and ten years later 54 of the original 62 were still in service, indicating a remarkably low attrition rate for a trainer. After the RAAF introduced all-through jet training with the Macchi MB-326H (see page 155), the Winjeel (aboriginal for "young eagle") remained in service for student pilot flight assessment, initial Army student pilot training and for flying by members of the RAAF Academy. The Winjeel was expected to be phased out during 1975, after nearly 20 years service, in favour of the Aerospace Airtrainer (see page 159).

DASSAULT-BREGUET/DORNIER ALPHA JET

The Franco-German Alpha Jet, illustrated by the photograph above and the general arrangement silhouette (left) is scheduled to enter service during 1977

Country of Origin: France/Federal Germany.
Type: Two-seat light tactical aircraft and trainer.
Power Plant: Two 2,960 lb st (1 345 kg) SNECMA/Turboméca Larzac 04 turbofans.
Performance (Estimated): Max speed (clean) 633 mph (1 020 km/h) at sea-level, 580 mph (933 km/h) at 36,090 ft (11 000 m); max range (with external fuel) 1,243 mls (2 000 km) at 485 mph (780 km/h); max endurance, 2 hr 35 min; typical endurance (low-level training mission without external fuel) 1 hr 40 min; service ceiling, 45,930 ft (14,000 m).
Weights: Empty equipped, 6,945–7,275 lb (3 150–3 300 kg); normal loaded (trainer), 9 744 lb (4 420 kg); max take-off (trainer), 10,542 lb (4 782 kg); max take-off (strike), 15,430 lb (7 000 kg).
Accommodation: Two in tandem, with dual control (trainer only).
Armament: (Strike variant only) one 30-mm DEFA cannon with 150 rounds in pod beneath fuselage and four underwing stores stations each capable of carrying one 250-lb (113,4-kg), 500-lb (227-kg) or 1,000-lb (454-kg) bomb, a 600-lb (272-kg) cluster dispenser or a pod containing 36 2·75-in (7-cm) rockets.
Status: First flight made at Istres on 26 October 1973, by first of four prototypes. First production deliveries scheduled for 1976 against planned orders of 200 each for *l'Armée de l'Air* and *Luftwaffe*.
Notes: The Alpha Jet was chosen for joint development by France and Germany in July 1969, after evaluation of proposals from Breguet, Nord, Dornier and MBB. It was designed to meet requirements for an advanced trainer for the *Armée de l'Air* and a light tactical strike aircraft for the *Luftwaffe*. Four prototypes were to be built of which the first two (01, for testing in France and 02 for testing in Germany) were basic models to test the overall handling, performance, systems, etc, the third (03, tested in France) was to represent the production standard strike variant and the fourth (04, tested in Germany) was to represent the production trainer. Production is shared equally between the French and German aircraft industries, with final assembly lines at Toulouse in France and at Oberpfaffenhofen, near Munich, in Germany. The Alpha Jet is scheduled to attain service status with both French and German air arms in 1977, replacing the Lockheed T-33As and Mystère IVAs currently used for training by the former, and for close air support, battlefield reconnaissance, anti-helicopter operations and tactical training by the latter.

DE HAVILLAND D.H.115 VAMPIRE

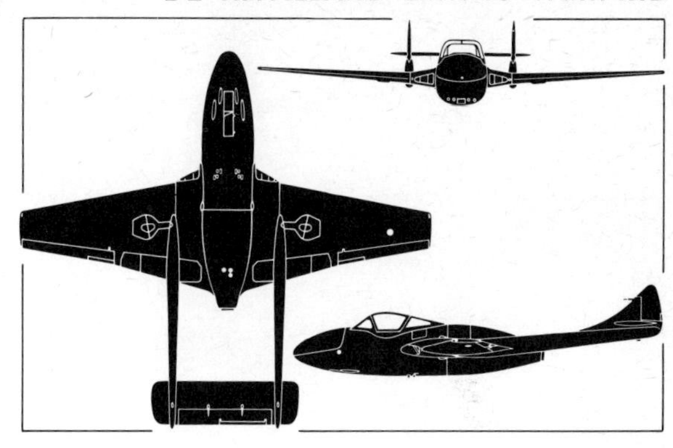

The Vampire two-seat trainer—the T Mk 55 version being illustrated by the photograph above and silhouette—is still included in the inventories of several air forces

Country of Origin: United Kingdom.
Type: Advanced trainer.
Power Plant: One 3,500 lb st (1 588 kgp) de·Havilland Goblin 35 turbojet.
Performance (T Mk 11 : clean) : Max speed, 538 mph (866 km/h) at sea level, 549 mph (885 km/h) at 20,000 ft (6 100 m) and 521 mph (839 km/h) at 40,000 ft (12 200 m); normal cruising speed, 403 mph (649 km/h) at 40,000 ft (12 200 m); initial rate of climb, 4,500 ft/min (22,8 m/sec); time to climb to 40,000 ft (12 200 m), 16·3 min; still air range, 623 mls (955 km) at 300 mph (480 km/h) at 20,000 ft (6 100 m) and 853 mls (1 370 km) at 403 mph (650 km/h) at 40,000 ft (12 200 m).
Weights: Empty, 7,380 lb (3 347 kg); loaded (clean), 11,150 lb (5 060 kg); loaded (two 100 Imp gal/455 1 drop tanks), 12,920 lb (5 860 kg).
Dimensions: Span, 38 ft 0 in (11,59 m); length, 34 ft 6½ in (10,51 m); height, 6 ft 2 in (1,88 m); wing area, 261 sq ft (24,25 m²).
Accommodation: Two side-by-side, full dual controls.
Armament: (T Mk 11) two or (T Mk 55) four 20 mm Hispano 404 cannon and provision for eight 25-lb (11,3 kg) or 60-lb (27,2 kg) rockets and two 500-lb (227-kg) bombs.
Status: Prototype D.H.100 Vampire single-seat fighter first flown on 20 September 1943; first D.H.115 Vampire Trainer flown on 15 November 1950; first production T Mk 11 flown on 1 December 1951. Production totals, 731 T Mk 11 and T Mk 55; 73 T Mk 22; 36 T Mk 33; 5 T Mk 34; 68 T Mk 35.
Notes: A two-seat training version of the Vampire fighter was adopted by the RAF as its standard advanced trainer from 1951, designated Vampire T Mk 11. The T Mk 22 was similar for use by the Royal Navy and the T Mk 55 was for export, some 20 different countries putting this variant into service. In addition to production in the UK, the Vampire trainer was produced in Australia for the RAAF (Vampire T Mk 33 and T Mk 35) and for the RAN (T Mk 34); and in India, Hindustan Aircraft Ltd assembled 55 T Mk 55s from British-built components. Principal users of two-seat Vampires in 1973 included the SAAF, Rhodesian Air Force, Austrian Air Force, Chilean Air Force, Swiss Air Force and Venezuelan Air Force. Some single-seat Vampires are also in use as advanced trainers in some of these countries.

DE HAVILLAND DHC-1 CHIPMUNK

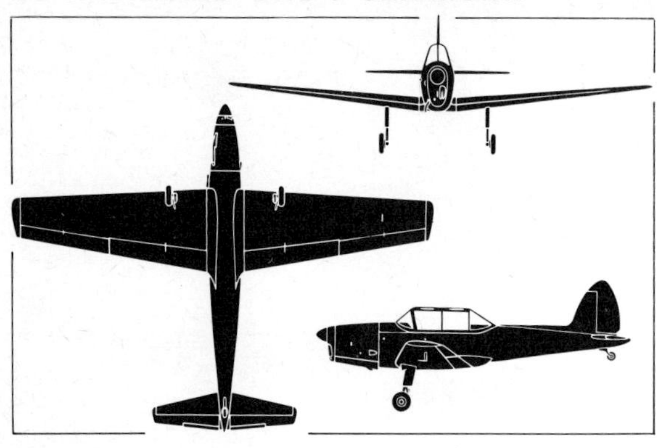

Both photograph above and general arrangement silhouette (left) depict the Chipmunk T Mk 20, the former illustrating an example flown by the Royal Danish Air Force

Country of Origin: Canada.
Type: Primary Trainer.
Power Plant: One 145 hp de Havilland Gypsy Major 8 (military) or Gipsy Major 10 Mk 2 (civil) four-cylinder inverted in-line engine.
Performance: Max speed, 138 mph (222 km/h) at sea level, cruising speed, 119 mph (191 km/h) at sea level; initial rate of climb, 840 ft/min (4·27 m/sec); service ceiling, 15,800 ft (4 820 m); range, 280 mls (445 km) at 116 mph (187 km/h) at 5,000 ft (1 525 m).
Weights: Empty, 1,425 lb (647 kg); normal loaded, 2,014 lb (914 kg); max take-off, 2,100 lb (953 kg).
Dimensions: Span, 34 ft 4 in (10,46 m); length, 25 ft 5 in (7,75 m); height, 7 ft 0 in (2,13 m); wing area, 172 sq ft (15,98 m²).
Accommodation: Two in tandem, full dual controls.
Status: Prototype first flown 22 May 1946; production deliveries began 1947. First British production aircraft completed September 1949. Production totals: DHC-1B-1, 157; DHC-1B-2, 60; T Mk 10, 740; T Mk 20, 231; Mk 21, 29; by OFEMA in Portugal, 60.
Notes: The Chipmunk was the first original design by the Canadian associate of the de Havilland company, which had built Mosquitoes during World War II and chose the primary trainer as a suitable type of aircraft with which to enter the market post-war. Initial sales were of the DHC-1B-1 version, and included 22 for the Egyptian Air Force and 18 for the Royal Thai Air Force. The RCAF also bought 8, followed by 60 of the DHC-1B-2 version with a one-piece blown canopy over the cockpits, which was designated Chipmunk T Mk 30. In 1948 the Chipmunk was adopted for service with the RAF, initially with Volunteer Reserve units and later with other organisations such as the University Air Squadrons. A production line was established at the Chester factory of the parent company, where 889 were completed, plus another 111 at the Hatfield works. The majority of these were Chipmunk T Mk 10s for the RAF, but examples were exported for military use to Denmark, Ceylon, Iraq, Portugal, Thailand, Burma, Eire, Syria, Egypt, Colombia, Uruguay, Saudi Arabia and the Lebanon, ex-RAF Chipmunks being supplied to Ghana, Jordan, Kenya and Zambia. Air forces of several of these nations continued to operate Chipmunks in 1973, while about 100 remained in service with the RAF, RN and British Army. The Chipmunk was being replaced by the Bulldog in University Air Squadrons in 1973–74.

FMA IA 35 HUANQUERO/PANDORA

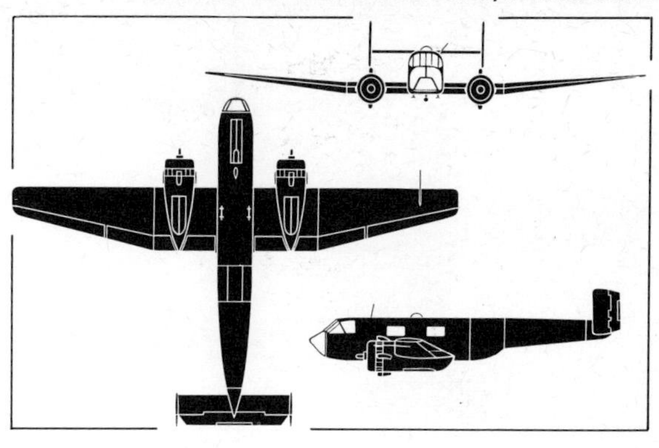

The photograph above and the general arrangement silhouette (right) illustrate the Type IA advanced pilot and navigational trainer of the FMA Huanquero

Country of Origin: Argentina.
Type: General-purpose military aircraft.
Power Plant: Two 620 hp IA 19R El Indio radial engines.
Performance: Max speed, 225 mph (362 km/h) at 9,840 ft (3 000 m); max cruising speed, 217 mph (350 km/h) at 9,840 ft (3 000 m); economical cruising speed, 200 mph (320 km/h) at 9,840 ft (3 000 m); initial rate of climb, 984 ft/min (5,0 m/sec); service ceiling, 21,000 ft (6 400 m); range with max fuel, 975 mls (1 570 km).
Weights: Empty equipped, 7,700 lb (3 500 kg); typical take-off weight, 12,540 lb (5 700 kg).
Dimensions: Span, 64 ft 3 in (19,60 m); length, 45 ft 10 in (13,98 m).
Accommodation: Two pilots and radio operator, plus other personnel appropriate to mission, eg, instructor and four pupils in Type IA navigation trainer; seven passengers in Type II transport or four stretcher patients and one attendant in Type III ambulance.
Armament (Type IB only): Two 12·7-mm Browning machine guns with 534 rpg and four wing strong points, two of which can carry bombs up to 220 lb (100 kg) in weight, and two of which each carry four 2·25-in rockets.
Status: Prototype first flown on 21 September 1953. First production aircraft flown on 29 March 1957. A total of 47 built for use by *Fuerza Aerea Argentina* (Argentine Air Force), production completed.
Notes: The Huanquero was developed at the Argentine's state-owned aircraft manufacturing organisation, the *Fabrica Militar de Avions*, a part of the Dinfia directorate at Cordoba. Work on this twin-engined general purpose aircraft began in 1950 and a prototype known as the IA 35 flew in 1953. After successful testing, the Argentine Air Force ordered 100, but production was terminated after a total of only 47 had been built, in three principal versions. These were the Type IA advanced pilot and navigation trainer, the Type IB weapons trainer and the Type IV photographic survey model. About 35 examples of these variants remained in service in 1973, these serving temporarily in the attack rôle with the *II Grupo Exploración y Ataque* until replaced by helicopters. Other variants produced were the Type III ambulance and the Type II light transport (prototype only, with tip tanks). One production Huanquero was converted as a VIP transport named the Constancia.

FOKKER S-11 AND S-12 INSTRUCTOR

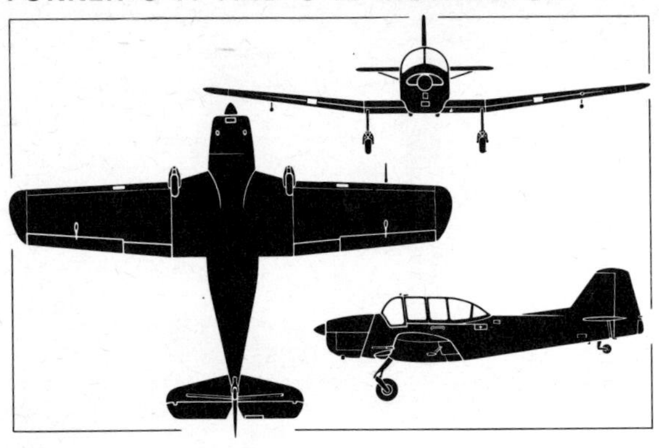

The S-11 Instructor, illustrated by the photograph above and the general arrangement silhouette (left), was being phased out of the Dutch Air Force and Naval Air Arm late 1973

Country of Origin: Netherlands.
Type: Primary Trainer.
Power Plant: One 190 hp Lycoming O-435A six-cylinder horizontally-opposed engine.
Performance: Max speed, 130 mph (209 km/h) at sea level; cruising speed, 102 mph (164 km/h) at sea level; time to climb to 3,280 ft (1 000 m), 5·6 min; service ceiling, 12,600 ft (3 850 m); endurance, 3 hrs 35 min; range 430 mls (695 km).
Weights: Empty equipped, 1,785 lb (810 kg); normal take-off weight, 2,425 lb (1 100 kg).
Dimensions: Span, 36 ft 1 in (11,0 m); length, 26 ft 8 in (8,18 m); height, 7 ft $10\frac{1}{2}$ in (2,70 m); wing area, 199 sq ft (18,5 m²).
Accommodation: Two in tandem, dual controls.
Status: Prototype (S-11) first flown on 18 December 1947. Production deliveries began in 1949, comprising 40 to Royal Netherlands AF, 41 to Israeli Defence Force, 150 to Italian Air Force (built by Macchi in Italy) and 100 to Brazilian Air Force (built in Brazil). First S-12 flown in 1948, production total 50 (built in Brazil).
Notes: The S-11 Instructor was one of the first post-war products of the Fokker company, designed for military or civil use as a primary trainer. In this rôle, it was adopted by the Royal Netherlands AF and about 10 were still in use in 1973 at the Gilze Rizen base to give a short period of *ab initio* training to military pilots before they are sent to Canada for a two-year training course to wings standard, the Instructor-equipped unit being known as the S.11 Introduction Flight. The Dutch naval air component, the *Marineluchtvaartdienst* was also using a few S.11 Instructors for pilot training with No 9 Squadron during 1973. The other major customer for S-11s built in Holland was the Israeli Defence Force, but the largest order came from the Italian Air Force and was fulfilled through a licence production arrangement between Fokker and Macchi, the Italian-built aircraft being designated M.416. To meet a Brazilian Air Force order for 100 S-11s, Fokker set up a company in Brazil, the Fokker Industria Aeronautica SA, and the latter company also built 50 S.12s after a single prototype had been built in the Netherlands. The S-12 differed from the S-11 only in having a nosewheel undercarriage. One S-12 was converted by the Brazilian PAR to the FG-8 Guanabara four-seat executive transport. This featured a retractable undercarriage, wing-tip fuel tanks and a 250 hp Lycoming O-540-AIA engine.

FUJI T1

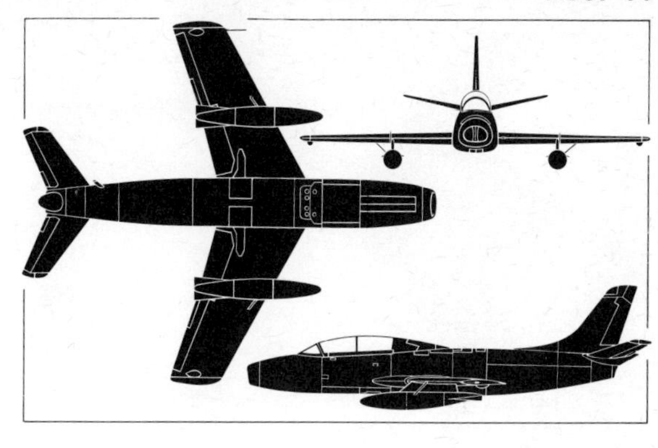

The Fuji T1, illustrated by the photograph above and the general arrangement silhouette (right), has now been in service with the Air Self-Defence Force for a decade

Country of Origin: Japan.
Type: Basic trainer.
Power Plant: One (T1A) 3,990 lb st (1 810 kgp) Rolls-Royce Bristol Orpheus 805, or (T1B) 2,645 lb st (1 200 kgp) Ishikawajima-Harima J3-IHI-3 or (T1C) 3,085 lb st (1 400 kgp) J3-IHI-7 turbojet.
Performance (T1A): Max speed (clean) 575 mph (925 km/h) at 20,000 ft (6 100 m); max speed with external tanks, 510 mph (825 km/h) at 25,000 ft (7 600 m); cruising speed, 357 mph (575 km/h) at 10,000 ft (3 050 m) and 385 mph (620 km/h) at 30,000 ft (9 150 m); initial rate of climb, 6,500 ft/min (33 m/sec); service ceiling, 47,244 ft (14 400 m); max range (clean) 805 mls (1 300 km); max range (external fuel) 1,210 mls (1 950 km).
Performance (T1B): Max speed, 518 mph (834 km/h) at 19,685 ft (6 000 m); cruising speed, 357 mph (575 km/h); initial rate of climb, 4,724 ft/min (24,0 m/sec), service ceiling 39,370 ft (12 000 m).
Weights (T1A): Empty, 6,078 lb (2 755 kg); normal take-off, 9,502 lb (4 310 kg); max overload, 11,005 lb (5 000 kg).
Weights (T1B): Empty, 6,261 lb (2 840 kg); normal take-off, 9,678 lb (4 390 kg).
Armament: Provision for one 0·50-in gun in nose and underwing provision on two strong points for two Sidewinder AAMs, four HVARs, rocket pods or bombs up to total of 1,500 lb (680 kg).
Accommodation: Two in tandem, dual controls.
Status: First of two prototypes flown on 8 January 1958; first T1B flown on 17 May 1960; first T1C flown in April 1965. Production totals: two prototypes, four pre-production, 40 T1A and 20 T1B, all for JASDF; production ended 1963.
Notes: This basic jet trainer was designed to replace North American T-6 Texans in service with the JASDF, and was Japan's first post-war jet aircraft. The initial version was designated T1F2 by Fuji, with imported Orpheus engines, and entered service with the designation T1A. Delivery of two batches of 20 each was completed by July 1962. A batch of 20 T1F1s with the indigenous J3 engine was delivered in 1962/63, with the JASDF designation T1B. One prototype was later re-engined with the uprated J3-IHI-7 as the T1F3 and service aircraft retrospectively fitted with this engine were designated T1C. Late in 1973, Fuji was proposing a revised version of the basic T1 with new fuselage and an Adour engine.

GRUMMAN TF-9J COUGAR

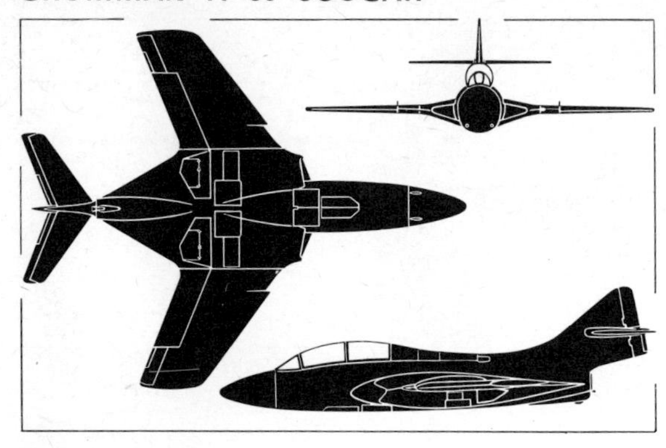

The TF-9J Cougar, illustrated by the photograph above and the general arrangement silhouette (left), remains the standard US Navy advanced trainer

Country of Origin: USA.
Type: Advanced trainer.
Power Plant: One 7,200 lb st (3 266 kgp) Pratt & Whitney J48-P-8A turbojet.
Performance: Max speed, 705 mph (1 135 km/h) at sea level; time to climb to 40,000 ft (12 192 m), 8·5 min; service ceiling, 50,000 ft (15 240 m); range, 600 mls (966 km).
Weights: Normal take-off, 20,600 lb (9 344 kg).
Dimensions: Span, 34 ft 6 in (10,52 m); length, 44 ft 5 in (13,54 m); height, 12 ft 3 in (3,73 m).
Accommodation: Two in tandem, dual controls.
Armament: Two 20 mm cannon in forward fuselage. Provision on underwing strong points for two 1,000-lb (454-kg) bombs or six rockets or four AAMs.
Status: Prototype (XF9F-6) first flown on 20 September 1951. Production deliveries began November 1952. First F9F-8 flown on 18 December 1953; first F9F-8P flown on 21 August 1955; first F9F-9T flown on 4 April 1956. Production deliveries of F9F-8T began in mid-1956. Production quantities: F9F-6, 706; F9F-7, 168; F-9F-8, -8B and -8P, 712; F9F-8T, 399. Production ended December 1959.
Notes: Grumman's Cougar was a swept-wing version of the company's first jet fighter, the F9F Panther. The F9F-6, -7 and -8 versions were single-seat fighters used by the USN and USMC, later re-designated F-9F, F-9H and F-9J respectively, while the missile-armed F9F-8B became the AF-9J and the camera-equipped F9F-8P became RF-9J. A considerable number of these fighters were converted as target drones designated QF-9F, QF-9G and QF-9J while the drone directors were DF-9F, and some of these still serve as missile targets. The two-seat F9F-8T, now the TF-9J, embodied a 34-in (86·3-cm) extension of the fuselage nose to accommodate the additional cockpit. It has been standard equipment at US Navy advanced flying training schools since 1955 and continues to be so used in a three-aircraft training syllabus which also embraces Beech T-34B (see page 162) in the primary training rôle, and the T-2 Buckeye (see page 191) for the basic training phase. Two TF-9Js were supplied to the Argentine Navy, in addition to 10 of the straight-wing single-seat F9F-2 Panthers, but the latter have now been withdrawn from service. Two single-seat F-IIA Tigers were in use in 1974 for in-flight reverse thrust research.

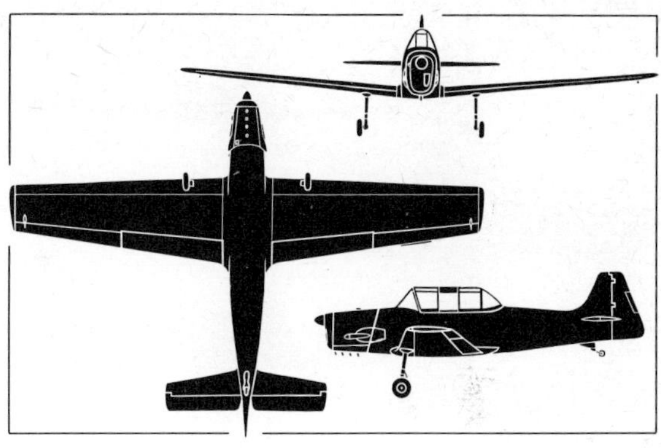

The HT-2 primary trainer, depicted by the photograph (above) and the general arrangement silhouette (right), was the first Indian aircraft of indigenous design to be built in quantity

Country of Origin: India.
Type: Primary trainer.
Power Plant: One 155 hp Blackburn Cirrus Major III four-cylinder in-line engine.
Performance: Max speed, 130 mph (209 km/h); max cruising speed, 115 mph (185 km/h); initial rate of climb, 800 ft/min (4,1 m/sec); service ceiling, 14,500 ft (4 420 m); max range, 350 mls (560 km); endurance, 3·5 hrs.
Weights: Empty, 1,540 lb (702 kg); max take-off, 2,240 lb (1 018 kg).
Dimensions: Span, 35 ft 2 in (10,72 m); length, 24 ft 8½ in (7,53 m); height, 8 ft 11 ins (2,72 m); wing area, 173·4 sq ft (16,0 m²).
Accommodation: Two in tandem, dual controls.
Status: Prototype first flown on 5 August 1951; second prototype flown on 19 February 1952. Production total 169, completed in 1958.
Notes: Although Hindustan Aircraft Ltd (now Hindustan Aeronautics) had built and flown a troop-carrying glider of original design in 1942, the design and construction of powered aircraft by the company was not authorised by the Indian government until September 1948. A programme was then drawn up covering work on three types—the HT-2 primary trainer, HT-10 advanced trainer (with turboprop engine) and HT-11 basic trainer intended to replace the Percival Prentice then in service with the Indian Air Force. Of this trio, only the first was destined to be completed, the prototype making its first flight in 1951 with a 145 hp DH Gipsy Major 10 engine. Early test flying indicated that the HT-2 was underpowered in this form and the second prototype was fitted with a 155 hp Blackburn Cirrus III engine, this power plant being selected for the production version which was ordered on behalf of the Indian Air Force, Indian Navy and local flying clubs. Of entirely conventional all-metal construction, the HT-2 entered service in 1955 and remains the standard primary trainer used by the Indian Air Force and Navy. Single examples were presented by the Indian government to Indonesia and Singapore, but the only export order was received from Ghana Air Force, which purchased 12 and used them until replaced by Bulldogs in 1973. No requirement currently exists for a replacement for the HT-2 in Indian service, and it is likely to be retained for the remainder of the present decade.

HAL HJT-16 KIRAN

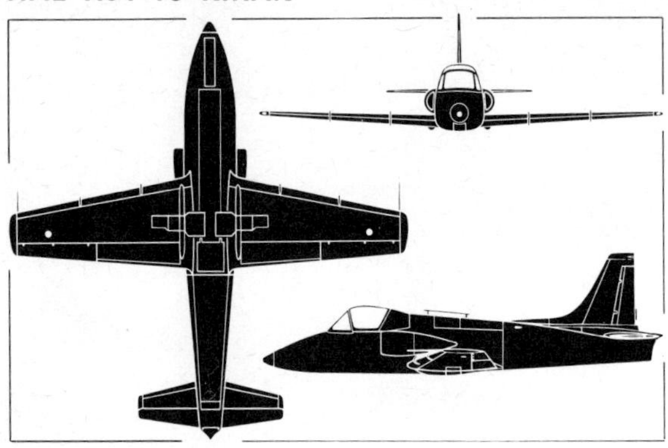

Illustrated by the photograph (above) and the general arrangement silhouette (left), the HJT-16 Kiran serves in relatively small numbers with the Indian Air Force and naval air component

Country of Origin: India.
Type: Basic trainer.
Power Plant: One 2,500 lb st (1 135 kg) Rolls-Royce Bristol Viper ASV.11 turbojet.
Performance: Max speed, 432 mph (695 km/h) at sea level and 427 mph (688 km/h) at 30,000 ft (9 150 m); initial rate of climb, 4,774 ft/min (24,1 m/sec); time to reach 30,000 ft (9 144 m), 10 min; service ceiling, 42,000 ft (12 800 m); range, 465 mls (748 km) at 265 mph (426 km/h) at 30,000 ft (9 144 m).
Weights: Empty, 5,644 lb (2 560 kg); normal take-off, 7,936 lb (3 600 kg); max overload, 9 259 lb (4 200 kg).
Dimensions: Span, 35 ft 1¼ in (10,70 m); length, 34 ft 9 in (10,60 m); height, 11 ft 11 in (3,64 m); wing area, 204·5 sq ft (19,00 m²).
Accommodation: Two side-by-side, with full dual control.
Status: Prototype first flown 4 September 1964; second prototype flown August 1965; pre-production deliveries began March 1968.
Notes: Design of the Kiran (Ray of Light) began at the HAL plant in Bangalore in 1961 by a team led by Dr V M Ghatage, with the object of developing a basic jet trainer to replace the licence-built Vampires.

Fully pressurised, the HJT-16 is comparable with other basic trainers in the same class produced in other countries, and is equipped with Martin-Baker zero-zero ejection seats. The undercarriage is by Dowty Rotol and much of the equipment is supplied by British companies but the structure is wholly Indian in design and manufacture. When the design was launched, it was intended that the ultimate production version should be powered by the HAL-developed HJE-2500 turbojet but this plan was later abandoned and all production aircraft are powered by the locally-built Viper engine. A pre-production batch of 24 Kirans was built, the first six of these being handed over to the Indian Air Force in March 1968, and production was continuing in 1973 against an order for 36 aircraft. Seven examples have been delivered to the Indian Navy and are operated by INAS 550 and INAS 551. During the course of 1973, the Kiran entered service with the Indian Air Force Academy at Dundigal, its operation having been previously confined to the Flying Instructors School at Tambaram. Follow-up orders for a total of a further 100 Kirans were expected to be placed on behalf of the Indian Air Force and some effort was being made to obtain export orders for the trainer during 1973.

HAWKER SIDDELEY DOMINIE (AND HS.125)

The Dominie T Mk 1, illustrated above and by the general arrangement silhouette (right), serves with RAF Training Command in the navigational training rôle

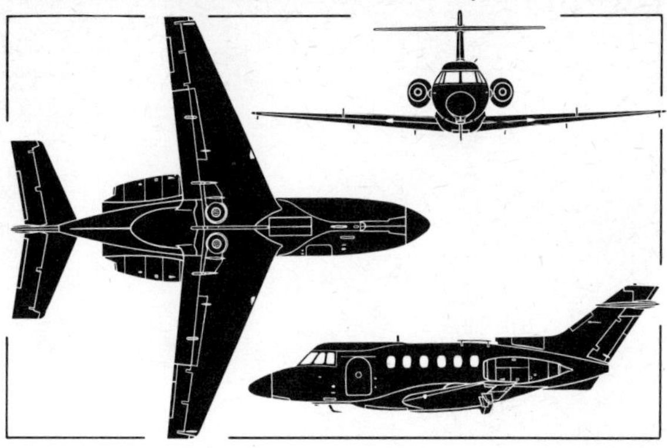

Country of Origin: United Kingdom.
Type: Navigation trainer and communications aircraft.
Power Plant: Two 3,000 lb st (1 360 kgp) Rolls Royce Bristol Viper 301 turbojets.
Performance: Max speed, 500 mph (805 km/h) at 30,000 ft (9 145 m); typical operating speed, 368 mph (593 km/h) at 35,000 ft (10 680 m); initial rate of climb, 4,000 ft/min (20,3 m/sec); service ceiling, 40,000 ft (12 200 m).
Weights: Basic weight, 10,100 lb (4 581 kg); radio and navaids, 1,390 lb (631 kg); max take-off, 21,200 lb (9 615 kg); max landing, 19,550 lb (8 865 kg).
Dimensions: Span, 47 ft 0 in (14,33 m); length, 47 ft 5 in (14,45 m); height, 16 ft 6 in (5,03 m); wing area, 353 sq ft (32,8 m²).
Accommodation: Two pilots with dual control (provision for single-pilot operation); four seats in cabin for instructor, two pupils and supernumerary crew.
Status: Prototype D.H.125 first flown on 13 August 1962. First HS.125 Srs 2 Dominie flown on 1 December 1964. Production of 20 Dominies for RAF completed in 1966; production of HS.125s for military and civil use continuing in 1973.

Notes: The Hawker Siddeley (de Havilland) 125 business jet was selected in September 1962 to meet RAF needs for a new navigation trainer, and a batch of 20 was produced. These are operated primarily from No 1 Air Navigation School at Stradishall, others being used by the College of Air Warfare at Manby. Installation of Decca Type 62 Doppler aerials under the fuselage results in the Dominie having a longer wing-to-fuselage fairing than the standard commercial model (see *Civil* volume) and this version was the first to introduce the small ventral fins fitted also on later HS.125s. In addition to the Dominie, the RAF uses HS.125s in the communications rôle, having acquired four Srs 400s and two Srs 600s, with the official designation of HS.125 CC Mk 1 and CC Mk 2 respectively. Several other air forces have also bought the HS.125 for use in the communications rôle, for checking and calibrating radio and radar aids and for VIP transportation. Among these users are the South African Air Force, which has acquired seven to which the name Mercurius has been applied, the Brazilian Air Force, which has bought 11, and the Royal Malaysian Air Force. The Ghana Air Force and the air component of the Argentine Navy have acquired one HS.125 each.

HISPANO HA-200 SAETA AND HA-220 SUPER SAETA

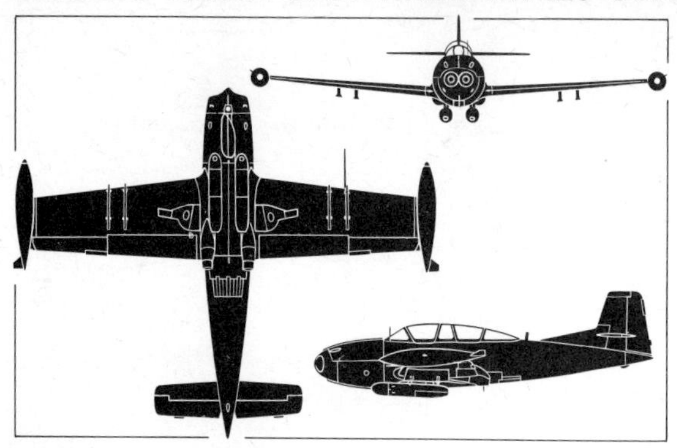

The photograph above depicts the HA-200D Saeta and the general arrangement silhouette (left) illustrates the more powerful, single-seat HA-220 Super Saeta

Country of Origin: Spain.
Type: Basic trainer and light attack aircraft.
Power Plant: (HA-200A, B & D) Two 880 lb st (400 kgp) Turboméca Marboré II turbojets or (HA-220) two 1,058 lb st (480 kgp) Marboré VI turbojets.
Performance (HA-200D): Max speed, 404 mph (650 km/h); max cruising speed, 330 mph (530 km/h); initial rate of climb, 2,755 ft/min (14,0 m/sec); service ceiling, 39,360 ft (12 000 m); range with max fuel, 930 mls (1 500 km).
Performance (HA-220): Max speed, 435 mph (700 km/h) at 23,000 ft (7 000 m); normal cruising speed, 354 mph (570 km/h) at 19,700 ft (6 000 m); initial rate of climb, 3,346 ft/min (17,0 m/sec); service ceiling, 42,650 ft (13 000 m); range with max fuel, 1,055 mls (1 700 km).
Weights (HA-200D): Empty equipped, 4,035 lb (1 830 kg); max take-off, 7,385 lb (3 350 kg); max landing, 5,840 lb (2 650 kg).
Weights (HA-220): Basic empty, 4,894 lb (2 220 kg); max take-off, 8,157 lb (3 700 kg).
Dimensions: Span, 34 ft 2 in (10,42 m); span over tip tanks, 35 ft 10 in (10,93 m); length, 29 ft 5 in (8,97 m); height, 9 ft 4 in (2,85 m); wing area, 187·2 sq ft (17,40 m²).
Accommodation (HA-200): Two in tandem, full dual controls; (HA-220) Pilot only.
Armament: Two 7·7-mm Breda machine guns in upper front fuselage; two strong points under fuselage with 375-lb (170-kg) capacity each and four underwing strong points with 551-lb (250-kg) capacity each.
Status: First prototype flown on 12 August 1955; first HA-200A flown on 11 October 1962; first HA-200B flown on 21 July 1960; first HA-200D flown in April 1965; first HA-220 flown 25 April 1970. Production quantities: two prototypes; HA-200, 10; HA-200A, 30; HA-200D, 55; HA-220, 25; plus 90 HA-200B Al-Kahira built under license in Egypt.
Notes: The HA-200 is used by the Spanish Air Force as a basic jet trainer, the E-14A, while the similar HA-200D, with underwing armament provision, is used in the light strike rôle as the C-10B. In the more powerful HA-220, the place of the second seat is taken by an extra fuel tank; this type operates as the C-10C. Five of the pre-production HA-200s were modified to HA-200B with a single 20-mm Hispano-Suiza cannon armament.

IPD-6201 UNIVERSAL

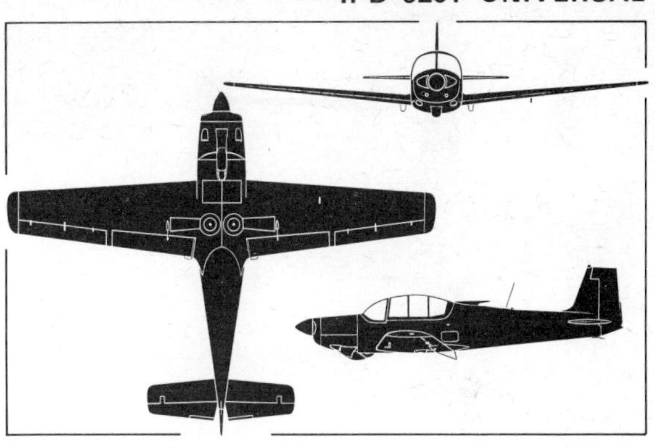

The T-25 Universal, illustrated by the photograph above and the general arrangement silhouette (right), is replacing the T-6 Texan as a basic trainer in the Brazilian Air Force

Country of Origin: Brazil.
Type: Basic trainer.
Power Plant: One 300 hp Lycoming IO-540-K1D5 six-cylinder horizontally-opposed engine.
Performance: Max speed, 195 mph (315 km/h) at sea level, cruising speed, 172 mph (278 km/h) at sea level, initial rate of climb, 1,380 ft/min (7,0 m/sec); service ceiling, 19,700 ft (6 000 m); range with max payload, 447 mls (720 km); range with max fuel, 975 mls (1 570 km).
Weights: Empty equipped, 2,425 lb (1 100 kg); max take-off and landing, 3,748 lb (1 700 kg).
Dimensions: Span, 36 ft 1 in (11,0 m); length, 28 ft 2½ in (8,60 m); height, 9 ft 9¾ in (3,0 m); wing area, 185,14 sq ft (17,20 m²).
Accommodation: Two seats side-by-side, with full dual control, and optional third seat to rear.
Status: Prototype first flown on 29 April 1966; first production model flown on 7 April 1971. Production deliveries began late summer of 1971; 150 in production for Brazilian Air Force.
Notes: The Universal was developed by the Neiva company to meet a Brazilian Air Force requirement for a basic trainer which could replace the North American T-6 Texan and the Fokker S-11/S-12 Instructor. Unlike the types it had to replace, the new trainer was required to provide side-by-side seating, with provision for a third occupant behind the two pilots. After the prototype had been tested in 1966/67, the Brazilian Air Force ordered the Universal into production and assigned it the designation T-25. Production is shared between the two Neiva factories, at Botucatú and San José dos Campos, with final assembly at the latter and sub-contract manufacture of components by the Avitec company. The Universal is of all-metal construction, with a single-spar wing, split flaps and slotted ailerons. The undercarriage retracts hydraulically, the main wheels having disc brakes and the nose wheel being steerable. Fuel is carried in six metal cells in the wings. For weapons training the Universal may have two wing-mounted 7·62-mm machine guns and underwing attachment points for bombs or rockets. Proposals have been made for a low-cost counter-insurgency derivative, the Turbo-Universal, with a Pratt & Whitney PT6A turboprop. Neiva was working in 1973 on the Bi-Universal, a six-seat light transport and communications aircraft using many of the components of the T-25 and powered by two Lycoming engines of 300 hp each.

LOCKHEED T-33A SHOOTING STAR

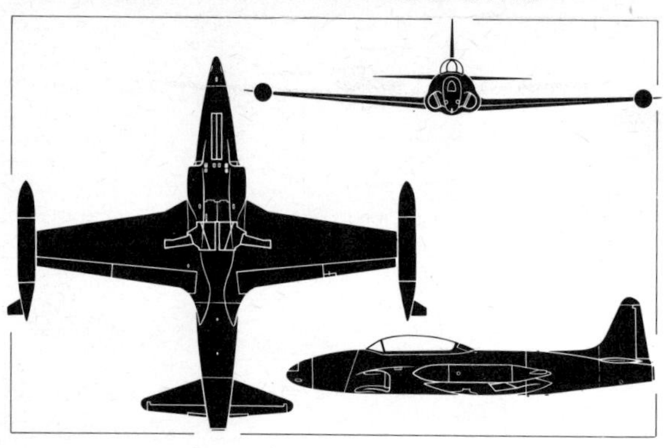

The T-33A, illustrated by the general arrangement silhouette (left) and by the photograph (above) which depicts it in Spanish service, remains one of the most widely used of trainers

Country of Origin: USA.
Type: Advanced trainer.
Power Plant: One 5,200 lb st (2 360 kgp) Allison J33-A-35 turbojet or (T-33A-N) 5,100 lb st (2 313 kgp) Rolls-Royce Nene 10 turbojet.
Performance: Max speed, 600 mph (960 km/h) at sea level, 543 mph (874 km/h) at 25,000 ft (7 620 m); initial rate of climb, 5,525 ft/min (28,0 m/sec); service ceiling, 47,500 ft (14 480 m); range, 1,345 mls (2 165 km); endurance, 3·12 hrs.
Weights: Empty, 8,084 lb (3 667 kg); normal take-off, 11,695 lb (5 305 kg); max overload, 14,442 lb (6 551 kg).
Dimensions: Span, 38 ft 10½ in (11,85 m); length 37 ft 9 in (11,48 m); height, 11 ft 8 in (3,55 m); wing area, 237 sq ft (22,01 m^2).
Armament: Two 0·50-in (12,7 mm) M-3 machine guns.
Accommodation: Two in tandem, full dual controls.
Status: Prototype (TF-80C) first flown on 22 March 1948. Production of approximately 5,800 by Lockheed completed between 1948 and 1959. Under license, Kawasaki built 210 in Japan and Canadair built 656 in Canada.

Notes: This two-seat training version of the F-80 Shooting Star (the USAF's first operational jet fighter) was evolved in 1947 to meet Air Force requirements for an advanced jet trainer. The modifications comprised lengthening the fuselage to provide space for a second seat and extending the cockpit canopy to cover both occupants. Initial production aircraft were designated TF-80C for the USAF and TV-2 for the US Navy before becoming T-33A and T-33B respectively. Of the production total, the Navy received 649, and 1,058 were procured by USAF for supply to foreign nations through MAP, some of these being armed for close support and interdiction with guns, rockets or bombs and designated AT-33A. Also included in the MAP-supplied batches were 85 RT-33As with lengthened, camera-carrying nose and special equipment replacing the rear seat. For service with the RCAF, Canadair built 656 Silver Star Mk 3s (CL-30), after a trial installation of the Nene powerplant in a Lockheed airframe as the Silver Star Mk 2. Twenty standard T-33As served with the RCAF as Silver Star Mk 1s. T-33A "T-Birds" remain in service with many air forces, including those of Belgium, Brazil, Canada, Chile, Colombia, Denmark, Ethiopia, France, Greece, Iran, Italy, Japan, Pakistan and Portugal.

MIKOYAN-GUREVICH MIG-15UTI

The MiG-15UTI, illustrated by the photograph above and by the general arrangement silhouette (right), is widely used by Soviet bloc and other air forces

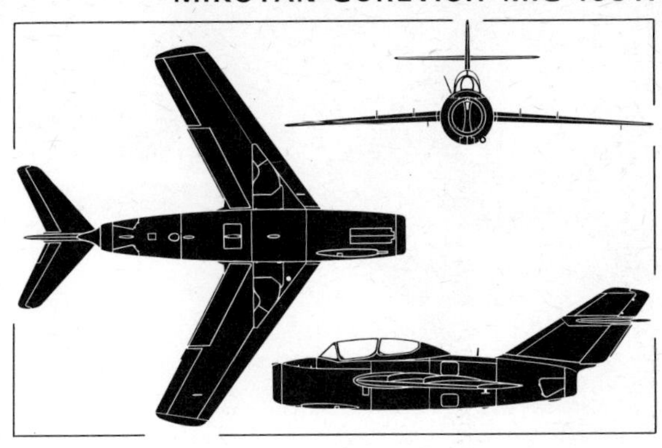

Country of Origin: USSR.
NATO Code Name: *Midget.*
Type: Advanced trainer.
Power Plant: One 5,952 lb (2 700 kg) Klimov VK-1 turbojet.
Performance: Max speed (clean), 630 mph (1 015 km/h) at sea level, (with two 54·4 Imp gal/247,5 l slipper tanks), 559 mph (900 km/h); initial climb (clean), 10,235 ft/min (52 m/sec); time to 16,405 ft (5 000 m), 2·6 min, to 32,810 ft (10 000 m), 6·8 min; service ceiling, 47,980 ft (14 625 m); range at 16,405 ft (5 000 m), (clean) 422 mls (680 km), (with slipper tanks) 590 mls (950 km); range at 32,810 ft (10 000 m), (clean) 590 mls (950 km), (with slipper tanks, 885 mls (1 424 km).
Weights: Empty, 8,818 lb (4 000 kg); loaded (clean), 10,692 lb (4 850 kg); max take-off, 11,905 lb (5 400 kg).
Dimensions: Span, 33 ft 0⅞ in (10,08 m); length, 32 ft 11¼ in (10,04 m); height, 12 ft 1⅝ in (3,70 m); wing area, 221·74 sq ft (20,6 m²).
Accommodation: Pupil and instructor in tandem.
Armament: One 23-mm NS-23 cannon with 80 rounds or one 12,7-mm UBK-E machine gun with 150 rounds offset to port in fuselage nose.
Status: Prototype flown as conversion of single-seat MiG-15 fighter in 1948 with production deliveries commencing in following year. Licence manufacture subsequently undertaken in Czechoslovakia as the CS 102, in Poland as the SBLim-1 and in China.
Notes: Still widely used in the advanced training rôle, the MiG-15UTI was derived from the initial production version of the MiG-15 interceptor and was originally powered by an RD-45F turbojet rated at 5,005 lb (2 270 kg) and, later, the improved RD-45FA, the more powerful VK-1 eventually being standardised together with other improvements introduced by the MiG-15bis interceptor. Some Polish-built SBLim-1 trainers were fitted with the complete power plant, rear fuselage and tail assembly of obsolete LIM-2 (MiG-15bis) fighters to conform with the late production Soviet MiG-15UTI. Although largely supplanted by the MiG-21U (*Mongol*), the MiG-15UTI remains in service in substantial numbers with the Soviet Air Forces, and the air arms of Afghanistan, Algeria, Cuba, Egypt, Guinea, Iraq, Khmer, Korea (North), Mali, Somalia, Southern Yemen, Syria, Uganda and North Yemen, the Chinese-built version serving with China, Pakistan and North Vietnam.

MITSUBISHI T-2

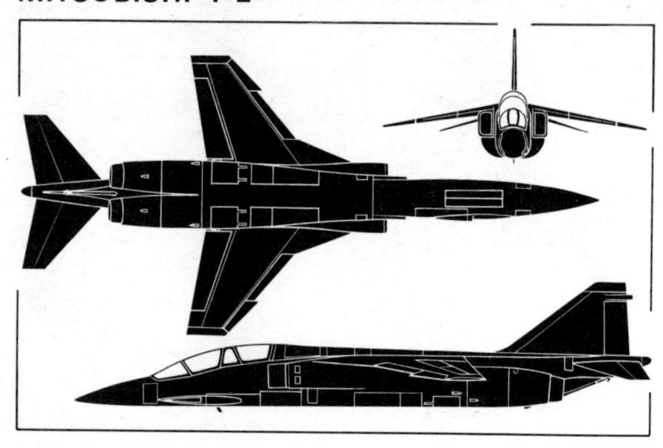

The Mitsubishi T-2A, illustrated by the photograph above and by the general arrangement silhouette (left), will enter Air Self-Defence Force service during 1975

Country of Origin: Japan.
Type: Advanced trainer and strike aircraft.
Power Plant: Two 4,600 lb st (2 086 kgp) dry and 6,950 lb st (3 150 kgp) with reheat Rolls-Royce Turboméca RB.172 Adour 801 turbofans.
Performance (Estimated): Max speed, 1,056 mph (1 700 km/h) or Mach 1·6 at 40,000 ft (12 190 m); service ceiling, 50,000 ft (15 250 m); max ferry range, 1,610 mls (2 595 km).
Weights: Approximate gross weight, 21,000 lb (9 525 kg).
Dimensions: Span, 25 ft 11 in (7,90 m); length, 58 ft 4¾ in (17,80 m); height, 14 ft 9¼ in (4,50 m); wing area, 228·2 sq ft (21,2 m²).
Accommodation: Two in tandem, full dual controls.
Armament: Provision for one 20-mm rotary cannon in lower front fuselage; provision for external stores on fuselage centre line and wing pylons and at wing tips.
Status: First of two prototypes flown on 20 July 1971. First pre-production T-2 flown in 1972. Production deliveries scheduled to begin in 1974.
Notes: Mitsubishi was selected in September 1967 to proceed with development of an advanced trainer with strike potential, intended to replace Lockheed T-33 trainers and North American F-86F Sabre close-support fighters in service with the JASDF. Final assembly of the T-2A is undertaken by Mitsubishi at Nagoya, with Fuji, Kawasaki and Shin Meiwa contributing as sub-contractors. Production of up to 80 T-2As is projected, of which 59 were included in the 1972–75 defence programme, with deliveries to the first JASDF training squadron scheduled to be formed during 1975. Planning in 1973 anticipated the re-equipment of two F-86F squadrons with the T-2A, these units then being transferred from the Central Air Command to the Air Training Command. In addition, a single-seat close support fighter variant, the Mitsubishi F-1 (FS-T-2Kai) is planned for introduction in 1975. The F-1, for which the JASDF has a projected requirement for 126 of which 68 are included in the 1972–75 programme, will have a gross weight of 30,865 lb (14 000 kg) and a combat radius of 345 mls (556 km) carrying eight 500 lb (227-kg) bombs. In addition to a 20-mm rotary-barrel cannon, the F-1 will have air-to-air and air-to-ground radar and an integrated data computer, and will be able to carry four IR-homing AAMs for the secondary intercept mission.

MORANE-SAULNIER MS-760 PARIS

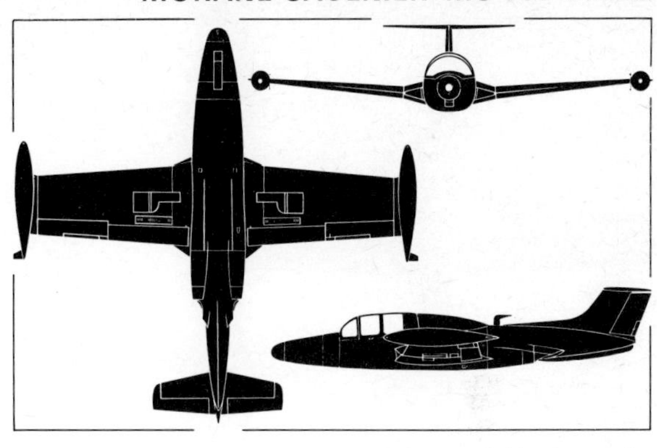

The photograph above illustrates an MS-760A Paris I of the Argentine Air Force, this version of the Paris also being depicted by the general arrangement silhouette (right)

Country of Origin: France.

Type: Basic trainer, photo-survey and liaison aircraft.

Power Plant: Two (MS-760A) 880 lb st (400 kgp) Turboméca Marboré IIC or (MS-760B) 1,058 lb st (480 kgp) Marboré VI turbojets.

Performance: (MS-760B) Max speed, 432 mph (695 km/h) at 25,000 ft (7 600 m); max cruising speed, 393 mph (633 km/h) at 16,400 ft (5 000 m); economical cruising speed, 342 mph (550 km/h); initial rate of climb, 2,460 ft/min (12,5 m/sec); service ceiling, 39,370 ft (12 000 m); max range, 1,080 mls (1 740 km).

Weights: Empty, equipped, 4,557 lb (2 067 kg); max take-off, 8,642 lb (3 920 kg).

Dimensions: Span, 33 ft 3 in (10,15 m); length, 33 ft 7 in (10,24 m); height, 8 ft 6 in (2,60 m); wing area, 193·7 sq ft (18 m²).

Accommodation: Side-by-side seats with full dual control, with provision for two passenger seats in rear when used in training rôle.

Armament: Optional armament may comprise (for training) two 7·5 mm machine guns plus four 3·5-in rockets or two 112-lb (50-kg) bombs or (for light tactical rôle, when flown as single-seater) one 30-mm cannon and two 7·5-mm machine guns, plus eight 112-lb (50-kg) and two 260-lb (118-kg) bombs.

Status: Prototype (MS-760-01) first flown on 29 July 1954; first production model (MS-760A) flown on 27 February 1958; first MS-760B flown on 12 December 1960. Production totals: one prototype, 150 MS-760A and 63 MS-760B.

Notes: The Paris I was built by the original Morane-Saulnier company as a development of the MS-755 Fleuret, one of the first light aircraft to be powered by jet engines. It was adopted by the French *Armée de l'Air* for use in the communications rôle and for continuation training, and a small batch was supplied to the *Aéronavale* for the same purpose. For weapon training, the rear seats were usually removed, and for aerobatic flying, the tip tanks were detached; up to 900 lb (408 kg) of external stores could be carried if the aircraft was used for light attack duties. Export orders for the Paris were obtained from the Argentine Air Force and the Brazilian Air Force, in both cases final assembly lines being established in the countries concerned. Argentina assembled 48 MS-760A Paris Is and Brazil assembled 48 MS-760B Paris IIs. One Paris was obtained by the Paraguayan Air Force.

NORTH AMERICAN T-6 TEXAN

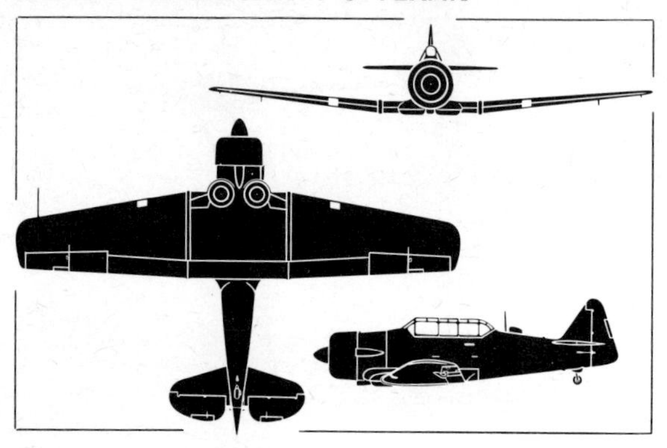

The photograph above depicts a T-6G Texan of the Portuguese Air Force and the general arrangement silhouette (left) illustrates the T-6J (Harvard Mk 4) built by CCF

Country of Origin: USA.
Type: Basic trainer and light attack aircraft.
Power Plant: One 550 hp Pratt & Whitney R-1340-AN-1 radial
Performance: (T-6G) Max speed, 212 mph (341 km/h) at 5,000 ft (1 524 m); max cruising speed, 170 mph (274 km/h); economical cruising speed, 146 mph (235 km/h); initial rate of climb, 1,643 ft/min (8,3 m/sec); service ceiling, 24,750 ft (7 338 m); normal range, 870 mls (1 400 km).
Weights: Empty, 4,271 lb (1 938 kg); max take-off, 5,617 lb (2 546 kg).
Dimensions: Span, 42 ft 0¼ in (12,80 m); length, 29 ft 6 in (8,99 m); height, 11 ft 8½ in (3,56 m); wing area, 253·7 sq ft (23,56 m²).
Accommodation: Two in tandem, dual controls.
Armament (Optional): Provision for one flexibly-mounted 0·30-in machine gun in rear cockpit and one 0·30-in fixed forward-firing in wing root; rockets, light bombs or gun pods under wings for weapon-training or light counter-insurgency rôle.
Status: Prototype (NA-26) flown in 1937. Production and re-manufacture of variants of basic design continued until 1952. Production of all variants (including fixed-undercarriage types) totalled more than 17,000 in the USA plus more than 4,500 under licence in Canada, Australia, Sweden and Japan. Production continued in Canada until 1954 as the T-6J for the USAF and Harvard Mk 4 for the RCAF.
Notes: The T-6 Texan or (British name) Harvard is a member of the famous North American family of trainers that began in April 1935 with the NA-16. Initial versions had fixed undercarriage legs but a redesign of the basic airframe in 1937 produced the NA-26 prototype with retractable undercarriage and more powerful engine. Production derivatives of the NA-26 were widely used throughout World War II, and thousands of examples that became surplus at the end of the war were distributed world wide. A modernisation programme for 2,068 USAF T-6s during 1949–50 gave these a new lease of life and in 1973 many hundreds were still in service with more than 30 air forces. Primarily used as trainers, T-6s were also adapted for light attack duties in some countries, notably Brazil, which maintained several front-line squadrons equipped with these armed versions, and Spain, where the armed version carries the Air Force nomenclature C.6, while the training variant is the E.16.

NORTH AMERICAN T-28 TROJAN

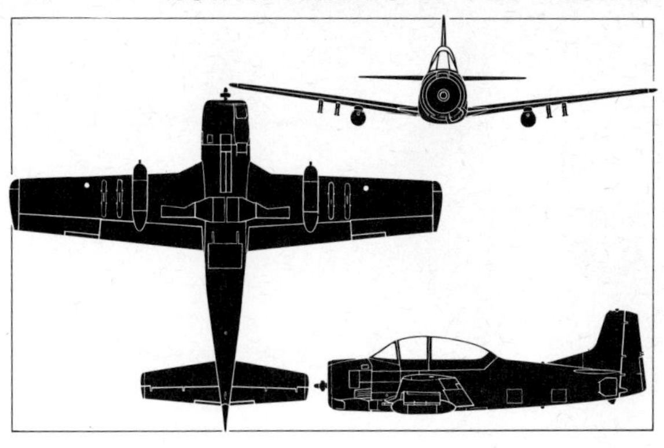

Both photograph (above) and general arrangement silhouette (right) depict the T-28D, the former illustrating an example serving with the Imperial Ethiopian Air Force

Country of Origin: USA.
Type: Basic trainer and light attack aircraft.
Power Plant: One (T-28A) 800 hp Wright R-1300-1A or (T-28B and C) 1,425 hp Wright R-1820-86 or (T-28D) 1,300 hp R-1820-56S radial engine.
Performance: (T-28B) Max speed, 346 mph (557 km/h) at 18,000 ft (5 486 m); normal cruising speed, 219 mph (352 km/h); initial rate of climb, 3,830 ft/min (1,9 m/sec); service ceiling 37,000 ft (11 278 m); max range, 907 mls (1 460 km).
Performance: (T-28D) Max speed, 352 mph (566 km/h) at 18,000 ft (1 167 m); max range, 1,184 mls (1 820 km) cruising at 203 mph (327 km/h).
Weights: (T-28D) Empty, 6,512 lb (2 953 kg); loaded, 8,118 lb (3 682 kg).
Dimensions: Span, 40 ft 7½ in (12,37 m); length, 32 ft 10 in (10,00 m); height, 12 ft 8 in (3,86 m); wing area, 271·1 sq ft (25,3 m²).
Accommodation: Two in tandem.
Armament: (T-28D only) Six underwing strong points carry two 0·50-in (12,7-mm) gun pods and assorted bombs, rockets, etc to a total weight of 4,000 lb (1 815 kg).
Status: First of two prototypes (XT-28) flown on 26 September 1949; first T-28B flown in 1954; first T-28C flown on 19 September 1955; first T-28D flown in 1961; first T-28D-5 flown in 1965; first of three YAT-28Es flown 15 February 1963. Production totals, XT-28, 2; T-28A, 1,194; T-28B, 489; T-28C, 299. (Other variants were conversions.) Production completed in 1957.
Notes: The NA-159 design was an attempt to produce a post-war successor to the NA-16/T-6 family of trainers (see page 184). The T-28A was built for the USAF and the T-28B and T-28C were for the US Navy, the latter having an arrester hook for dummy deck-landing approaches. Through MAP and other programmes, T-28As were supplied to several overseas air forces, particularly in South America and South-East Asia, and the T-28D conversion was evolved with a heavy underwing armament, several hundred surplus T-28As being converted by North American and Fairchild, with the final batches by the latter company being designated AT-28Ds. A similar conversion scheme was undertaken by the Sud company in France, which delivered 245 modified T-28As to the *Armée de l'Air* as Fennecs for service in North Africa.

NORTHROP T-38 TALON

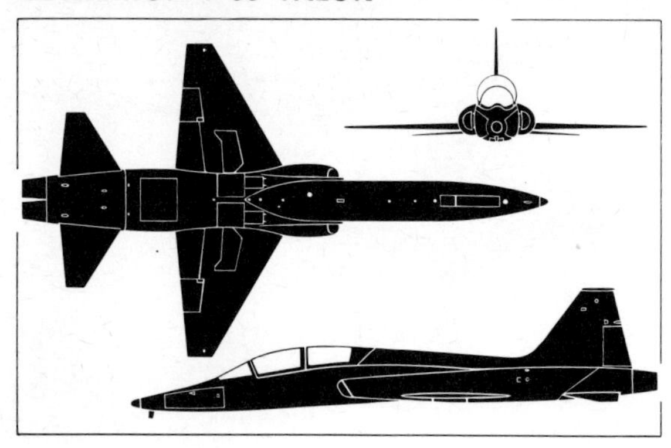

The T-38A Talon, illustrated by the photograph above and the general arrangement silhouette (left), was the first supersonic advanced trainer to attain production status

Country of Origin: USA.
Type: Advanced Trainer.
Power Plant: Two 3,850 lb st (1 748 kgp) with reheat General Electric J85-GE-5 turbojets.
Performance: Max speed, Mach 1·23 at 36,000 ft (11 000 m) at half fuel weight; max cruising speed, 630 mph (1 104 km/h) at 40,000 ft (12 200 m); economical cruising speed, 594 mph (956 km/h) above 40,000 ft (12 200 m); initial rate of climb at half fuel weight, 30,000 ft/min (152,4 m/sec); service ceiling at half fuel weight, 53,600 ft (16 335 m); range with max fuel, 20 min reserve, two crew, 1,093 mls (1 759 km).
Weights: Max take-off and landing, 12,093 lb (5 485 kg); max zero fuel, 7,662 lb (3 475 kg).
Dimensions: Span, 25 ft 3 in (7,70 m); length, 46 ft 4½ in (14,13 m); height, 12 ft 10½ in (3,92 m); wing area, 170 sq ft (15,80 m²).
Accommodation: Two in tandem, full dual control.
Status: First YT-38 prototype flown on 10 April 1959; first production T-38A flown in May 1960. Production of 1,187 T-38As completed in January 1972.
Notes: The T-38A Talon was an outgrowth of the Northrop programme to develop a lightweight supersonic fighter, the N-156F Freedom Fighter, which eventually went into production as the F-5 (see page 35). Design of the similar N-156T trainer proceeded in parallel with that of the N-156F for two years as a private venture, until the USAF drew up a requirement for a supersonic basic/advanced trainer and ordered prototypes of the Northrop design. The first contract, issued in May 1956, was for six YT-38s, the first two of which were flown with non-afterburning YJ85-GE-1 engines pending availability of the production standard J85-GE-5s. The T-38A Talon entered service with the USAF on 17 March 1961 and has been in continuous large-scale service at USAF training bases ever since. A series of contracts kept the Talon in production until the end of 1971, the production total including 46 T-38As which were funded by the *Luftwaffe* but retained in the US in USAF markings for use at training bases for *Luftwaffe* personnel. This was the only non-USAF purchase of T-38As, but 24 examples were acquired through USAF by NASA for use as space flight readiness trainers and five were acquired by the US Navy. The Talon was the first production trainer designed specifically to reproduce the flying characteristics of operational supersonic aircraft.

PIAGGIO P.148

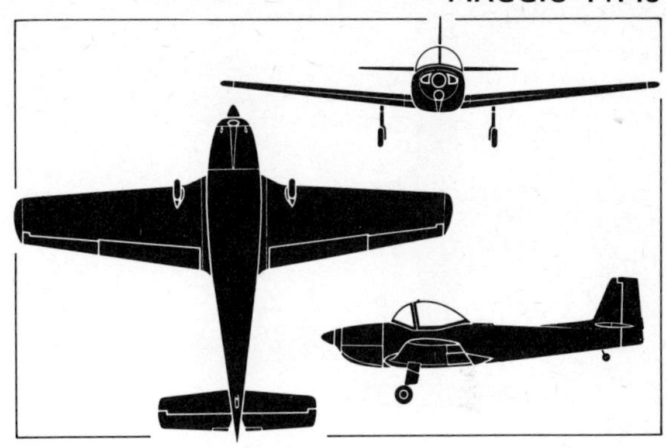

The P.148, illustrated by the photograph above and the general arrangement silhouette (right), serves with the Italian, Congolese and Somali air arms

Country of Origin: Italy.
Type: Primary trainer.
Power Plant: One 190 hp Lycoming O-435-A six cylinder horizontally-opposed engine.
Performance: Max speeds, 144 mph (232 km/h) at sea level, 137 mph (220 km/h) at 6,560 ft (2 000 m) and 123 mph (198 km/h) at 13,120 ft (4 000 m); cruising speed, 125 mph (201 km/h) at 2,950 ft (900 m); time to climb to 3,280 ft (1 000 m) 4½ min and to 9,840 ft (3 000 m) 19 min; service ceiling, 14,436 ft (4 400 m); range, 500 mls (800 km); endurance 4 hrs.
Weights: Empty, 1,930 lb (876 kg); max take-off (two-seater), 2,645 lb (1 200 kg); max take-off (three-seater) 2,822 lb (1 280 kg).
Dimensions: Span, 36 ft 6 in (11,12 m); length, 27 ft 8 in (8,44 m); height, 7 ft 10½ in (2,40 m); wing area, 203 sq ft (18,81 m²).
Accommodation: Two pilots side by side; optional third seat behind.
Status: Prototype first flown on 12 February 1951. Production total 100, completed 1955.
Notes: One of the first post-war designs of the Piaggio company, the P.148 was designed to meet an Italian Air Force requirement for a primary trainer, fully aerobatic and with a cockpit layout similar to the basic arrangement adopted by the USAF and RAF. It was adopted for use at primary training schools alongside the Italian-built version of the Fokker Instructor, an initial order for 50 being placed and later doubled. Following the precedent set in the RAF with the Percival Prentice, the P.148 provided space for a third occupant behind the instructor and pupil; although in practice this third seat was seldom used after the P.148 entered service. After the Italian Air Force adopted an "all-through" jet training sequence using the AerMacchi MB.326, the P.148s were withdrawn from service and under Italian technical aid programmes 12 were supplied to the Congolese (Kinshasa) Air Force, now the *Force Aérienne Zaïroise*, and a few others to the Somali Air Force, and the P.148 remained in service with both air arms in 1973. However, in 1970 the Italian Air Force decided, in common with most other air forces, to re-introduce a primary training stage on piston-engined aircraft as a preliminary to the jet training sequence, and the P.148s were then restored to service. Many of the structural components of the P.148 were utilised by the P.149 (see page 188) four/five-seat trainer and communications aircraft.

PIAGGIO P.149D

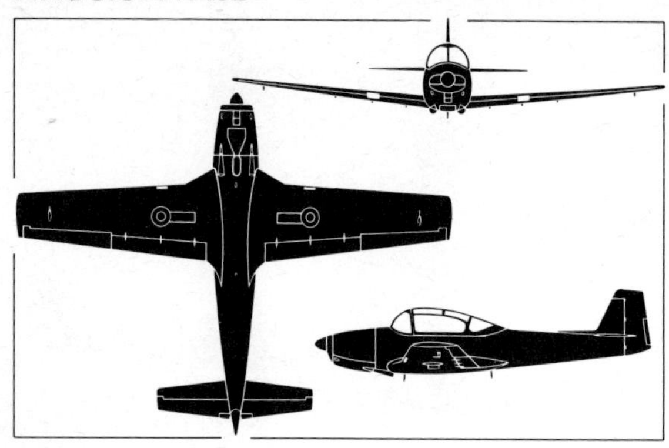

Both photograph (above) and general arrangement silhouette (left) illustrate the P.149-D, the former depicting an ex-Luftwaffe aircraft of the Nigerian Air Force

Country of Origin: Italy.
Type: Basic trainer and communications aircraft.
Power Plant: One 270 hp Lycoming GO-480-B1A6 six-cylinder horizontally-opposed engine.
Performance: Max speeds, 192 mph (304 km/h) at sea level and 177 mph (285 km/h) at 6,560 ft (2 000 m); cruising speed, 165 mph (266 km/h) on 67 per cent power at 7,500 ft (2 300 m) and 145 mph (234 km/h) on 56 per cent power at 10,800 ft (3 300 m); initial rate of climb, 980 ft/min (5 m/sec); service ceiling, 19,800 ft (6 050 m); range with full tanks, 30-min reserve, 680 mls (1 090 km) at 10,800 ft (3 300 m).
Weights: Empty, 2,557 lb (1 160 kg); max take-off weight, 3,704 lb (1 680 kg).
Dimensions: Span, 36 ft 6 in (11,12 m); length, 28 ft 9½ in (8,80 m); height, 9 ft 6 in (2,90 m); wing area, 203 sq ft (18,81 m²).
Accommodation: Two pilots side-by-side with full dual control for training rôle, with seats in rear for two-three passengers when used as a communications aircraft.
Status: Prototype first flown on 19 June 1953. First delivery from Italian production line, May 1957; first delivery from German production line, October 1957. Production totals, two prototypes and 72 by Piaggio; 190 by Focke-Wulf, completed 1959.
Notes: Piaggio developed the P.149 as a civil four-seat light aircraft using many structural components of the P.148 (see page 187) including the wing. Swissair acquired five for its pilot training programme, but major production of the type began only after a modified version, with a one-piece rearward sliding hood, had been adopted by the *Luftwaffe*. Piaggio supplied 72 complete aircraft to Germany, and a licence plus 16 sets of assemblies to the Focke-Wulf company. The latter built a total of 190, including the 16 supplied by Piaggio. Designated P.149-D, the *Luftwaffe* version was adopted for both basic training and liaison tasks, subsequently serving at the three *Luftwaffe* "C" schools for piston-engined aircraft pilots, supplanting the North American T-6 Texan, and normally flown as a two-seater. Although no longer used in the training rôle in Germany, many remain in *Luftwaffe* service, and others have been transferred from the latter service to the Air Forces of Nigeria, Tanzania and Uganda under German military assistance programmes. The P.149-Ds serving with the Nigerian Air Force were scheduled to be replaced by the Scottish Aviation Bulldog from late 1973.

PILATUS P.3

The P-3, illustrated in service with the Brazilian Air Force above, has been the standard Swiss Air Force (Schweizerische Flugwaffe) basic trainer for almost 20 years

Country of Origin: Switzerland.
Type: Basic trainer.
Power Plant: One 260 hp Lycoming GO-435-C2A six-cylinder horizontally-opposed engine.
Performance: Max speed, 193 mph (310 km/h) at 6,560 ft (2 000 m); max cruising speed, 170 mph (275 km/h); economical cruising speed, 157 mph (252 km/h); initial rate of climb, 1,378 ft/min (7,0 m/sec); service ceiling, 18,040 ft (5 500 m); max range, 465 mls (750 km).
Weights: Empty equipped, 2,447 lb (1 110 kg); normal loaded, 3,120 lb (1 415 kg); max take-off, 3,307 lb (1 500 kg).
Dimensions: Span, 34 ft 1 in (10,40 m); length, 28 ft 8 in (8,75 m); height, 10 ft 0 in (3,05 m); wing area, 177 sq ft (16,5 m²).
Accommodation: Two in tandem, full dual control.
Armament (Optional): Provision for underwing armament comprising one 7·9 mm gun pod with 180 rounds, one carrier for two practice bombs and two rocket launchers, plus a gun camera.
Status: First of two prototypes flown on 3 September 1953. Production batch of 72 supplied to Swiss Air Force and six supplied to Brazil. Production completed.

Notes: Development of the P-3 was initiated in Switzerland in 1951 to provide the *Schweizerische Flugwaffe* (Swiss Air Force) with a replacement for its North American T-6 intermediate trainers. Following successful testing of two prototypes, the type was ordered into production and a batch of 72 was built. They have served since 1955 at Flying Schools to provide basic training after an *ab initio* stage has been completed on Bücker Bü 131 biplanes. The performance and characteristics of the P.3, despite its low power, are such that pupils can progress straight from this type to Vampire Trainers and single-seat Vampires and Venoms. Provision is made in the P.3 design for underwing stores, allowing its use as a weapons trainer. Apart from the Swiss Air Force, the only customer for the P.3 was the Brazilian Navy, which bought six for service in an observation and liaison rôle. These were later transferred to the Brazilian Air Force and were still in service in 1973. In an attempt to modernize the P.3 design, Pilatus fitted a 550 shp Pratt & Whitney PT6A-20 turboprop in one of the prototypes which, re-designated PC-7, first flew on 12 April 1966. In addition to serving with the Swiss 1st Stage Flying School, the P-3 is operated by the *Leichte Fliegerstaffeln* in the liaison rôle.

POTEZ/AEROSPATIALE CM 170 MAGISTER

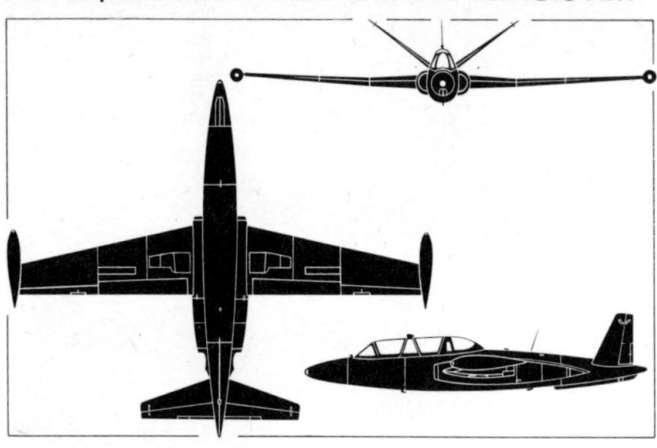

The CM 170-1 Magister is illustrated by the photograph above and by the general arrangement silhouette (left), the former depicting an example operated by the Algerian Air Force

Country of Origin: France.
Type: Basic trainer.
Power Plant: Two (CM 170-1 and CM 175) 880 lb st (400 kgp) Turboméca Marboré IIA or (CM 170-2 Super Magister) 1,058 lb st (480 kgp) Marboré VIC turbojets.
Performance (CM 170-1): Max speed, 403 mph (650 km/h) at sea level and 444 mph (715 km/h) at 29,520 ft (9 000 m); initial rate of climb, 2,950 ft/min (15,0 m/sec); service ceiling, 36,090 ft (11 000 m); range (with tip tanks) 735 mls (1 200 km); endurance at 29,520 ft (9 000 m) 2 hr 40 min.
Performance (CM 170-2): Max speed, 435 mph (700 km/h) at sea level and 463 mph (745 km/h) at 29,520 ft (9 000 m); initial rate of climb, 3,540 ft/min (18 m/sec); service ceiling 44,300 ft (13 500 m); range, 775 mls (1 250 km); endurance 2 hr 50 min.
Weights: Empty equipped (CM 170-1) 4,740 lb (2 150 kg); (CM 170-2) 5,093 lb (2 310 kg); normal take-off, 6,835 lb (3 100 kg); max take-off, 7,055 lb (3 200 kg).
Dimensions: Span, 37 ft 5 in (11,40 m); span over tip tanks, 39 ft 10 in (12,15 m); length, 33 ft 0 in (10,06 m); height, 9 ft 2 in (2,80 m); wing area, 186·1 sq ft (17,30 m²).

Accommodation: Two in tandem, full dual controls.
Armament: Two 7·5-mm or 7·62-mm fixed machine guns in nose, underwing strong points for bombs, rockets or missiles.
Status: First of three prototypes flown on 27 June 1951; first pre-production aircraft flown on 7 July 1954; first full production aircraft flown on 29 February 1956; first CM 175 Zephyr flown on 30 May 1959. Production total 916 comprising: prototypes, 3; pre-production, 10; *l'Armée de l'Air* 387; CEV, 5; *Aéronavale* 32; Brazil, 7; Germany, 250; Israel, 52; Austria, 18; Finland, 82; Belgium and Netherlands, 48; Cambodia, 4; Congo (Leopoldville) 6; Lebanon, 4. Production was also undertaken in Finland (Valmet OY built 62), Israel (IAI built 36) and Germany (188 by Flugzeug Union Sud).
Notes: The CM 170 was originally designed to a French Air Force specification, winning a contract for prototype construction in June 1951 and entering service as the standard basic trainer at French flying schools. Of the 400 delivered to the French Air Force, 130 were to CM 170-2 Super Magister standard, as were seven supplied to Brazil. Morocco received eight from the *Armée de l'Air*. The *Aéronavale* CM 175 Zephyr had an arrester hook.

ROCKWELL (NORTH AMERICAN) T-2 BUCKEYE

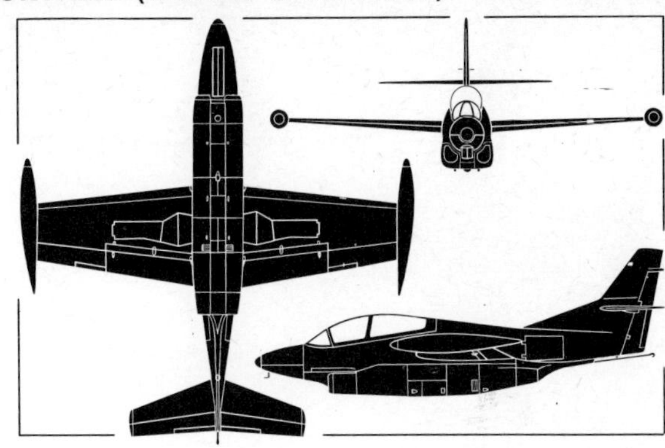

The photograph above depicts a T-2B Buckeye of the US Navy and the general arrangement silhouette (right) illustrates the externally similar T-2C

Country of Origin: USA.
Type: Basic trainer (and light attack aircraft).
Power Plant: Two (T-2B) 3,000 lb st (1 360 kgp) Pratt & Whitney J60-P-6 or (T-2C and T-2D) two 2,950 lb st (1,339 kgp) General Electric J85-GE-4 turbojets.
Performance: (T-2C) Max level speed, 522 mph (840 km/h) at 25,000 ft (7 620 m); initial rate of climb, 6,200 ft/min (31,5 m/sec); service ceiling, 40,414 ft (12 320 m); max range, 1,047 mls (1 685 km).
Weights: Empty, 8,115 lb (3 680 kg); max take-off, 13,179 lb (5 977 kg).
Dimensions: Span (over tip tanks) 38 ft 1½ in (11,62 m); length, 38 ft 3½ in (11,67 m); height, 14 ft 9½ in (4,51 m); wing area, 255 sq ft (23,69 m²).
Accommodation: Two in tandem, full dual controls.
Armament: (Optional) one store station beneath each wing with capacity of 320-lb (145-kg) each.
Status: First production model T2J-1 (no prototypes built) first flown on 31 January 1958; first T2J-2 (T-2B) conversion flown on 30 August 1962; first production T-2B flown on 21 May 1965; first T-2C prototype flown on 17 April 1968; first production T-2C flown on 10 December 1968. Production quantities, T-2A, 217; T-2B, 97; T-2C, 183 ordered by 1973; T-2D, 12.
Notes: The North American NA-241 design met a US Navy requirement for a basic jet trainer, and was based on the wing of the FJ-1 Fury jet fighter, with the control system used in the T-28C and a single 3,400 lb st (1 540 kgp) Westinghouse J34-WE-36 or -48 engine. Entering production in 1957 as the T2J-1, it was in service by mid-1959, finally retiring from USN service early in 1973, by which time the designation had changed to T-2A. The T-2B differed from the T-2A primarily in having two J60 engines, and the T-2C followed on, with another change of power plant. The first export order came from Venezuela in 1972, for the T-2D version, and these were delivered to the Venezuelan Air Force Academy in 1973. The first production T-2B, converted to T-2C standard was later fitted with a NASA designed supercritical wing, with which it flew in November 1970. During the course of 1973, Rockwell was offering potential Buckeye customers an armaments package to modify the aircraft to attack configuration with two 750-lb (340-kg) and four 500-lb (227-kg) wing ordnance stations.

ROCKWELL (NORTH AMERICAN) T-39 SABRELINER

The photograph above depicts a CT-39E of the US Navy and the general arrangement silhouette (left) illustrates the T-39A staff transport and combat readiness trainer of the USAF

Country of Origin: USA.
Type: Crew trainer and communications aircraft.
Power Plant: (T-39A to D) Two 3,000 lb st (1 360 kgp) Pratt & Whitney J60-P-3 or (CT-39E) two 3,300 lb st (1 497 kgp) Pratt & Whitney JT12A-8 turbojets.
Performance: (T-39A) Max speed, 540 mph (869 km/h) at 36,000 ft (10 973 m); high-speed cruise, 502 mph (808 km/h) at 43,500 ft (13 250 m); long-range cruise, 475 mph (764 km/h) at 44,000 ft (13 400 m); service ceiling, over 45,000 ft (13 716 m); max range, with reserves, 1,950 mls (3 130 km).
Weights: Basic operating weight, 9,257 lb (4 200 kg); max take-off, 17,760 lb (8 055 kg); max landing, 13,000 lb (5 900 kg); max zero fuel, 10,984 lb (4 982 kg).
Dimensions: Span, 44 ft 5 in (13,53 m); length, 43 ft 9 in (13,33 m); height, 16 ft 0 in (4,88 m); wing area, 342·6 sq ft (31,83 m²).
Accommodation: Two pilots as combat readiness trainer; up to eight passengers as transport; two pilots, instructor and three pupils as specialised trainer.
Status: Prototype (NA-246) first flown on 16 September 1958; first production T-39A flown on 30 June 1960; deliveries of T-39B began in February 1961 and of T-39D in August 1963. Production totals: NA-246, one; T-39A, 143; T-39B, 6; T-39D, 42; CT-39E, 9; CT-39G, 5 ordered up to 1973.
Notes: North American built the NA-246 as a private venture to meet the requirements of the USAF's UTX (utility and "combat readiness" trainer) specification. The prototype was powered by 2,500 lb st (1 134 kgp) General Electric J85 turbojets but when the USAF selected the type for production a switch was made to the more powerful J60 engines. The principal version is the T-39A, staff transport, communications aircraft and combat readiness trainer. USAF also procured six T-39Bs with radar and Doppler equipment to train F-105 crews and three T-39As were later converted to T-39Fs to train electronic warfare officers for the special "Wild Weasel" F-105Gs. To train F-8 and F-4 crews in the use of intercept radar, the USN bought 42 aircraft similar to the T-39B; originally T3J-1s, they are now T-39Ds. The USN has also bought, for "rapid response airlift", nine examples of the commercial Sabreliner 40 with uprated engines, designated CT-39E, and five Sabre 60s with lengthened fuselage (see *Civil* volume) which carry the designation CT-39G.

SAAB 91 SAFIR

The Saab 91D, illustrated in Finnish service by the photograph above and by the general arrangement silhouette (right), was the definitive production version of the Safir

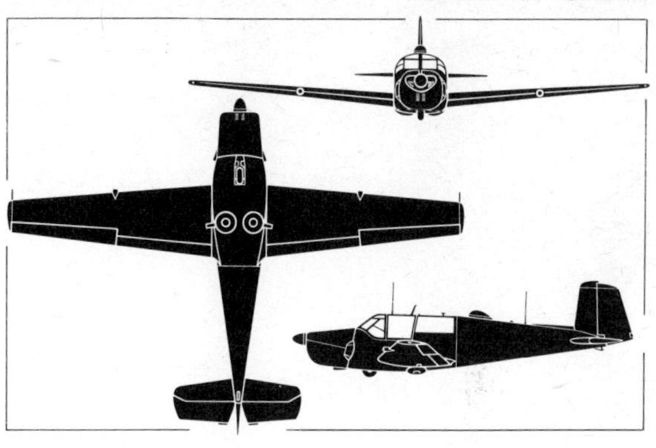

Country of Origin: Sweden.
Type: Basic trainer and communications aircraft.
Power Plant: One (918 and C) 190 hp Lycoming O-435-A six-cylinder horizontally-opposed or (91D) 180 hp Lycoming O-360-AIA four-cylinder horizontally-opposed engine.
Performance (918): Max speed, 171 mph (275 km/h); max cruising speed, 152 mph (244 km/h); economical cruising speed, 144 mph (232 km/h); initial climb, 1,140 ft/min (5,8 m/sec); service ceiling, 20,500 ft (6 248 m); max range, 670 mls (1 078 km).
Performance (91D): Max speed, 165 mph (265 km/h) at sea level; max cruising speed, 146 mph (235 km/h); economical cruising speed, 136 mph (220 km/h); initial rate of climb, 800 ft/min (4,1 m/sec); service ceiling, 16,400 ft (5 000 m); max range, 660 mls (1 062 km).
Weights (91B): Empty equipped, 1,590 lb (720 kg); max take-off, 2,686 lb (1 218 kg).
Weights (91D): Empty equipped, 1,570 (710 kg); max take-off, 2,660 lb (1 205 kg); max aerobatic weight, 2,320 lb (1 050 kg).
Dimensions: Span, 34 ft 9 in (10,60 m); length (91B) 26 ft 0 in (7,92 m), (91D) 26 ft 4 in (8,03 m); height, 7 ft 2⅔ in (2,20 m); wing area, 146,3 sq ft (13,60 m²).
Accommodation: Two pilots side-by-side, with full dual control, plus one (91A, B and D) or two (91C) passenger seats in rear.
Status: Prototype first flown on 20 November 1945; first 91B flown on 18 January 1949; first 91C flown in September 1953; first 91D flown in 1957. Total production for military and civil use approximately 320, including 10 91A for Swedish Air Force (Tp 91), 16 91A to Ethiopian AF, 75 91B to Swedish AF (Sk 50B); 18 91B to Ethiopian AF; 25 91B to R Norwegian AF; 14 91C to Swedish AF (Sk 50C); 14 91C to Ethiopian AF; 24 91D to Austrian AF; 35 91D to Finnish AF and 15 91D to Tunisia.
Notes: Saab completed the prototype Saab 91 Safir in 1945 as a two-three-seat lightplane for the civil market, with a 130 hp DH Gipsy Major 1C engine. In production guise as the Saab 91A with a 145 hp Gipsy Major 10, the Safir soon attracted military as well as civil interest. A switch to the Lycoming engine was made in the Saab 91B, which was adopted as the standard primary trainer for the Swedish Air Force and built under licence by De Schelde in Holland. Saab resumed production with the four-seat 91C in 1955 and concluded with the 91D, production ending in 1966.

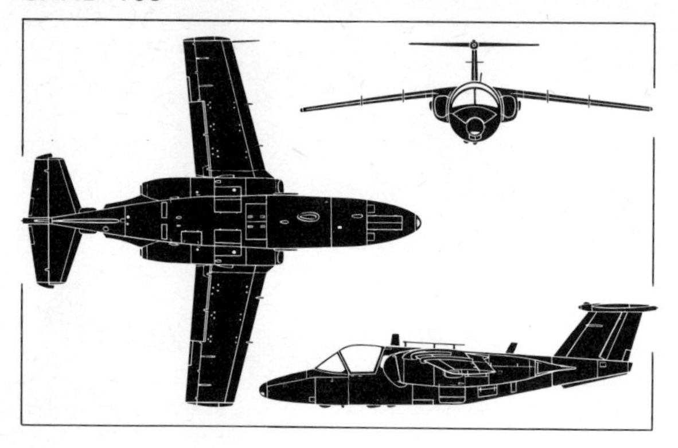

Both photograph above and general arrangement silhouette depict the Saab 105Ö as operated by the Austrian Air Force, this being more powerful than the version used in Sweden

Country of Origin: Sweden.

Type: Basic trainer and light attack aircraft.

Power Plant: Two (Sk 60) 1,640 lb st (743 kg) Turboméca Aubisque turbofans or (Saab-105XT, -105XH and -105G) 2,850 lb st (1 293 kgp) General Electric J85-GE-17B turbojets.

Performance (SK 60 at gross weight of 8,820 lb/4 000 kg): Max speed, 447 mph (720 km/h) at sea level and 475 mph (765 km/h) at 20,000 ft (6 000 m); cruising speed, 435 mph (700 km/h) at 20,000 ft (6 000 m) and 426 mph (685 km/h) at 30,000 ft (9 000 m); initial rate of climb, 3,440 ft/min (17,5 m/sec); service ceiling, 39,400 ft (12 000 m); range, 1,106 mls (1 780 km).

Performance (105XT at max weight): Max speed, 603 mph (970 km/h) at sea level and 544 mph (875 km/h) at 33,000 ft (10 000 m); time to 33,000 ft (10 000 m) 4–5 mins; range at 43,000 ft (13 100 m) at 435 mph (700 km/h) with 20-min reserve, 1,423 mls (2 290 km).

Weights (Sk 60): Empty, 5,534 lb (2 510 kg); max take-off according to rôle—aerobatic, 8,380 lb (3 800 kg), utility, 8,930 lb (4 050 kg) and ground attack, 9,920 lb (4 500 kg).

Weights (105XT): Empty, 6,173 lb (2 800 kg); normal take-off (trainer) 10,174 lb (4 615 kg); max take-off (armed) 14,330 lb (6 500 kg).

Dimensions: Span, 31 ft 2 in (9,50 m); length (Sk 60), 34 ft 5 in (10,50 m); length (105XT), 35 ft 5¼ in (10,80 m); height 8 ft 10 in (2,70 m); wing area, 175 sq ft (16,3 m²).

Accommodation: Two side-by-side, dual controls. Optional provision for two additional seats if ejection seats not fitted.

Armament: Wing stressed for six strong points to carry (Sk60B) up to 1,543 lb (700 kg) or (105XT) up to 4,410 lb (2 000 kg) total stores.

Status: Two prototypes first flown on 29 June 1963 and 17 June 1964 respectively. First production Sk 60A flown on 27 August 1965; prototype Sk 60C flown on 18 January 1967; 105XT flown on 29 April 1967; 105G flown on 26 May 1972. Production of 150 for Swedish Air Force completed; 40 for Austrian Air Force.

Notes: The multi-purpose Saab 105 was adopted by the Royal Swedish Air Force as its standard basic trainer (Sk 60A), for weapons training and light strike duties (Sk 60B) and for reconnaissance (Sk 60C). The Austrian Air Force's Saab 105Ös are similar to the 105XT prototype described above.

SAAB SUPPORTER

The photograph above depicts the Saab Supporter primary trainer, the general arrangement silhouette (right) illustrating the light attack version of the same basic design

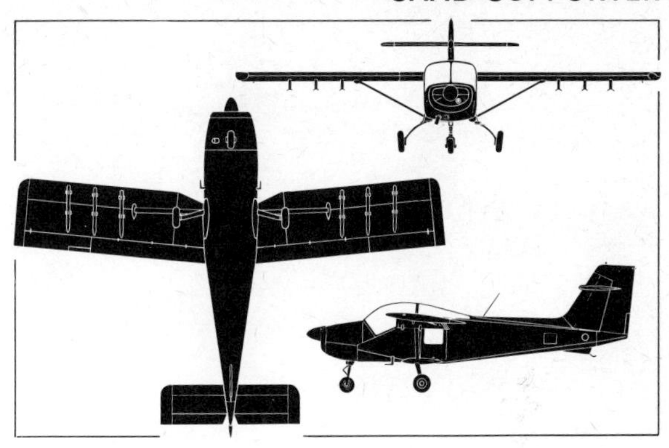

Country of Origin: Sweden.
Type: Primary trainer and light attack aircraft.
Power Plant: One 200 hp Avco-Lycoming IO-360-A1B four-cylinder horizontally-opposed engine.
Performance: (MFI-15 at 2,200 lb/1 000 kg weight; Max speed, 154 mph (248 km/h) at sea level; cruising speed, 137 mph (220 km/h) at 75 per cent power at sea level; initial rate of climb 1,050 ft/min (5,4 m/sec); service ceiling, 17,000 ft (5 190 m); endurance on max fuel, 65 per cent power, 10 per cent reserve, 5 hr 10 min.
Performance: (MFI-17 at 1,819 lb/825 kg weight); Max speed, 163 mph (262 km/h) at sea level; cruising speed, 148 mph (238 km/h) at 75 per cent power at sea level; initial rate of climb, 1,550 ft/min (7,9 m/sec); service ceiling, 22,800 ft (6 950 m); endurance on max fuel, 10 per cent reserve, 4 hr 45 min.
Weights: Empty equipped, 1,323 lb (600 kg); take-off weights according to category—normal (MFI-15), 2,200 lb (1 000 kg), (MFI-17) 2,095 lb (950 kg); utility, 1,985 lb (900 kg); aerobatic, 1,819 lb (825 kg); restricted, 2,423 lb (1 100 kg).
Dimensions: Span 28 ft 6½ in (8,7 m); length, 23 ft 0 in (7,0 m); height, 8 ft 6¼ in (2,6 m); wing area, 127 sq ft (11,80 m²).

Accommodation: Two side-by-side, with dual controls; optional third seat to rear.
Armament (Supporter): Six underwing hardpoints of which two inboard stressed for 220 lb (100 kg) and remainder for 110 lb (50 kg) each.
Status: Prototype (MFI-15) flown on 11 July 1969 with 160 hp engine and on 26 February 1971 with 200 hp engine. First MFI-17 flown early summer 1972. Production deliveries of Safari and Supporter began mid-1973.
Notes: Designed as a primary trainer and for miscellaneous military liaison duties, the MFI-15 first flew with a 160 hp engine, low-mounted tailplane and tail-wheel undercarriage. It was subsequently re-engined with the standard 200 hp engine and a high tailplane position was adopted. A nosewheel undercarriage is available on both the Safari (MFI-15) and the Supporter (MFI-17) as an alternative to the tailwheel type. The Supporter is the light attack version of the same design, the airframes being identical and both types being built on the same assembly line. The Supporter is fully aerobatic and stressed for maximum limiting load factors of +6g to −3g.

SCOTTISH AVIATION BULLDOG

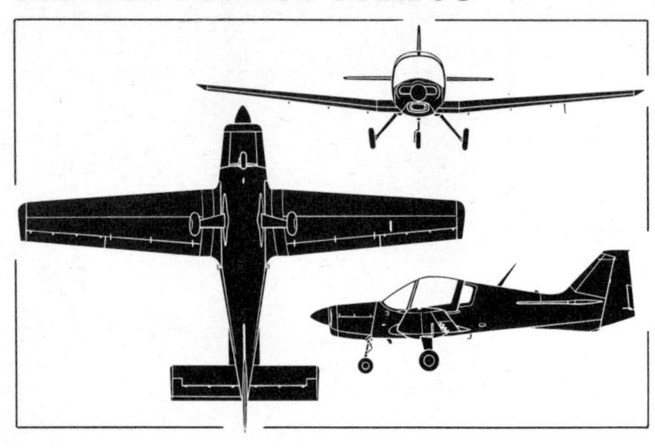

The photograph above depicts a Bulldog T Mk 1 of the RAF, the general arrangement silhouette (left) being applicable to all production models

Country of Origin: United Kingdom.
Type: Primary trainer.
Power Plant: One 200 hp Avco-Lycoming IO-360-AIB6 four-cylinder horizontally-opposed engine.
Performance: Max speed, 150 mph (241 km/h) at sea level; max cruising speed, 138 mph (222 km/h) at 75 per cent power at 4,000 ft (1 220 m); economical cruising speed, 121 mph (194 km/h) at 55 per cent power at 4,000 ft (1 220 m); initial rate of climb, 1,034 ft/min (5,25 m/sec); service ceiling, 16,000 ft (4 880 m); range with max fuel, no reserves, 55 per cent power, 621 mls (1 000 km).
Weights: Empty equipped, 1,430 lb (649 kg); max take-off (non-aerobatic), 2,350 lb (1 066 kg); max take-off (aerobatic, Srs 100) 2,150 lb (975 kg); max take-off (aerobatic, Srs 120) 2,238 lb (1 015 kg).
Dimensions: Span, 33 ft 0 in (10,06 m); length, 23 ft 3 in (7,08 m); height, 7 ft 5¾ in (2,28 m); wing area, 129·4 sq ft (12,02 m²).
Accommodation: Two side-by-side, full dual control.
Armament (Optional): Four hard points in wings to carry pylons for light HE bombs, practise bombs, machine gun pods, air-to-surface missiles, etc.
Status: Prototype (by Beagle) first flown on 19 May 1969; second prototype (by Scottish Aviation) flown on 14 February 1971. First production Srs 100 flown on 22 July 1971; first Model 102 flown on 25 November 1971; first Model 103 flown on 25 April 1972; first Srs 120 flown on 30 January 1973. Production of 100 completed in January 1973, with 138 more then on order.
Notes: The Bulldog originated as a military derivative of the Beagle Pup light aircraft (see *Civil* volume). Deliveries of the initial Series 100 comprised 58 to the Swedish Air Force (Model 101, Swedish designation Sk 61), 20 to the Swedish Army (Model 101), 15 to the Royal Malaysian Air Force (Model 102); 5 for the Kenya Air Force (Model 103). The Series 120 is restressed to permit semi-aerobatic manoeuvres at the full gross weight. Orders for this variant include 132 for the RAF as Bulldog T Mk 1 (Model 121) the first of which was delivered to the CFS at Little Rissington in April 1973, six for the Ghana Air Force (Model 122), delivered in June 1973, and 20 for the Nigerian Air Force (Model 123), ordered in May 1973 for delivery in 1973/74. The RAF Bulldogs are to be used primarily by the 16 University Air Squadrons.

SCOTTISH AVIATION JETSTREAM T MK 1

Illustrated by the photograph above and the general arrangement silhouette (right), the Jetstream T Mk 1 was entering service with the RAF as a crew trainer late 1973

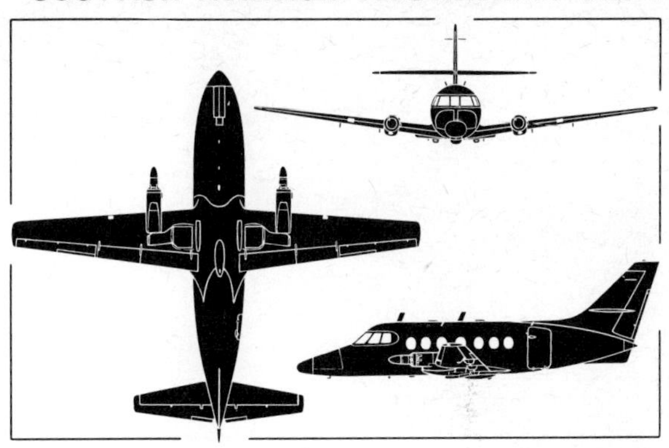

Country of Origin: United Kingdom.
Type: Crew trainer and communication aircraft.
Power Plant: Two 996 eshp Turboméca Astazou 16 turboprops.
Performance: Max speed, 285 mph (459 km/h) at 12,000 ft (3 657 m); max cruising speed, 278 mph (448 km/h) at 12,000 ft (3 657 m) and 254 mph (409 km/h) at 22,000 ft (6 705 m); initial rate of climb, 2,500 ft/min (12,7 m/sec); service ceiling, 26,000 ft (7 928 m); range with max fuel, 1,380 mls (2 224 km).
Weights: Typical empty equipped weight, 8,741 lb (3 973 kg); max take-off weight, 12,500 lb (5 670 kg); max zero fuel weight 12,250 lb (5 556 kg).
Dimensions: Span, 52 ft 0 in (15,85 m); length, 47 ft $1\frac{1}{2}$ in (14,37 m); height, 17 ft $5\frac{1}{2}$ in (5,32 m); wing area, 270 sq ft (25,08 m²).
Accommodation: Normal flight crew of three (two pilots and supernumerary), with up to 18 passenger seats in cabin as required.
Status: Prototype (HP Jetstream 1) first flown on 18 August 1967; first Mk 3 prototype flown on 21 November 1968. Production deliveries began in 1969. First Scottish Aviation Jetstream T Mk 1 flown on 13 April 1973; 26 ordered by RAF.

Notes: The Jetstream originated as the Handley Page HP.137, a low-priced third-level airliner and executive aircraft that was in production at the Radlett factory when the company went bankrupt in March 1970. About 36 had been completed and some 20 delivered, these being of the Mk 1 variety with Astazou XIV engines. Under development were the Mk 2 with Astazou XVI engines and the Mk 3 with Garrett AiResearch TPE-331-3A-301W engines as specified by the USAF for its C-10A version of the Jetstream. Design and production rights for the Jetstream were acquired by Jetstream Aircraft Ltd, which continued development of the Mk 2, and were then purchased by Scottish Aviation, together with all existing production components and assemblies. From Scottish Aviation, the RAF ordered 26 Jetstreams to fill a requirement for a multi-engined pilot trainer. Included in this batch of aircraft are a number which had been completed or part-assembled by Handley Page, while the balance of the RAF order is made up of completely new airframes. The first Jetstream T Mk 1 was handed over to the RAF on 26 June 1973, the type subsequently entering service with No 5 Flying Training School at Oakington, replacing the Varsity. Commercial model (see *Civil* volume) is the Jetstream 200.

SIAI-MARCHETTI SF.260MX

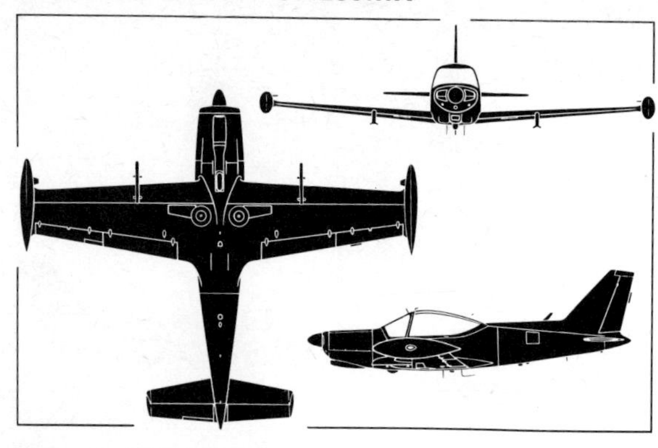

The armed version of the SF 260M primary trainer, the SF 260W Warrior, is illustrated by the photograph above and by the general arrangement silhouette (left)

Country of Origin: Italy.
Type: Primary trainer and light attack aircraft.
Power Plant: One 260 hp Avco Lycoming O-540-E4A5 six-cylinder horizontally-opposed engine.
Performance: Max speed, 211 mph (340 km/h) at sea level; max cruising speed, 186 mph (300 km/h) at 75 per cent power at 4,925 ft (1 500 m); initial rate of climb, 1,493 ft/min (7,6 m/sec); service ceiling, 16,400 ft (5 000 m); range with max fuel, 894 mls (1 440 km).
Weights: Empty equipped, 1,587 lb (720 kg); max take-off and landing, 2,998 lb (1 360 kg).
Dimensions: Span, 26 ft 11¾ in (8,25 m); span over tip-tanks 27 ft 4¾ in (8,35 m); length, 23 ft 3½ in (7,10 m); height, 7 ft 11 in (2,41 m); wing area, 108·5 sq ft (10,10 m²).
Accommodation: Two seats side-by-side with full dual controls and third seat behind.
Armament (SF.260W only): One strong point in each wing with max capacity of 330 lb (150 kg); typical loads can include two Matra gun pods each containing two 7·62-mm guns; two rocket pods or two 110-lb (50-kg) or two 264·5-lb (120-kg) bombs.

Status: Prototype (as Frati F.250) flown on 15 July 1964. First SF.260MX flown in September 1969. First SF.260W Warrior flown in May 1972. Production quantities include 36 SF.260M for Belgian Air Force; 12 SF.260MC for Zaïre Air Force; 16 SF.260MS for Singapore Air Defence; 8 SF.260MZ for Zambian Air Force, 32 SF.260MP and 16 SF.260WP Warrior for Philippine Air Force and 12 SF.260MT for Royal Thai Air Force.
Notes: The SF.260MX is the generic designation for the export military version of the SF.260 three-seat high performance lightplane (see *Civil* volume). A number of significant structural alterations were introduced by SIAI-Marchetti in the design of the original SF.260M, and a further stage in the development of the type was reached in 1972 with the introduction of the SF.260W Warrior. This differs from the SF.260M primarily in having a strengthened wing with two pylons for the carriage of assorted weapon loads, as indicated above, and entered production in 1973, the first export order for this type having been received from the Philippine Air Force which is to use the Warrior in the weapons training and counter-insurgency rôles, complementing the North American T-28 used by Philippine Army and Constabulary support units.

SOKO G2-A GALEB AND JASTREB

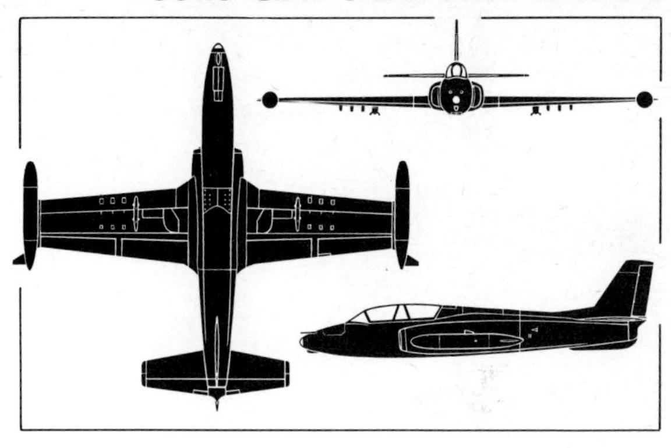

The Galeb, illustrated by the photograph above and the general arrangement silhouette (right), was the first jet aircraft of Yugoslav design to attain production status

Country of Origin: Yugoslavia.
Type: Basic trainer.
Power Plant: One 2,500 lb st (1 134 kgp) Rolls-Royce Viper II Mk 22-6 turbojet.
Performance: Max speed, 470 mph (756 km/h) at sea level and 505 mph (812 km/h) at 20,350 ft (6 200 m); max cruising speed 453 mph (730 km/h) at 19,680 ft (6 000 m); initial rate of climb, 4,500 ft/min (22,8 m/sec); service ceiling, 39,375 ft (12 000 m); max range, 770 mls (1 240 km) at 29,520 ft (9 000 m) with full tip tanks; max endurance at 23,000 ft (7 000 m) 2 hr 30 min.
Weights: Empty equipped, 5,775 lb (2 620 kg); max take-off weights according to mission: aerobatic (clean), 7,438 lb (3 374 kg); basic trainer (clean), 7,690 lb (3 488 kg); navigation trainer (with tip-tanks) 8,439 lb (3 828 kg); weapons trainer, 8,792 lb (3 988 kg); strike, 9,210 lb (4 178 kg).
Dimensions: Span, 34 ft 4½ in (10,47 m); span over tip-tanks, 38 ft 1½ in (11,62 m); length, 33 ft 11 in (10,34 m); height, 10 ft 9 in (3,28 m); wing area, 209·14 sq ft (19,43 m²).
Accommodation: Two in tandem, with full dual control.
Armament: Two 0·50-in (12,7-mm) machine guns in nose and underwing pylons for two 110-lb (50-kg) or 220-lb (100-kg) bombs and four 57-mm or two 127-mm rockets.
Status: First of two prototypes flown in May 1961; first pre-production example flew in February 1963 and deliveries to Yugoslav AF began in 1965.
Notes: Design of the Galeb (Gull) began in 1957 and after prototypes had been evaluated the type became the first jet-powered aircraft of Yugoslavian design to enter production. In addition to the Yugoslav Air Force, which has adopted the type as its standard trainer, the Galeb is used by the Zambian Air Force. A version powered by the 3,395 lb st (1 540 kgp) Viper 532 was designated Galeb-3 and made its first flight on 19 August 1970, but did not enter production. The Yugoslav Air Force also has in service a single-seat light attack version of the same design, known as the Jastreb (Hawk). This is powered by the 3,000 lb st (1 360 kgp) Viper 531 and has three nose-mounted 0·5-in (12,7-mm) machine guns and eight underwing strong points, with a wide stores capability. Provision is made for day and night reconnaissance equipment, but apart from local structural strengthening, the airframe of the Jastreb is essentially similar to that of the Galeb.

WSK-MIELEC TS-11 ISKRA

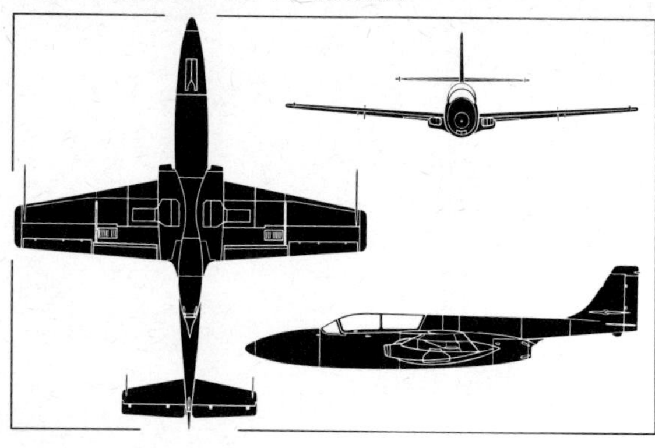

The TS-11 Iskra, illustrated by the photograph above and the general arrangement silhouette (left), was designed to compete with the Czechoslovak L 29 Delfin

Country of Origin: Poland.
Type: Basic trainer.
Power Plant: One 2,205 lb st (1 000 kgp) OKL SO-1 turbojet.
Performance: Max speed, 447 mph (720 km/h) at 16,000 ft (5 000 m); normal cruising speed, 373 mph (600 km/h); initial rate of climb, 3,150 ft/min (16,0 m/sec); service ceiling, 41,000 ft (12 500 m); range with max fuel, 907 mls (1 460 km).
Weights: Empty, 5,423 lb (2 460 kg); normal take-off, 8,068 lb (3 660 kg); max loaded, 8,377 lb (3 800 kg); max landing weight, 7,716 lb (3 500 kg).
Dimensions: Span, 33 ft $0\frac{1}{4}$ in (10,07 m); length, 36 ft $10\frac{3}{4}$ in (11,25 m); height, 10 ft 8 in (3,25 m); wing area, 188,37 sq ft (17,5 m^2).
Accommodation: Two in tandem, full dual controls.
Armament (Iskra 100 only): One 23-mm fixed forward-firing cannon in nose, with gun camera, and four wing strong-points each stressed to carry 110-lb (50-kg) load.
Status: Prototype first flown on 5 February 1960. Production deliveries to Polish Air Force (the only user) began in March 1963.
Notes: The TS-11 was designed in the Polish State Aircraft Factories to meet the requirement for a standardised jet basic trainer for use by the Warsaw Pact countries. After an evaluation programme using four prototypes each powered by the 1,700 lb st (771 kgp) HO-10 turbojet, the TS-11 was placed second to the Czech L-29 Delfin (see page 156). Nevertheless, the Polish Air Force decided to adopt the home design rather than the L-29 for its own use and production began at the WSK (Transport Equipment Manufacturing Centre) at Mielec in 1962. A few pre-production aircraft were completed with the HO-10 engine but the more powerful SO-1 was then introduced and has been the standard engine for the production run of several hundred TS-11s. A variant for weapon training has also been produced, with a nose gun and wing pylons, and is designated Iskra 100 (the name Iskra means Spark). A TS-11 has established speed records for aircraft of its weight class including 521·33 mph (839 km/h) over a straight course and 454·03 mph (730,701 km/h) over a 500-km (806-mls) closed circuit. The TS-11 is used to provide an all through jet training syllabus at the Polish Air Force's Zwirko and Wigura Officers' Flying School, carrying pupils to wings standard. Pupils subsequently convert to the SBLim-2 version of the MiG-15UTI (see page 181) for advanced flying.

The original production version of the Yak-18, still widely used, is illustrated by the photograph above and by the general arrangement silhouette (right)

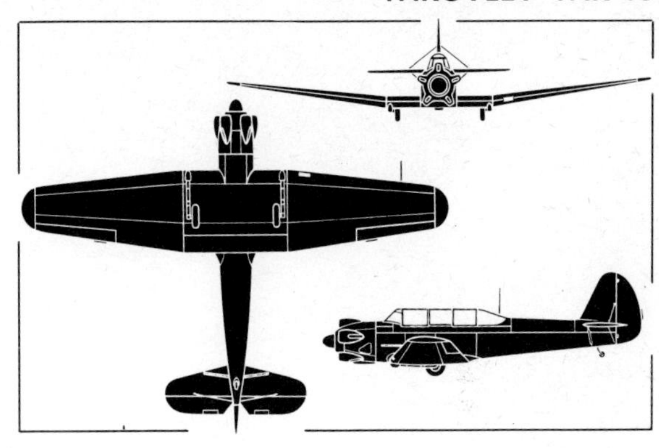

Country of Origin: USSR.
NATO Code Name: *Max*
Type: Primary trainer and aerobatic monoplane.
Power Plant: One (Yak-18 and -18U) 160 hp M-11FR or (Yak 18A, -18P, -18PM and -18PS) 300 hp Ivchenko AI-14RF nine-cylinder radial engine.
Performance (Yak-18): Max speed, 153 mph (248 km/h); max cruising speed, 133 mph (215 km/h); initial rate of climb, 689 ft/min (3,5 m/sec); service ceiling, 13,130 ft (4 000 m); range 630 mls (1 015 km).
Performance (Yak-18A): Max speed, 163 mph (263 km/h); initial rate of climb, 1,043 ft/min (5,3 m/sec); service ceiling, 16,600 ft (5 060 m); range 560 mls (900 km).
Weights (Yak-18): Empty, 1,799 lb (816 kg); loaded, 2,469 lb (1 120 kg).
Weights (Yak-18A): Empty, 2,259 lb (1 025 kg); loaded, 2,900 lb (1 316 kg).
Dimensions (Yak-18): Span, 33 ft 9½ in (10,3 m); length, 26 ft 6 in (8,07 m); wing area, 183 sq ft (17 m²).
Dimensions (Yak-18A): Span, 34 ft 9¼ in (10,6 m); length, 28 ft 0 in (8,53 m); height, 11 ft 0 in (3,35 m); wing area, 183 sq ft (17 m²).
Accommodation: Two in tandem, dual controls.
Status: Prototype first flown 1946; production deliveries began 1947; Yak-18U first appeared in 1955; Yak-18A introduced in 1957. Total of 6,760 Yak-18 series aircraft manufactured.
Notes: Derived from a prototype built by the Yakovlev bureau in 1938 (the AIR-19 or Ya-19), the Yak-18 entered production in 1946 and quickly became the standard primary trainer in the Soviet Air Force. It remained operational in the same capacity in 1973—in its developed Yak-18A version—being used to provide *ab initio* training at both military and civil schools in the Soviet Union before student pilots progressed to the L-29 jet basic trainer (see page 156). The original Yak-18 had a tailwheel undercarriage but this was changed to a nosewheel version in the Yak-18U and the Yak-18A, the latter having an uprated engine. Several single-seat versions of the Yak-18 have been evolved for aerobatic competitions flying, all with tricycle undercarriage and AI-14RF engine, and designated Yak-18P or -18PM. In addition to the Soviet Union, some Warsaw Pact nations also still have Yak-18 trainers in use.

AÉROSPATIALE ALOUETTE II

The photograph above depicts an SE 313B Alouette II of France's Aéronavale. Externally this differs little from the SA 315B Lama and the SA 318C Alouette II

Country of Origin: France.
Type: Five-seat light utility helicopter.
Power Plant: One (SA 318C) 523 shp Turboméca Astazou IIA or (SA 315B) 870 shp Artouste IIB turboshaft.
Performance: (SA 318C) Max speed, 127 mph (205 km/h) at sea level; max cruise, 112 mph (180 km/h); max inclined climb, 1,396 ft/min (7,1 m/sec); hovering ceiling (IGE), 5,085 ft (1 550 m), (OGE), 2,960 ft (900 m); range with max fuel, 447 mls (720 km), with max payload 62 mls (100 km).
Weights: (SA 318C) Empty, 1,961 lb (890 kg); max take-off, 3,630 lb (1 650 kg).
Dimensions: Rotor diam, 33 ft 5 in (10,20 m); fuselage length, 31 ft 11¾ in (9,75 m), height, 9 ft (2,75 m).
Notes: The SE 313B Alouette II was first flown on 12 March 1955, powered by a 360 shp Artouste IIC6 turboshaft. Production totalled 923. The SA 318C differed in having an Astazou engine and about 1,300 have been bought by operators all over the world. The SA 315B Lama, first flown on 17 March 1969, has the dynamics system of the Alouette III, and is in production by HAL in India as the Cheetah for the Indian Air Force.

AÉROSPATIALE ALOUETTE III

The photograph above depicts an SA 316 Alouette III of the Search-and-Rescue Flight of the Royal Netherlands Air Force with pontoons partly inflated

Country of Origin: France.
Type: Seven-seat light utility helicopter.
Power Plant: One (SA 316C) 870 shp Turboméca Artouste IIID or (SA 319B) 789 shp Astazou XIV turboshaft.
Performance: (SA 319B) Max speed, 136 mph (220 km/h) at sea level; max cruise, 122 mph (197 km/h); max inclined climb, 853 ft/min (4,32 m/sec); hovering ceiling (IGE), 5,740 ft (1 750 m); range with six passengers, 375 mls (605 km).
Weights: (SA 319B) Empty, 2,403 lb (1 090 kg); max take-off, 4,960 lb (2 250 kg).
Dimensions: Rotor diam, 36 ft 1¾ in (11,02 m); fuselage length, 32 ft 10¾ in (10,03 m), height, 9 ft 10 in (3,0 m).
Notes: The SA 316 Alouette III was a development of the Alouette II, first flown on 28 February 1959, with 870 shp Artouste IIIB turboshaft. Initial production model was SA 316A, followed by SA 316B with higher weights and SA 316C with Artouste IIID engine. The SA 319B (data above) has an Astazou engine and increased range. Orders for all versions exceed 1,000 and licence production includes 80 by HAL in India, 50 by IAR in Roumania and 60 in Switzerland.

AÉROSPATIALE SA 321 SUPER FRELON

The photograph above depicts a pre-production example of the SA 321G Super Frelon which serves in the anti-submarine rôle with Flottille 32F of the Aéronavale

Country of Origin: France.
Type: Medium transport and multi-purpose helicopter.
Power Plant: Three 1,550 shp Turboméca Turma IIICB turbo-shafts.
Performance: Max speed, 149 mph (240 km/h) at sea level; max cruise, 143 mph (230 km/h); max inclined climb, 1,495 ft/min (7,6 m/sec); hovering ceiling (IGE), 7,380 ft (2 250 m), (OGE), 1,804 ft (550 m); range with 5,511-lb (2 500-kg) payload and 20 min reserves, 404 mls (650 km).
Weights: Empty, 14,420 lb (6,540 kg); max take-off, 27,557 lb (12 500 kg).
Dimensions: Rotor diam, 62 ft 0 in (18,90 m); fuselage length, 63 ft $7\frac{3}{4}$ in (19,40 m); height, 21 ft $10\frac{1}{4}$ in (6,66 m).
Notes: Derived from the smaller Frelon prototypes, the Super Frelon first flew on 7 December 1962. The initial production model was the SA 321G amphibious model for the *Aéronavale* with Sylph radar in outrigger floats and up to four torpedoes and other ASW stores; 17 were built for *Flotilla* 32F. Non-amphibious transport versions have been delivered to Israel (SA 321K), South Africa (SA 321L) and Libya.

AÉROSPATIALE SA 330 PUMA

The photograph above depicts an SA 330E Puma HC Mk 1 of No 33 Squadron, RAF Support Command, the first RAF unit to operate this tactical medium transport helicopter

Country of Origin: France.
Type: Medium transport helicopter.
Power Plant: Two 1,320 shp Turboméca Turmo IIIC4 turbo-shafts.
Performance: Max speed, 174 mph (280 km/h) at sea level; max cruise, 165 mph (265 km/h); max inclined climb, 1,400 ft/min (7,1 m/sec); hovering ceiling (IGE), 9,186 ft (2 800 m), (OGE), 6,233 ft (1 900 m); max range, 390 mls (630 km).
Weights: Empty, 7,561 lb (3 430 kg); max take-off, 14,110 (6 400 kg).
Dimensions: Rotor diam, 49 ft $2\frac{1}{2}$ in (15,00 m); fuselage length, 46 ft $1\frac{1}{2}$ in (14,06 m); height, 16 ft $10\frac{1}{2}$ in (5,14 m).
Notes: Subject of an Anglo-French joint production agreement, the SA 330 first flew on 15 April 1965. Production versions are the SA 330B for the French Army (130 ordered); SA 330C military export versions (ordered by Portuguese, South African, Zaïre Republic, Abu Dhabian, Algerian and Ivory Coast air arms); SA 330E for the RAF (40 delivered from Westland assembly line, first flown on 25 November 1970) and SA 330F civil version for 15–17 passengers with 1,385 shp Turmo IVA engines.

AÉROSPATIALE/WESTLAND SA 341 GAZELLE

The photograph above depicts the third pre-production SA 341 Gazelle serving as a development aircraft for the SA 341B Gazelle AH Mk 1 for the British Army

Country of Origin: France.
Type: Five-seat light utility helicopter.
Power Plant: One 592 shp Turboméca Astazou IIIN turboshaft.
Performance: Max speed, 165 mph (265 km/h) at sea level; max cruise 149 mph (240 km/h) max inclined rate, 1,214 ft/min (6,16 m/sec); hovering ceiling (IGE), 10,170 ft (3 100 m), (OGE), 8,530 ft (2 600 m); max range, 403 mls (650 km).
Weights: Empty, 1,873 lb (850 kg); max take-off, 3,747 lb (1 700 kg).
Dimensions: Rotor diam, 34 ft $5\frac{1}{2}$ in (10,50 m); fuselage length, 31 ft $2\frac{3}{4}$ in (9,52 m); height, 10 ft $3\frac{1}{2}$ in (3,14 m).
Notes: The SA 340 prototype first flew on 7 April 1967 and in SA 341 production form was adopted for joint development and production in France and Britain. First production examples flew in France on 6 August 1971 and in Britain on 31 January 1972. Variants are the SA 341B, Gazelle AH Mk 1 for British Army; SA 341C, HT Mk 2 for RN; SA 341D HT Mk 3 for RAF; SA 341E HCC Mk 4 for RAF; SA 341F for French Army; SA 341G civil version and SA 341H export model, also licence-built in Yugoslavia by SOKO with deliveries commencing late 1973.

BELL 47

The photograph above depicts a Bell 47G-3B-1 of the Royal New Zealand Air Force, the primary task of which is Army support. The Bell 47 is employed by many air forces

Country of Origin: USA.
Type: Three-seat light utility and training helicopter.
Power Plant: One (OH-13G & TH-13M) 200 hp Franklin 6V-200-C32 or (OH-13H and UH-13H) 240 hp Lycoming VO-435-AIA, AIB or AID or (OH-13S and TH13T) 260 hp Lycoming TVO-435-AIA flat six engine.
Performance: (OH-135) Max speed, 105 mph (169 km/h) at sea level; max cruise, 86 mph (138 km/h) at 5,000 ft (1 524 m); initial climb, 1,190 ft/min (6,05 m/sec); hovering ceiling (IGE), 18,000 ft (5 486 m); range, 324 mls (521 km).
Weights: Empty, 1,936 lb (877 kg); max loaded, 2,850 lb (1 293 kg).
Dimensions: Rotor diameter, 37 ft $1\frac{1}{2}$ in (11,31 m); fuselage length, 32 ft 7 in (9,93 m); overall height, 9 ft $3\frac{3}{4}$ in (2,82 m).
Notes: First flown in prototype form on 8 December 1945, the Bell 47 has a longer production span than any other helicopter and over 5,000 examples have been built. Production was being phased out in 1973 both by Bell and the European licencee, Agusta; the type had also been built by Kawasaki in Japan and Westland in the UK.

BELL MODEL 204B IROQUOIS

The photograph above depicts a UH-1B Iroquois of the Royal Australian Air Force's No 5 Squadron which serves in the battlefield support rôle

Country of Origin: USA.
Type: Ten-seat utility helicopter.
Power Plant: One 1,100 shp Lycoming T5311A turboshaft.
Performance: (At 8,500 lb/3 855 kg) Max speed, 120 mph (193 km/h) at sea level); max cruise, 110 mph (177 km/h); max inclined climb, 1,400 ft/min (7,1 m/sec); hovering ceiling (IGE), 10,000 ft (3 050 m), (OGE), 4,500 ft (1 370 m); max range, 392 mls (630 km).
Weights: Empty, 4,600 lb (2 086 kg); max take-off, 9,500 lb (4 309 kg).
Dimensions: Rotor diam, 48 ft 0 in (14,63 m); fuselage length, 40 ft 4 in (12,31 m); height, 12 ft $8\frac{1}{2}$ in (3,87 m).
Notes: First flown on 22 October 1956, the Bell UH-1 is usually known by its unofficial name "Huey". Versions include UH-1B and C for Army, UH-1E for USMC assault support, UH-1F missile-site support and TH-1F trainer for USAF with 1,272 shp T58-GE-3 engine, and HH-1K rescue, TH-1L Seawolf trainer and UH-1L utility models for USN. Fuji builds Model 204B in Japan and Agusta the AB-204B in Italy, plus a special ASW version for Italian Navy.

BELL MODEL 205 IROQUOIS

The photograph above depicts a UH-1H of the US Army. This 10-seat utility helicopter has been licence built in Germany, Japan and on Taiwan in its UH-1D version

Country of Origin: USA.
Type: Fifteen-seat utility helicopter.
Power Plant: One 1,400 shp Lycoming T5313A turboshaft.
Performance: (At 9,500 lb/4 309 kg) Max speed, 127 mph (204 km/h) at sea level; max cruise, 111 mph (179 km/h) at 8,000 ft (2 440 m); max inclined climb, 1,680 ft/min (8,53 m/sec); hovering ceiling (IGE), 10,400 ft (3 170 m), (OGE), 6,000 ft (1 830 m); range 344 mls (553 km) at 8,000 ft (2 440 m).
Weights: Empty equipped, 5,082 lb (2 305 kg); normal take-off, 9,500 lb (4 309 kg).
Dimensions: Rotor diam, 48 ft 0 in (14,63 m); fuselage length, 41 ft 6 in (12,65 m); height, 14 ft 6 in (4,42 m).
Notes: The Model 205 first flew on 16 August 1961 and was produced initially for the US Army as the UH-1D (1,000 shp T53-L-11 engine) and UH-1H (data above). The Canadian Armed Forces' CUH-1Hs are similar. Dornier built 352 UH-1Ds for the German Army, a similar model is built in Taiwan for the Nationalist Chinese Air Force and Agusta builds the AB-205 variant in Italy. Total production exceeds 6,000. The Model 214A HueyPlus is similar with uprated engine and Iran has ordered 287 examples.

BELL MODEL 206 KIOWA

The photograph above depicts the OH-58A Kiowa light observation version of the Model 206 JetRanger supplied to the US Army. The US Navy trainer versions is the TH-57A SeaRanger

Country of Origin: USA.
Type: Five-seat light utility helicopter.
Power Plant: One 400 shp Allison 250-C20 turboshaft.
Performance: (At 3,000 lb/1 361 kg) Max cruise, 138 mph (219 km/h) at sea level, 142 mph (228 km/h) at 5,000 ft (1 524 m); hovering ceiling (IGE), 13,200 ft (4 023 m), (OGE), 8,700 ft (2 652 m); max inclined climb, 1,540 ft/min (7,82 m/sec); max range 436 mls (702 km) at 10,000 ft (3 048 m).
Weights: Empty, 1,455 lb (660 kg); max take-off, 3,000 lb (1 360 kg).
Dimensions: Rotor diam, 33 ft 4 in (10,16 m); fuselage length, 31 ft 2 in (9,50 m); height, 9 ft 6½ in (2,91 m).
Notes: The Model 206 design originated for a US Army requirement and prototypes were designated OH-4A (first flight 8 December 1962). Production models are designated OH-58A Kiowa and a training version for the US Navy is the TH-57A SeaRanger. Agusta builds the same type in Italy as the AB 206B-1, Commonwealth Aircraft assembles OH-58As for the Australian Army and Canada is another major user. The data refer to the export model; the OH-58A has slightly larger dimensions.

BELL MODEL 209 HUEYCOBRA

The photograph above depicts the twin-engined version of the Model 209 HueyCobra which, ordered in quantity by Iran, is used by the US Marine Corps as the AH-1J SeaCobra

Country of Origin: USA.
Type: Two-seat attack helicopter.
Power Plant: (AH-1G) One 1,400 shp Lycoming T53-L-13 turboshaft.
Performance: (AH-1G) Max speed, 219 mph (352 km/h) at sea level; max inclined climb, 1,580 ft/min (8 m/sec); hovering ceiling, (IGE), 9,900 ft (3 015 m); max range, 387 mls (622 km) at sea level.
Weights: Operational empty, 6,096 lb (2 765 kg); max take-off, 9,500 lb (4 309 kg).
Dimensions: Rotor diam, 44 ft 0 in (13,41 m); fuselage length, 44 ft 5 in (13,54 m); height, 12 ft 8 in (4,14 m).
Notes: The Model 209 was developed for armed missions by the US Army and first flight was made on 7 September 1965. The original Army version is designated AH-1G and has two 7·62-mm Miniguns in the nose turret and rockets or gun pods on the stub wings, but many are being converted to AH-1Q with TOW missile armament. The USMC variant is the twin-engined AH-1J with 1,800 shp T400-CP-400 coupled turboshaft and a three-barrel 20-mm cannon in the nose barbette. Iran has ordered 202 AH-1Js.

BELL MODEL 212

The photograph above depicts a CUH-1N of the Mobile Command of the Canadian Armed Forces, this being similar to the UH-1N of the USAF, the US Navy and the US Army

Country of Origin: USA.
Type: Fifteen-seat utility helicopter.
Power Plant: One 1,800 shp Pratt & Whitney PT6T-3 coupled turboshaft.
Performance: Max speed, 121 mph (194 km/h) at sea level; max inclined climb at 10,000 lb (4 535 kg), 1,460 ft/min (7.4 m/sec); hovering ceiling (IGE), 17,100 ft (5 212 m); (OGE), 9,900 ft (3 020 m); max range, 296 mls (476 km) at sea level.
Weights: Empty, 5,500 lb (2 495 kg); max take-off, 10,000 lb (4 535 kg).
Dimensions: Rotor diam, 48 ft $2\frac{1}{2}$ in (14,69 m); fuselage length, 42 ft $10\frac{1}{4}$ in (13,07 m); height, 14 ft $4\frac{3}{4}$ in (4,39 m).
Notes: Derived from the Model 205, the 212 differs primarily in having a twin-engined powerplant (coupled to a single gearbox to drive the main and tail rotors). Examples are used by the USAF, USN and US Army, all with the designation UH-1N, and by the Canadian Armed forces as CUH-1N. Export models for other military users are known as Model 212s. The same twin-engined power plant is used in the AH-1J version of the Model 209 HueyCobra. Licence manufacture is undertaken by Agusta in Italy.

BOEING VERTOL CH-46/107-II

The Kawasaki-built KV-107/II-5, illustrated above, is a long-range search and rescue version of the Model 107-II for Japan's Air Self-Defence Force

Country of Origin: USA.
Type: Medium transport helicopter.
Power Plant: Two 1,500 shp General Electric T58-GE-5 turboshafts.
Performance: (At 20,800 lb/9 434 kg) Max speed, 139 mph (224 km/h), max inclined climb, 1,920 ft/min (9,75 m/sec); hovering ceiling, (IGE), 10,000 ft (3 048 m), (OGE), 7,100 ft (2 165 m); range with 2,400 lb (1 088 kg) payload and 30 min reserves, 633 mls (1 020 km).
Weights: Empty equipped, 11,585 lb (5 240 kg); max take-off, 21,400 lb (9 706 kg).
Dimensions: Rotor diam (each), 50 ft 0 in (15,24 m); fuselage length, 44 ft 10 in (13,66 m); height, 16 ft $8\frac{1}{2}$ in (5,09 m).
Notes: The Model 107-II has been in production for 12 years and is licence-built by Kawasaki in Japan for Air, Ground and Maritime Self-Defence Forces and for export. Boeing built several hundred for the USMC (the CH-46A, D and F Sea Knight) and the USN (the UH-46A and D Sea Knight). Others have been supplied to Sweden and Canada. The Model 107-II accommodates three crew and 25 passengers.

BOEING VERTOL CH-47 CHINOOK

The CH-47C Chinook of the US Army, illustrated above, is the current production version of the Model 114 which is also licence built in Italy for the Italian and Iranian forces

Country of Origin: USA.
Type: Medium transport helicopter.
Power Plant: (CH-47C) Two 3,750 shp Lycoming T55-L-11 turboshafts.
Performance: (CH-47C at 33,000 lb/14 969 kg) Max speed, 190 mph (306 km/h) at sea level; average cruise, 158 mph (254 km/h); max inclined climb, 2,880 ft/min (14,63 m/sec); hovering ceiling (OGE), 14,750 ft (4 495 m); mission radius, 115 mls (185 km).
Weights: Empty, 20,378 lb (9 243 kg); max take-off, 46,000 lb (20 865 kg).
Dimensions: Rotor diam (each), 60 ft 0 in (18,29 m); fuselage length, 51 ft 0 in (15,54 m); height, 18 ft 7 in (5,67 m).
Notes: The Model 114 (first flown on 21 September 1961) is the standard medium transport helicopter of the US Army, operating under the designation CH-47A (T55-L-5 engines), CH-47B (T55-L-7) and CH-47C (data above). The RAAF, Spanish Air Force and South Vietnam Air Force have received Chinooks from the USA and the Italian Army and Iranian Army are supplied from the Elicotteri Meridionali line in Italy.

FAIRCHILD HILLER H-23

A small number of Hiller 12E helicopters serve with the Royal Navy in the training rôle, one of these being illustrated above, the US Army version being the OH-23G Raven

Country of Origin: USA.
Type: Three-seat light utility and observation helicopter.
Power Plant: One (OH-23D) 250 hp Lycoming VO-435-A1C or (OH-23G) 305 hp Lycoming VO-540-A1B flat-six engine.
Performance: (OH-23G) Max speed, 96 mph (154 km/h); cruising speed, 90 mph (145 km/h); initial climb, 1,290 ft/min (6,5 m/sec); hovering ceiling (IGE), 9,500 ft (2 895 m), (OGE), 5,800 ft (1 768 m); service ceiling, 15,200 ft (4 633 m); range, 250 mls (402 km).
Weights: Empty, 1,755 lb (794 kg); loaded, 2,800 lb (1 270 kg).
Dimensions: Rotors diameter, 35 ft $4\frac{3}{4}$ in (10,79 m); fuselage length, 28 ft 6 in (8,69 m); overall height, 9 ft $3\frac{1}{2}$ in (2,82 m).
Notes: The Hiller UH-12A was first flown in 1950 and provided the basis of Stanley Hiller's most successful helicopter design, more than 2,000 examples being built. Small batches were supplied to the RN for use as trainers (still in use in 1973) and to the Canadian Armed Forces, and a few examples to several other air forces but the US Army was the largest user. A four-seat derivative, the Model E4, entered service with the US Army in 1962 as the OH-23F for high-performance utility missions.

HUGHES OH-6/MODEL 500 CAYUSE

The Model 500M, essentially similar to the OH-6A Cayuse, is licence manufactured by Kawasaki for Japan's Ground Self-Defence Force, an example being illustrated above

Country of Origin: USA.
Type: Six-seat light utility helicopter.
Power Plant: One 317 shp Allison 250-C18A turboshaft.
Performance: Max speed, 152 mph (244 km/h) at 1,000 ft (305 m); range cruise, 138 mph (222 km/h) at sea level; max inclined climb, 1,700 ft/min (8,64 m/sec); hovering ceiling (IGE), 8,200 ft (2 500 m), (OGE), 5,300 ft (1 615 m); max range, 377 mls (589 km) at 4,000 ft (1 220 m).
Weights: Empty, 1,086 lb (492 kg); max take-off, 2,550 lb (1 157 kg).
Dimensions: Rotor diam, 26 ft 4 in (8,03 m); fuselage length, 23 ft 0 in (7,01 m); height, 8 ft $1\frac{1}{2}$ in (2,48 m).
Notes: The Model 500 (originally developed as Model 369) was winner of a US Army contest in 1965 for a light observation helicopter and 1,415 were built as OH-6A Cayuse with Allison T65-A-5A engine. Several armed forces have ordered the similar Model 500M (data above) and this type is built in Japan by Kawasaki for the Ground Self Defence Force (as the OH-6J) and by Nardi in Italy. The GSDF is to receive 165 OH-6Js. The Spanish Navy has 12 of a special ASW version, carrying two torpedoes.

HUGHES TH-55A OSAGE

The photograph above depicts the TH-55A Osage, a primary trainer variant of the Hughes Model 300 supplied in large numbers to the US Army

Country of Origin: USA.
Type: Three-seat light utility helicopter.
Power Plant: (Model 300C) One 190 shp Lycoming HIO-360-D1A four-cylinder horizontally-opposed engine.
Performance: (Model 300C) Max speed, 105 mph (169 km/h); max cruise, 100 mph (161 km/h) at 5,000 ft (1 525 m); max inclined climb, 1,100 ft/min (5,08 m/sec); hovering ceiling (IGE), 7,600 ft (2 316 m), (OGE), 5,200 ft (1 585 m); range, 255 mls (410 km).
Weights: Empty, 1,025 lb (465 kg); max take-off, 1,900 lb (861 kg).
Dimensions: Rotor diam, 26 ft 10 in (8,18 m); fuselage length, 23 ft 1 in (7,03 m); height, 8 ft $2\frac{3}{4}$ in (2,50 m).
Notes: The Hughes Tool Co developed its initial light helicopter prototype, the Model 269, in 1956, and production began in 1961. The US Army tested five YHO-2HUs in the observation rôle and subsequently, in 1964, selected the Model 269A as its standard primary helicopter trainer. Production of the training model, TH-55A, was completed in March 1969, 792 being delivered. Licence manufacture is undertaken in Japan by Kawasaki, 48 TH-55As being on order for the Ground Self-Defence Force.

KAMAN H-43 HUSKIE

The HH-43F Huskie, an example of which in USAF service is illustrated above, has been supplied for crash rescue duties to a half-dozen air forces

Country of Origin: USA.
Type: Crash rescue and utility helicopter.
Power Plant: One 825 hp Lycoming T53-L-11A turboshaft.
Performance: Max speed, 120 mph (193 km/h); cruising speed, 110 mph (177 km/h); inclined rate of climb, 1,800 ft/min (9,1 m/sec); service ceiling, 23,000 ft (7 010 m); hovering ceiling (IGE), 20,000 ft (6 100 m), (OGE), 16,000 ft (4 880 m); range, 500 mls (810 km).
Weights: Empty, 4,620 lb (2 095 kg); normal loaded, 6,500 lb (2 950 kg).
Dimensions: Rotor diam, 47 ft (14,33 m); fuselage length, 25 ft 2 in (7,67 m); height, 15 ft $6\frac{1}{2}$ in (4,73 m).
Notes: The USAF's HH-43B Huskie (often known as the *Pedro*) was developed from the US Navy HOK-1 observation helicopter to serve on crash rescue duties at all Air Force bases. The HH-43B had a T53-L-1B engine and was followed by the improved HH-43F. Examples were supplied through MAP to several nations including Burma, Colombia, Morocco, Iran, Pakistan and Thailand, and the US Navy used a single QH-43G drone for special duties. A 1,000-lb load of fire-fighting and rescue equipment is carried by the HH-43F.

KAMAN SH-2 SEASPRITE

Illustrated above is the SH-2F Seasprite of the US Navy, the latest in a progressive series of conversions that have transformed the Seasprite for the ASW and missile defence rôles

Country of Origin: USA.
Type: All-weather search and rescue helicopter.
Power Plant: Two 1,350 shp General Electric T58-GE-8F turbo-shafts.
Performance: (SH-2F) Max speed, 168 mph (270 km/h) at sea level; normal cruise, 150 mph (241 km/h); max inclined climb, 2,440 ft/min (12,4 m/sec); hovering ceiling (IGE), 18,600 ft (5 670 m), (OGE), 15,400 ft (4 695 m); max range, 445 mls (716 km).
Weights: (SH-2F) Empty, 7,500 lb (3 401 kg); normal take-off, 12,800 lb (5 805 kg); max overload, 13,300 lb (6 032 kg).
Dimensions: Rotor diam, 44 ft 0 in (13,41 m); fuselage length, 40 ft 6 in (12,3 m); height, 15 ft 6 in (4,72 m).
Notes: Original production of the Seasprite totalled 190, in UH-2A and UH-2B single-engined versions; 40 were converted to twin-engined UH-2C, six to armed HH-2C and about 70 to HH-2D. Starting in 1970, the US Navy has assigned the Seasprite force to the LAMPS (Light Airborne Multi-Purpose System) programme for anti-submarine warfare and missile defence, with 20 converted to SH-2Ds and 85 under modification to SH-2F.

KAMOV KA-25

The KA-25, illustrated above, is the standard Soviet Navy shipboard ASW helicopter, currently serving aboard the helicopter cruisers Moskva and Leningrad

Country of Origin: USSR.
NATO Code Name: *Hormone A.*
Type: Shipboard anti-submarine warfare helicopter.
Power Plant: Two 900 shp Glushenkov GTD-3 turboshafts.
Performance: (Estimated) Max speed, 130 mph (209 km/h); normal cruise, 120 mph (193 km/h); max range, 400 mls (644 km); service ceiling, 11,000 ft (3 353 m).
Weights: (Estimated) Empty, 10,500 lb (4 765 kg); max take-off, 16,500 lb (7 484 kg).
Dimensions: Rotor diam (each), 51 ft 7½ in (15,74 m); approx fuselage length, 35 ft 6 in (10,82 m).
Notes: The Ka-25 serves with the Soviet Navy primarily in the ASW rôle but is also employed in the utility and transport rôles. The ASW Ka-25 serves aboard the helicopter cruisers *Moskva* and *Leningrad* as well as with shore-based units. A search radar installation is mounted in a nose radome, but other sensor housings and antennae differ widely from helicopter to helicopter. Each landing wheel is surrounded by an inflatable pontoon surmounted by inflation bottles. Sufficient capacity is available to accommodate up to a dozen personnel.

MIL Mi-1

The first Soviet helicopter to be manufactured in series, the Mil Mi-1, an example in Polish service being illustrated above, has now been largely phased out in favour of the Mi-2

Country of Origin: USSR.
NATO Code Name: *Hare.*
Type: Four-seat military helicopter.
Power Plant: One 575 hp Ivchenko A1-26V piston radial.
Performance: Max speed, 111 mph (180 km/h); cruising speed, 84 mph (135 km/h); inclined rate of climb, 1,280 ft/min (6,5 m/sec); hovering ceiling, (OGE), 9,842 ft (3 000 m); range, 340 mls (550 km).
Weights: Empty, 4,035 lb (1 831 kg); normal take-off, 5,325 lb (2 416 kg).
Dimensions: Rotors diam, 47 ft 1 in (14,35 m); fuselage length, 39 ft 9 in (12,11 m); height, 10 ft 10 in (3,30 m).
Notes: First flown in 1948, the Mi-1 was the first Soviet helicopter to enter large-scale production, being used for both military and civil duties. Largely replaced now by the Mi-2, it remains in use with some smaller air arms as a communications and general-purpose type. The Mi-1 was built in Poland by the WSK works as the SM-1, and the Polish Air Force also has in service the SM-2, developed in Poland from the Mi-1 with a lengthened, five-seat front fuselage and a nose-door for stretcher loading.

MIL MI-2

The Mil Mi-2, although developed in the Soviet Union, has been manufactured almost exclusively in Poland and has been exported to the forces of a large number of countries

Country of Origin: USSR.
NATO Code Name: *Hoplite.*
Type: Light general-purpose helicopter.
Power Plant: Two 437 shp Izotov GTD-350 turboshafts.
Performance: Max speed, 130 mph (210 km/h) at 1,640 ft (500 m); max cruise, 124 mph (200 km/h); econ cruise, 118 mph (190 km/h); max inclined climb, 885 ft/min (4,5 m/sec); hovering ceiling (IGE), 6,550 ft (2 000 m), (OGE), 3,275 ft (1 000 m); range with max payload and 5 per cent reserves, 360 mls (580 km).
Weights: Operational empty, 5,180 lb (2 350 kg); normal take-off, 7,826 lb (3 550 kg); overload take-off, 8,157 lb (3 700 kg).
Dimensions: Rotor diam, 47 ft $6\frac{3}{4}$ in (14,50 m); fuselage length, 37 ft $4\frac{3}{4}$ in (11,40 m); height, 12 ft $3\frac{1}{2}$ in (3,75 m).
Notes: Manufacture of the Mi-2, developed in the Soviet Union, has been undertaken since 1966 at the WSK-Swidnik in Poland. The Mi-2 has been built in large numbers for both civil and military tasks, and has been exported widely. Accommodation may be provided for a single pilot and six to eight passengers, or up to 1,543 lb (700 kg) of freight can be carried. For the aeromedical rôle four stretchers and a medical attendant may be carried.

MIL MI-4

The Mil Mi-4 was exported more widely than the smaller Mi-1 and, despite its age, is still quite widely used, that illustrated above serving with the Czechoslovak Air Force

Country of Origin: USSR.
NATO Code Name: *Hound.*
Type: General purpose helicopter.
Power Plant: One 1,700 hp ASh-82V piston radial engine.
Performance: Max speed, 130 mph (210 km/h) at 5,000 ft (1 500 m); economical cruise, 99 mph (160 km/h); service ceiling, 18,000 ft (5 500 m); normal range, 155–250 mls (250–400 km).
Weights: Max take-off, 17,200 lb (7 800 kg).
Dimensions: Rotor diam, 68 ft 11 in (21,0 m); fuselage length, 55 ft 1 in (16,80 m); overall height, 17 ft 0 in (5,18 m).
Notes: Able to carry 14 troops or up to 3,835 lb (1 740 kg) of freight or vehicles, the Mi-4 is comparable to the Sikorsky S-55 in design concept and rôle, and several thousand are believed to have been produced to meet Soviet Army and Air Force requirements. The type remains in service as a transport and was also used by the Soviet Navy in an anti-submarine rôle. More than 20 other countries, including all the Soviet Bloc nations, received examples of the Mi-4 and it remains in service with several of these. A commercial version, the Mi-4P, accommodates 8–11 passengers and an agricultural version is known as the Mi-4S.

MIL MI-6

The Mil Mi-6 remains one of the largest of military helicopters and was, at one time, operated by the Egyptian Air Force, an example in Egyptian service being illustrated above

Country of Origin: USSR.
NATO Code Name: *Hook.*
Type: Heavy transport helicopter.
Power Plant: Two 5,500 shp Soloviev D-25V turboshafts.
Performance: (At 93,700 lb/42 500 kg) Max speed, 186 mph (300 km/h); max cruise, 155 mph (250 km/h); service ceiling, 14,750 ft (4 500 m); range with 17,640-lb (8 000-kg) payload, 385 mls (620 km), with 9,920-lb (4 500-kg) payload and external tanks, 620 mls (1 000 km).
Weights: Empty, 60,055 lb (27 240 kg); normal take-off, 89,285 lb (40 500 kg); max take-off (for VTO), 93,700 lb (42 500 kg).
Dimensions: Rotor diam, 114 ft 10 in (35,00 m); fuselage length, 108 ft $9\frac{1}{2}$ in (33,16 m); height, 32 ft 4 in (9,86 m).
Notes: First flown in 1957, the Mi6 has clam-shell doors and folding ramps to facilitate the loading of vehicles and bulky freight. Two heavy flying-crane helicopters have been evolved from the Mi-6, the Mi-10 (NATO *Harke*), flown in 1961 and the Mi-10K flown in 1965, these being almost identical to the Mi-6 above the line of the cabin. The Mi-6 has been supplied to the armed forces of North Vietnam, Egypt and Indonesia.

MIL MI-8

Currently the most widely used of Soviet military helicopters, the Mil Mi-8 serves with the armed forces of many countries, that illustrated belonging to the Pakistan Army

Country of Origin: USSR.
NATO Code Name: *Hip.*
Type: General-purpose transport helicopter.
Power Plant: Two 1,500 shp Izotov TB-2-117A turboshafts.
Performance: (At 24,470 lb/11 100 kg) Max speed, 155 mph (250 km/h); max cruise, 140 mph (225 km/h); hovering ceiling (IGE), 5,900 ft (1 800 m), (OGE), 2,625 ft (800 m); service ceiling, 14,760 ft (4 500 m); range with 6,615 lb (3 000 kg) of freight, 264 mls (425 km).
Weights: Empty (cargo), 15,787 lb (7 171 kg), (passenger), 16,352 lb (7 417 kg); normal take-off, 24,470 lb (11 100 kg); max take-off (for VTO), 26,455 lb (12 000 kg).
Dimensions: Rotor diam, 69 ft $10\frac{1}{4}$ in (21,29 m); fuselage length, 59 ft 7 in (18,17 m); height, 18 ft $4\frac{1}{2}$ in (5,60 m).
Notes: The Mi-8 has been in continuous production since 1964 for both civil and military tasks. As a freighter the Mi-8 will carry up to 8,818 lb (4 000 kg) of cargo, and military tasks include assault transport, search and rescue, and anti-submarine warfare. The Mi-8 is operated by several Warsaw Pact air forces, serving primarily in the support transport rôle.

MIL MI-12

While the status of the Mi-12, the world's largest helicopter, was uncertain late 1973, this type is obviously intended primarily for a military role

Country of Origin: USSR.
NATO Code Name: *Homer.*
Type: Heavy transport helicopter.
Power Plant: Four 6,500 shp Soloviev D-25VF turboshafts.
Performance: Max speed, 161 mph (260 km/h); cruise, 149 mph (240 km/h); range with max payload of 78,000 lb (35 380 kg), 310 mls (500 km); service ceiling, 11,500 ft (3 500 m).
Weights: Normal take-off, 213,848 lb (97 000 kg); max take-off, 231,485 lb (105 000 kg).
Dimensions: Rotor diam (each), 114 ft 9½ in (35,00 m); fuselage length, 121 ft 4 in (37,00 m); height, 41 ft 0 in (12,50 m).
Notes: First flown in the autumn of 1968 and currently the world's largest helicopter, the Mi-12 carries a crew of six of which the pilot, co-pilot, flight engineer and electrician are accommodated on the lower flight deck with the navigator and radio-operator on the upper deck. The Mi-12 was evidently designed to carry loads compatible with those carried by the fixed-wing An-22 transport and three prototypes have been built. Production was expected to commence during the course of 1972, but it was not thought to have entered V-VS service by late 1973.

SIKORSKY S-55/WHIRLWIND

The S-55 illustrated above is a Mitsubishi-built example operated by the Japanese Ground Self-Defence Force. S-55s (and the UK-built Whirlwind) remain in service with several air arms

Country of Origin: USA.
Type: General purpose helicopter.
Power Plant: One 800 hp Wright R-1300-3 piston radial engine.
Performance: Max speed, 112 mph (180 km/h) at sea level, cruising speed, 91 mph (148 km/h); inclined rate of climb, 1,020 ft/min (5,2 m/sec); hovering ceiling (IGE), 5,800 ft (1 770 m), (OGE), 2,300 ft (700 m); normal range, 360 mls (578 km).
Weights: Empty, 5,250 lb (2 381 kg); max take-off, 7,900 lb (3 583 kg).
Dimensions: Rotor diam, 53 ft (16,16 m); fuselage length, 42 ft 3 in (12,88 m); height, 13 ft 4 in (4,07 m).
Notes: The S-55 was one of the most-used of Sikorsky helicopters, production by the parent company totalling 1,281 in addition to 44 by Mitsubishi in Japan and over 400 by Westland in the UK, as the Whirlwind. Examples remain in service with a number of smaller air arms around the world, some being of Whirlwind Srs 3 type with a 1,050 shp Gnome turboshaft engine. While the Westland-built Series 1 Whirlwind remained essentially similar to the basic S-55, the Series 2 was re-engined with the 750 hp Leonides Major 755.

SIKORSKY S-58

The Sikorsky S-58 illustrated above is operated by a Belgian joint-service search-and-rescue flight. Many S-58s remain in service in the ASW and rescue rôles

Country of Origin: USA.
Type: General purpose 16–18 seat helicopter.
Power Plant: One 1,525 hp Wright R-1820-84 piston radial.
Performance: Max speed, 123 mph (198 km/h) at sea level; cruising speed, 98 mph (158 km/h); inclined rate of climb, 1,100 ft/min (5,6 m/sec); service ceiling, 9,000 ft (2 740 m); hovering ceiling (IGE), 4,900 ft (1 490 m), (OGE), 2,400 ft (730 m); range, 280 mls (450 km).
Weights: Empty, 8,275 lb (3 754 kg); max take-off weight, 14,000 lb (6 350 kg).
Dimensions: Rotor diam, 56 ft (17,07 m); fuselage length, 46 ft 9 in (14,25 m); height, 14 ft $3\frac{1}{2}$ in (4,36 m).
Notes: The first example of the S-58 flew on 8 March 1954 as the HSS-1, a US Navy anti-submarine helicopter, and the type subsequently appeared in several other versions for the US Navy, Marine Corps, Army and Coast Guard, with H-34 designations. Production totalled 1,821 and many remain in service with foreign services. The French Army and Navy received 166 from the Sud-Aviation production line and Westland built the similar Wessex (page 217).

SIKORSKY S-61B SEA KING

The photograph above depicts an SH-3G Sea King of US Navy Squadron HS-15, one of a number of versions of the S-61B built by the parent company and under licence in UK, Italy and Japan

Country of Origin: USA.
Type: Anti-submarine, air-sea rescue and transport helicopter.
Power Plant: Two 1,400 shp General Electric T58-GE-10 turbo-shaft engines.
Performance: (SH-3D) Max speed, 155 mph (266 km/h) at sea level, initial rate of climb, 2,200 ft/min (11,2 m/sec); hovering ceiling, (OGE), 8,200 ft (2 500 m); service ceiling, 14,700 ft (4 480 m); range 620 mls (1 000 km).
Weights: Empty, 12,087 lb (5 481 kg); max take-off, 20,500 lb (9 297 kg).
Dimensions: Rotor diam, 62 ft 0 in (18,90 m); fuselage length, 54 ft 9 in (16,69 m); height, 15 ft 6 in (4,72 m).
Notes: Basic Sea King is anti-submarine helicopter, used by USN, Argentina, Brazil and Spain; Agusta-built version serves with Italian and Iranian navies. Westland-built Sea Kings serve with the RN, Indian and Pakistan Navies. Similar S-61A transport is used by the Malaysian Air Force. Search-rescue models are used by Denmark and (Westland-built) by Norway and Germany. Westland is also producing a land-based troop transport model as the Commando, the first example of which flew on 12 September 1973.

SIKORSKY S-61R

The HH-3E illustrated above is a variant of the S-61R for the USAF Aerospace Rescue and Recovery Service, featuring armour, self-sealing tanks and a refuelling probe

Country of Origin: USA.
Type: Amphibious transport and rescue helicopter.
Power Plant: (CH-3E) Two 1,500 shp General Electric T58-GE-5 turboshafts.
Performance: (CH-3E at 21,247 lb/9 635 kg) Max speed, 162 mph (261 km/h) at sea level; range cruise, 144 mph (232 km/h); max inclined climb, 1,310 ft/min (6,6 m/sec); hovering ceiling, (IGE), 4,100 ft (1 250 m); range with 10 per cent reserves, 465 mls (748 km).
Weights: (CH-3E) Empty, 13,255 lb (6 010 kg); normal take-off, 21,247 lb (9 635 kg); max take-off, 22,050 lb (10 000 kg).
Dimensions: Rotor diam, 62 ft 0 in (18,90 m); fuselage length, 57 ft 3 in (17,45 m); height, 18 ft 1 in (5,51 m).
Notes: Based on the S-61 Sea King (page 215), the S-61R has many design changes, including a rear-loading ramp and retractable tricycle undercarriage. Primary customer was USAF, which bought CH-3C with T58-GE-1 engines, CH-3E (data above) and HH-3E with armour for rescue mission in hostile environment. Similar search-rescue HH-3F Pelican is used by US Coast Guard and built for Italian Air Force by Agusta.

SIKORSKY CH-54 TARHE

The CH-54B Tarhe crane-type helicopter of the US Army illustrated above is a more powerful derivative of the CH-54A with a new rotor system

Country of Origin: USA.
Type: Heavy flying-crane helicopter.
Power Plant: (CH-54B) Two 4,800 shp Pratt & Whitney T37-P-700 turboshafts.
Performance: (CH-54A at 38,000 lb/17 237 kg) Max speed, 127 mph (204 km/h) at sea level; max cruise, 109 mph (175 km/h); max inclined climb, 1,700 ft/min (8,64 m/sec); hovering ceiling (IGE), 10,600 ft (3 230 m), (OGE), 6,900 ft (2 100 m); range, 253 mls (407 km).
Weights: (CH-54A) Empty, 19,234 lb (8 724 kg); max take-off, (CH-54A) 42,000 lb (19 050 kg); (CH-54B), 47,000 lb (21 320 kg).
Dimensions: Rotor diam, 72 ft 0 in (21,95 m); fuselage length, 70 ft 3 in (21,41 m); height, 25 ft 5 in (7,75 m).
Notes: The S-64 Skycrane is used by the US Army in the heavy lift rôle and may be fitted with a 15,000-lb (6 800-kg) hoist or an all-purpose pod which can accommodate 45 troops or 24 stretchers. Initial production version was the CH-54A with 4,500 shp T73-P-1 engines (data above). CH-54B had uprated engines and other new features.

SIKORSKY CH-53 SEA STALLION

The Sikorsky S-65 is licence-built in Federal Germany as the CH-53DG and has been exported to Austria as the S-65Ö, an example of the latter being illustrated above

Country of Origin: USA.
Type: Heavy assault transport helicopter.
Power Plant: Two 3,925 shp General Electric T64-GE-413 turbo-shafts.
Performance: (CH-53D) Max speed, 196 mph (315 km/h) at sea level; max cruise, 173 mph (278 km/h); max inclined climb, 2,180 ft/min (11,08 m/sec); hovering ceiling (IGE), 13,400 ft (4 080 m), (OGE), 6,500 ft (1 980 m); range, 257 mls (413 km).
Weights: (CH-53D) Empty, 23,485 lb (10 653 kg); normal take-off, 36,400 lb (16 510 kg); 41,000 lb (19 050 kg).
Dimensions: Rotor diam, 72 ft 3 in (22,02 m); fuselage length, 67 ft 2 in (20,47 m); height, 24 ft 11 in (7,60 m).
Notes: US Navy has CH-53A assault transport and RH-53D special mine-sweeping version. US Marine Corps version is CH-53D and USAF uses similar HH-53B and HH-53C, respectively with T64-GE-3 and -7 engines for heavy lift. German Army has bought 165 CH-53D (G), mostly built in Germany, and Austria has two S-65-Oe. Prototypes of an improved version with three T64 engines, the YCH-53E, were ordered in 1973 with first flight scheduled for April 1974.

WESTLAND WESSEX

Derived from the Sikorsky S-58, the Wessex differs from its US progenitor in many respects, the photograph above illustrating a Wessex HC Mk 2 of the RAF

Country of Origin: United Kingdom.
Type: Anti-submarine and transport helicopter.
Power Plant: One (HAS Mk 3) Rolls-Royce Napier Gazelle 165 turboshaft.
Performance: (HAS Mk 3) Max speed, 132 mph (213 km/h) at sea level; cruising speed for ranges, 68 mph (111 km/h); anti-submarine cruising speed, 115 mph (185 km/h); endurance at zero datum (50/50 hover and cruise) 83 min.
Weights: (HAS Mk 3) Empty, 9,350 lb (4 241 kg); gross, 13,500 lb (6 123 kg).
Dimensions: Rotor diam, 56 ft (17,07 m); fuselage length, 48 ft $4\frac{1}{2}$ in (14,74 m); height, 14 ft 5 in (4,39 m).
Notes: The Wessex is a development of the Sikorsky S-58, from which it differs in powerplant and equipment. First flight with a Gazelle turboshaft was made on 17 May 1957 and production models for the RN were the Wessex HAS Mk 1 and improved HAS Mk 3 with dorsal radome. RAF versions are HC Mk 2 with twin Gnome engine and HU Mk 5 commando assault transport. R Australian Navy uses HAS Mk 31 and others were exported to Iraq, Ghana and Brunei.

WESTLAND WASP (AND SCOUT)

The Westland Scout and Wasp are essentially similar helicopters developed for the British Army and British Navy respectively, an example of the Wasp ASW helicopter appearing above

Country of Origin: United Kingdom.
Type: Five/six-seat general-purpose and anti-submarine warfare helicopter.
Power Plant: One 710 shp Rolls-Royce Bristol Nimbus 503 turbo-shaft.
Performance: Max speed, 120 mph (193 km/h) at sea level; max cruise, 110 mph (177 km/h); max inclined climb, 1,440 ft/min (7,4 m/sec); hovering ceiling (IGE), 12,500 ft (3 810 m), (OGE), 8,800 ft (2 682 m); max range with standard fuel, 303 mls (488 km).
Weights: Empty, 3,452 lb (1 566 kg); max take-off, 5,500 lb (2 495 kg).
Dimensions: Rotor diam, 32 ft 3 in (9,83 m); fuselage length, 30 ft 4 in (9,24 m); height, 11 ft 8 in (3,56 m).
Notes: The Wasp HAS Mk 1 is standard anti-submarine equipment aboard RN frigates, usually flown by a solo pilot and carrying two 270-lb (122-kg) torpedoes. Wasps have also been supplied to Brazil, New Zealand, the Netherlands and South Africa. The British Army Scout AH Mk 1 is similar but has a skid landing gear, Nimbus 101 or 102 engine and different interior.

WESTLAND/AÉROSPATIALE WG.13 LYNX

Several variants of the Lynx are currently under development for the British services and the French Navy, that illustrated above being a Lynx AH Mk 1 of the British Army

Country of Origin: United Kingdom.
Type: Multi-purpose and transport helicopter.
Power Plant: Two 900 shp Rolls-Royce BS.360-07-26 turbo-shafts.
Performance: (General purpose versions) Max speed, 207 mph (333 km/h); cruise, 176 mph (284 km/h) at sea level; max inclined climb, 2,800 ft/min (14,2 m/sec); hovering ceiling, (OGE), 12,000 ft (3 650 m); range (with 10 passengers), 173 mls (278 km), (with cargo and full tanks), 489 mls (788 km).
Weights: Operational empty, 5,532–6,125 lb (2 509–2 778 kg); max take-off, 8,000 lb (3 620 .g); overload, 9,105 lb (4 130 kg).
Dimensions: Rotor diam, 42 ft 0 in (12 80 m); fuselage length, 38 ft $3\frac{1}{4}$ in (11,66 m); height, 12 ft 0 in (3,66 m).
Notes: Lynx is an Anglo-French project, developed in several versions for the British Army (AH Mk 1), Royal Navy (HR Mk 2) and French *Aéronavale*. First of 12 prototypes flew on 21 March 1971 and first production orders were placed in May 1973. Total procurement is expected to include 80 for *Aéronavale* and 277 for the British services. The first example in French Navy configuration flew on 6 July 1973.

INDEX OF AIRCRAFT TYPES

ACKNOWLEDGEMENTS

The authors wish to record their thanks to the following sources of copyright photographs appearing in this volume:

Stephen P Peltz, pages 7, 8, 10, 11, 13, 14, 18, 20, 21, 34, 37, 48, 78, 90, 115, 130, 131, 132, 136, 138, 158, 167, 172, 176, 189, 190, 203, 205, 210, 215
Image in Industry, page 9
Ministry of Defence, pages 23, 26, 52, 67, 86, 133, 217
William Strandberg, page 25
Capt Keith Sissons, page 28
John Fricker, pages 29, 213
Robert C. Mikesh, page 47
Claus Haugebo, pages 57, 180
Blandin-Régnier, pages 64, 149
Robert J Ruffle, page 106
Howard Levy, pages 108, 109, 148, 151
A Reinhard, page 120
Mick West, pages 122, 208
W Gysin-Aegerter, page 145
Ronaldo S Olive, page 150